Vaccines: What's Happening

VACCINES: WHAT'S HAPPENING

Edited by:

ShivSanjeevi Sripathi

Delve
PUBLISHING

www.delvepublishing.com

Vaccines: What's Happening

Shiv Sanjeevi Sripathi

Delve Publishing
224 Shoreacres Road
Burlington, ON L7L 2H2
Canada
www.delvepublishing.com
Email: orders@arclereducation.com

ISBN: 978-1-77469-407-7 (Hardcover)

DECLARATION

ABOUT THE EDITOR

ShivSanjeevi Sripathi completed his Masters in Biotechnology from Mumbai University in 2008. He was awarded for academic excellence in both his Bachelors and Masters for securing second rank in Mumbai University in 2006 and first rank in his college: Kishinchand chellaram College. For his Masters he secured first rank in his college KET's V.G.Vaze College. He qualified CSIR and NET and and TOEFL in September 2008. He then worked on a stem cell project at the Specialized Centre for Cell Based Therapy (SCCT),KEM Hospital at Mumbai on a project entitled, : Isolation & detection of stem cells from Human Umbilical cord/ amniotic membrane" following which he worked at Junior Research Fellow at Microbiology & Cell Biology Department, Indian Institute of Science, Bangalore from August 2009 to November 2011. The field of work involved cloning of cell wall genes and transcription factors in *E.coli & M.smegmatis*. As a writer, he has authored and co-authored 25 books with Delve and Arcler International Press. He loves to read and share on interesting aspects of life sciences in books. In his free time he loves to travel and explore ancient customs and traditions.

TABLE OF CONTENTS

LIST OF CONTRIBUTORS

ShivSanjeevi Sripathi

Stefan H. E. Kaufmann
Author information Article notes Copyright and License information Disclaimer
Max Planck Institute for Infection Biology, Berlin, Germany
Hagler Institute for Advanced Study, Texas A&M University, College Station, TX, United States

Charles Ndawula, Jr.
National Agricultural Research Organization, Entebbe, Wakiso 256, Uganda
National Livestock Resources Research Institute, Vaccinology Research Programme, Nakyesasa, Wakiso 256, Uganda

Tiza Ng'uni
KwaZulu-Natal Research Institute for Tuberculosis and HIV (K-RITH), Nelson R. Mandela School of Medicine, University of KwaZulu-Natal, Durban, South Africa

Caroline Chasara
KwaZulu-Natal Research Institute for Tuberculosis and HIV (K-RITH), Nelson R. Mandela School of Medicine, University of KwaZulu-Natal, Durban, South Africa

Zaza M. Ndhlovu
KwaZulu-Natal Research Institute for Tuberculosis and HIV (K-RITH), Nelson R. Mandela School of Medicine, University of KwaZulu-Natal, Durban, South Africa
Ragon Institute of Massachusetts General Hospital, Massachusetts Institute of Technology, and Harvard University, Cambridge, MA, United States

Rekha R. Rapaka
Center for Vaccine Development and Global Health, University of Maryland School of Medicine, 685 West Baltimore Street, Baltimore, MD 21201, USA

Alan S. Cross
Center for Vaccine Development and Global Health, University of Maryland School of Medicine, 685 West Baltimore Street, Baltimore, MD 21201, USA

Monica A. McArthur
Center for Vaccine Development and Global Health, University of Maryland School of Medicine, 685 West Baltimore Street, Baltimore, MD 21201, USA

Chiranjib Chakraborty
Department of Biotechnology, School of Life Science and Biotechnology, Adamas University, Kolkata, India
Institute for Skeletal Aging & Orthopedic Surgery, Hallym University-Chuncheon Sacred Heart Hospital, Chuncheon, Gangwon-do, South Korea

Ashish Ranjan Sharma
Institute for Skeletal Aging & Orthopedic Surgery, Hallym University-Chuncheon Sacred Heart Hospital, Chuncheon, Gangwon-do, South Korea

Manojit Bhattacharya
Department of Zoology, Fakir Mohan University, Odisha, India

Sang-Soo Lee
Institute for Skeletal Aging & Orthopedic Surgery, Hallym University-Chuncheon Sacred Heart Hospital, Chuncheon, Gangwon-do, South Korea

Anke Osterloh
Research Center Borstel, Parkallee 22, 23845 Borstel, Germany

Alberto E. Tozzi
Unit of Telemedicine, IRCCS, Bambino Gesù Children's Hospital, Rome, Italy

Francesco Gesualdo
Unit of Telemedicine, IRCCS, Bambino Gesù Children's Hospital, Rome, Italy

Angelo D'Ambrosio
Unit of Telemedicine, IRCCS, Bambino Gesù Children's Hospital, Rome, Italy

Elisabetta Pandolfi
Unit of Telemedicine, IRCCS, Bambino Gesù Children's Hospital, Rome, Italy

Eleonora Agricola
Unit of Telemedicine, IRCCS, Bambino Gesù Children's Hospital, Rome, Italy

Pierluigi Lopalco
European Centre for Disease Prevention and Control, Stockholm, Sweden

LIST OF ABBREVIATIONS

ADS	Air-dried sheets
AFM	Atomic force microscopy
BIIR	Bromobutyl rubbers
CB	Carbon black
CNTs	Carbon nanotubes
XNBR	Carboxylated Nitrile Butadiene Rubber
CVD	Chemical vapor deposition
CSM	Chlorosulfonated Polyethylene
CSPE	Chlorosulfonated polyethylene
CRRC	Combat Rubber Raiding Craft
EPLED	Elastomeric polymer light-emitting device
EPDM	Ethylene Propylene Diene Monomer
EQE	External quantum efficiency
FWHM	Full width half maximum
GO	Graphene oxide
GNPs	Graphite nanoplatelets
HMPE	High-modulus polyethylene
HPPE	High-performance polyethylene
HNBR	Hydrogenated nitrile rubbers
ITO	Indium tin oxide
IPN	Interpenetrating polymer network
IIR	Isobutylene-Isoprene Rubber
LEDs	Light-emitting diodes
LCST	Lower critical solution temperature
MPR	Melt Processible Rubbers
MWCNTs	Multiwall carbon nanotubes
NR	Natural rubber
OENR	Oil-extended natural rubber

OM	Optical microscopy
PEC	Piezoelectric elastic composite
PCEAs	Polycarbonate esteramides
PVC	Polyvinyl chloride
PVDF	Polyvinylidene fluoride
PCE	Power conversion efficiency
RTD	Resistance temperature detector
RSS	Ribbed smoked sheets
SEM	Scanning electron microscopy
SBR	Styrene-Butadiene Rubber
SBS	Styrene-butadiene-styrene
SEBS	Styrene/ethylene-butylene copolymer
SEPS	Styrene/ethylene-propylene copolymer
SIS	Styrene-isoprene-styrene
TPO	Thermoplastic Olefin Elastomers
TPV	Thermoplastic vulcanizates
TPOs	Polyolefin thermoplastic elastomers
TEM	Transmission electron microscopy
UCST	Upper critical solution temperature

PREFACE

"I shall endeavour still further to prosecute this inquiry, an inquiry I trust not merely speculative, but of sufficient moment to inspire the pleasing hope of its becoming essentially beneficial to mankind"

-Edward Jenner (the famous smallpox vaccination)

Pandemics have not been recent. Apart from the Spanish flu and the recent COVID-19 pandemic, the ability of pathogens to target humanity is a repeated pattern over time. For instance, what was first reported in late 2019 and brought the globe to a standstill in 2020, COVID-19 is firmly etched in the minds of the survivors. Conditions considered "old scourges of humanity" like cholera and typhoid are reported as becoming more resistant to antimicrobials. Research has always pointed out vaccines as being the cornerstone approach of quelling epidemics and pandemics: on the lines of the adage "better safe than sorry".

Prior to Lady Mary Wortley Montagu and Edward Jenner, the practice of inoculating a disease was recorded in ancient India, China and Africa. This entailed the induction of immunity via inhalation of smallpox crusts or inserting them into small cuts. Now, there are routine immunization schedules for infants, children and adults. The recent MERSm Ebola, NIPAH and COVID-19 pandemics have ushered in a new era of vaccination with the latest being "selfies"after COVID shots!

"The likelihood of a severe breakthrough infection as tracked by the CDC after vaccination remains rare. Among 161 million Americans vaccinated, 4072 hospitalizations occurred due to COVID after vaccination (0.0025%) and 849 deaths occurred due to COVID after vaccination (0.0005%.)" (Posted July 28, 2021)

- Monica Gandhi, MD, MPH

Infectious Diseases doctor and Professor of Medicine, University of California, San Francisco

This book comes at this crucial juncture when our conquest of land and overpopulation is resulting in waves of epidemics. The book starts with groundwork entailing the impact of vaccines, their production and costs. it then discusses the development of vaccines in a traditional scenario over pandemic situations: the latter necessitating fast-tracking the development process. However, the safety of people must be kept in mind along with "ethical" systems of vaccine development and administration. This is followed by a discussion of adjuvants used and how attempts are on to boost the efficiency of available vaccine platforms. Finally, the scope of vaccines for neglected tropical diseases is also elucidated as backed by research.

"It is possible to create an epidemic of health which is self-organizing and self-propelling."

- Jonas Salk (developer of the first successful polio vaccine)

CHAPTER 1

Groundwork

"I hope that some day the practice of producing cowpox in human beings will spread over the world - when that day comes, there will be no more smallpox."

-Edward Jenner

Contents

1.1 AN ANCIENT TREND

The act of preventing disease via inoculation or injecting a healthy individual with a pathogen to cause a mild version of the disease dates back to 1000 BC in India, China, Turkey and Africa (Lahariya, 2014). Quoting from John Shoolbred, *Report on the progress of vaccine inoculation in Bengal, from the period of its introduction in November 1802 to the end of the year 1803*, Calcutta, Honorable Company's Press, 1804, Reprint: London, Blacks and Parry, 1805, p. 73:

"The operator takes a piece of cloth in his hand ... and with it gives a dry friction on the part intended for inoculation, for the space of eight or ten minutes; then with a small instrument he wounds by many slight touches, about the compass of a silver groat, just making the smallest appearance of blood. Then opening a linen double rag, (which he always keeps in a cloth round his waist,) he takes from thence a small pledget of cotton charged with the variolous matter, which he moistens with two or three drops of the Ganges water, and applies it to the wound, fixing it on a slight bandage, and ordering it to remain on for six hours without being moved, then the bandage to be taken off, and the pledget to remain until it falls off itself"

Subsequent work by Jenner, Pasteur and other "milestones" are depicted in table 1.1 below (© 2015 Hajj Hussein, Chams, Chams, El Sayegh, Badran, Raad, Gerges-Geagea, Leone and Jurjus. *This is an open-access article distributed under the terms of the Creative Commons Attribution License (CC BY). The use, distribution or reproduction in other forums is permitted, provided the original author(s) or licensor are credited and that the original publication in this journal is cited, in accordance with accepted academic practice. No use, distribution or reproduction is permitted which does not comply with these terms*):

Vaccine (year introduced)	Infectious agent	Kind of vaccine first introduced	Vaccine used in present	Time for vaccination	Need of booster	Efficacy
Smallpox (1798)	Variola virus	Live vaccinia virus	N/A	Stopped in 1972 after eradication	N/A	Global eradication
Anthrax (1881)	*Bacillus anthracis*	Live, attenuated	Cell-free filtrates of microaerophilic cultures of a toxigenic, non-encapsulated strainof *B. anthracis* V770-NP1-R	Pre-exposure in adults ≥18 years old; 5 shots over 18 months	Yes; annually	92.50%
Rabies (1884)	Rabies virus	Live, attenuated	Inactivated virus	Post-exposure; 4 doses (0, 3, 7, 14)	Not recommended	Inconclusive data
Typhoid (1896)	*Salmonella typhi*	Inactivated	Inactivated; live, attenuated	At risk population; inactivated: one dose; live, attenuated: 4 doses every other day	Yes if at risk; inactivated: every 2 years; live, attenuated: every 5 years	Varies with age; >50%
Cholera (1884–1896)	*Vibrio cholerae*	Live, attenuated	Oral, inactivated, killed whole cell of *V. cholerae*	At risk population; 2 doses 1 week apart	Yes if at risk; every 6 months	50–60%
Tuberculosis (1927)	*Mycobacterium tuberculosis*	Live, attenuated *Mycobacterium bovis*	N/A	Single dose for children	Not recommended	70–80%
Yellow fever (1935)	Yellow fever virus	Live, attenuated	Live, attenuated	Single dose ≥9 months old	Not recommended	Long term (80–100%)
Diphtheria and tetanus toxoids (1930s and 1940s) and acellular pertussis (dtap)[a] (1948)	*Corynebacterium diphtheria, Bordetella pertussis, Clostridium tetani*	Inactivated	Inactivated	2, 4, 6, 15–18 months, 4–6 years	Yes; Tdap: 11–12 years; if Tdap not received between 11–18 years, Tdap dose should be given then followed with Td booster doses every 10 years	80–85%
Poliovirus (1955)	Poliovirus	Inactivated poliovirus	Inactivated and oral, live attenuated	2, 4, 6–18 months	Yes; 4–6 years	IPV: ≥99% OPV: >95%
Influenza (1954–1955)	Influenza virus	Inactivated	Inactivated and live attenuated	Annually	Not recommended	Inactivated: 60% Live attenuated: >87%
Measles, Mumps, Rubella (1971)	Measles, mumps, rubella	Inactivated	Live, attenuated	12–15 months, 4–6 years	Not recommended	>95%
Meningococcal (1974)	*Neisseria meningitidis*	Polysaccharide	Conjugate	11-12 years	Yes; 16 years	>85%
Pneumococcal (1977)	*Streptococcus pneumoniae*	Polysaccharide	Polysaccharide-protein conjugate	2, 4, 6, 12–15 months	Yes; if at risk, ≥65 years old	Children: >90%; adults >65: 75%
Hepatitis B (1981)	Hepatitis B virus	Plasma derived	DNA recombinant	0, 1–2, 6–18 months	Not recommended	Long term (80–100%)
Haemophilus influenzae type b (Hib) (1985)	*Haemophilus influenzae* type B	Polysaccharide	Polysaccharide-protein conjugate	2, 4 months	Yes; 12–15months	95–100%
Japanese Encephalitis (1992)	Japanese encephalitis virus	N/A	Inactivated virus	Endemic countries: two-dose series 28 days apart	Yes; if risk of exposure	No data available
Varicella (1995)	Varicella zoster virus	Live	Live, attenuated	12–15 months, 4–6 years	Not recommended	>70%
Hepatitis A (1995)	Hepatitis A virus	Inactivated	Inactivated, whole virus	Two-dose series 6 months apart: 12–23 months old	Not recommended	94–100%
Rotavirus (1998)	Rotavirus	Rhesus-based tetravalent rotavirus	RV1: live, oral, attenuated, monovalent human RV5: live, oral, attenuated, pentavalent bovine-human reassortant	RV1: 2, 4 months RV5: 2, 4, 6 months	Not recommended	85–98%
Human papillomavirus (2006)	Human papillomavirus	DNA recombinant	DNA recombinant	Three-dose series starting at 11 years old: 0, 1–2, 6 months	Not recommended	>99%
Shingles (2006)	Varicella zoster virus	N/A	Live, attenuated	Single dose for people ≥60 years old	Not recommended	97.20%
H1N1 (2009)	Influenza virus type A	N/A	Inactivated virus	Two doses 4 weeks apart for children aged 6 months-9 years or one dose for adults and children >10 years old	N/A	>80%

[a]*Upper-case letters denote full-strength doses of diphtheria (D) and tetanus (T) toxoids and pertussis (P) vaccine. Lower-case "d" and "p" denote reduced doses of diphtheria and pertussis used in the adolescent/adult-formulations. The "a" in DTaP and Tdap stands for "acellular."*
Sources: Ref. (61, 62).

1.2 PRE AND POST-VACCINATION

A predominant impact of vaccines has been in the attenuation of maladies that target children. For 2009, the administration of vaccines for 13 diseases across an annual birth cohort was pitched to avert 20 million disease incidences and around 42000 mortalities in the US. A dip in 90% was recorded for infectious diseases of the early 1900s in the year 2017 in the US due to vaccines viz. polio vaccine, the DTaP (diphtheria, tetanus, and acellular pertussis) along with MMR (measles, mumps, and rubella) vaccines. This trend of lowered disease incidence as per the Centers for Disease Control and Prevention from the 20th century to 2017 is depicted in table 1.2 below (© 2020 Rodrigues and Plotkin. *This is an open-access article distributed under the terms of the Creative Commons Attribution License (CC BY). The use, distribution or reproduction in other forums is permitted, provided the original author(s) and the copyright owner(s) are credited and that the original publication in this journal is cited, in accordance with accepted academic practice. No use, distribution or reproduction is permitted which does not comply with these terms*):

Vaccine	Peak cases in prevaccine era (year)	Vaccine coverage in children 19–35 months old (% [95% CI])	Cases in 2017 (*n*)	Disease reduction (%)
Smallpox	110, 672 (1920)	–	0	100
Diphtheria	30, 508 (1936)	94.0 (93.3 – 94.7)	0	100
Measles (non-imported)	763, 094 (1958)	91.5 (90.6 – 92.3)	99	99.99
Mumps	212, 932 (1964)	91.5 (90.6 – 92.3)	6,109	97.13
Rubella	488, 796 (1964)	91.5 (90.6 – 92.3)	7	100.00
Congenital rubella syndrome	20, 000 (1964 – 65)	–	5	99.98
Pertussis	265, 269 (1934)	94.0 (93.3 – 94.7)	18,975	92.85
Polio (paralytic)	21, 269 (1952)	92.7 (91.9 – 93.5)	0	100
Tetanus	601 (1948)	94.0 (93.3 – 94.7)	33	94.51

Despite the lowering of vaccine-preventable disease burden across the world, certain challenges remain in low- and middle-income countries (LMIC) inclusive of infrastructure paucity, political issues, economic feasibility, program management and resistance.

Figure 1.1 below illustrates ***TOP***: the vaccine uptake across economies from the World Health Organization and UNICEF dataset "Coverage Estimates Series" for high-income (solid line), middle-income (dashed line), and low-income (dotted line) countries for the last two decades (© 2020 Rodrigues and Plotkin. *This is an open-access article distributed under the terms of the Creative Commons Attribution License (CC BY). The use, distribution or reproduction in other forums is permitted, provided the original author(s) and the copyright owner(s) are credited and that the original publication in this journal is cited, in accordance with accepted academic practice. No use, distribution or reproduction is permitted which does not comply with these terms*):

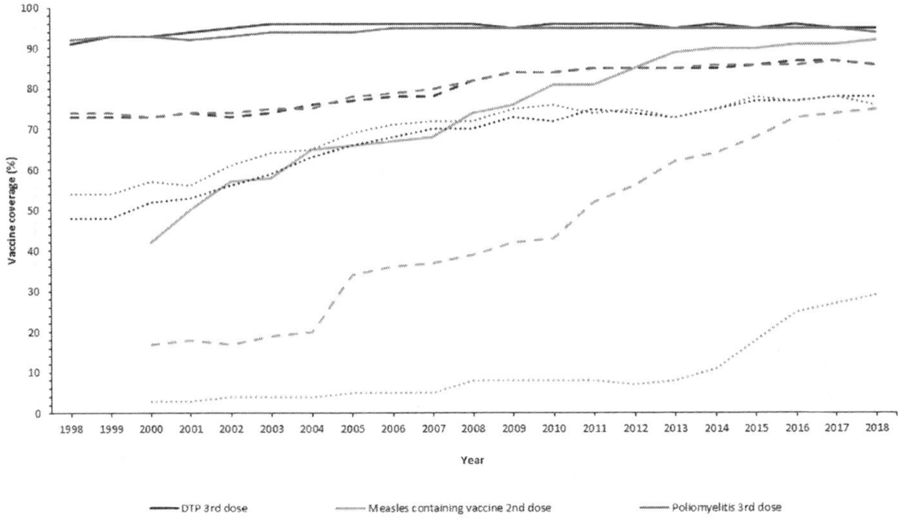

BOTTOM: Lowering of infectious diseases across the globe from the World Health Organization dataset "Reported cases of vaccine-preventable diseases" for the last two decades (© 2020 Rodrigues and Plotkin. *This is an open-access article distributed under the terms of the Creative Commons Attribution License (CC BY). The use, distribution or reproduction in other forums is permitted, provided the original author(s) and the copyright owner(s) are credited and that the original publication in this journal is cited, in accordance with accepted academic practice. No use, distribution or reproduction is permitted which does not comply with these terms*):

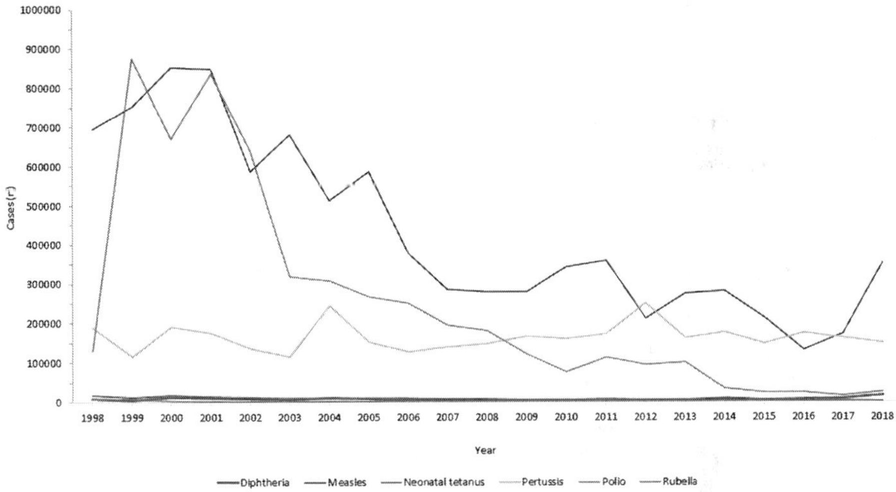

1.3 VACCINES: THE PROCESSES AND COSTS

Before we look at a snapshot of the process and costs, a research article by Bechini and team in *Vaccines* (2021) explored how much the general population was aware of the process of vaccine production. The cross-sectional study entailing the use of an online anonymous ad hoc questionnaire developed by experts employing the Google Modules application was available from January 2020 to May 2020. While quality controls during the vaccine production process saw good awareness in the participants with 90% aware of the importance of controls to circumvent contamination and 94% were aware that controls are carried out. Yet, knowledge gaps emerged in several spheres. For instance, 53% of participants felt that the quality controls of vaccines were at par with that of drugs-it is established that the former entails more rigorous assurance and controls. That quality control comprises 50–70% of the vaccine production process was not known by around 38% of the participants. Further, 55.6% of the participants were of the opinion that contamination cannot occur in vaccine production. Such gaps call for increased awareness programs in this regard to ensure more of the public is aware of the processes given the vital importance of vaccination.

Figure 1.2 below illustrates the opinions about vaccine production of 135 participants [88 females (65.2%) and 47 males (34.8%): mean age= 35.4 years; 34.8% were educated in or worked in healthcare] (Bechini A, Bonanni P, Zanella B, Di Pisa G, Moscadelli A, Paoli S, Ancillotti L, Bonito B, Boccalini S, 2021: © 2021 by the authors. Licensee MDPI, Basel, Switzerland. *This article is an open access article distributed under the terms and conditions of the Creative Commons Attribution (CC BY) license: https://creativecommons.org/licenses/by/4.0/)*:

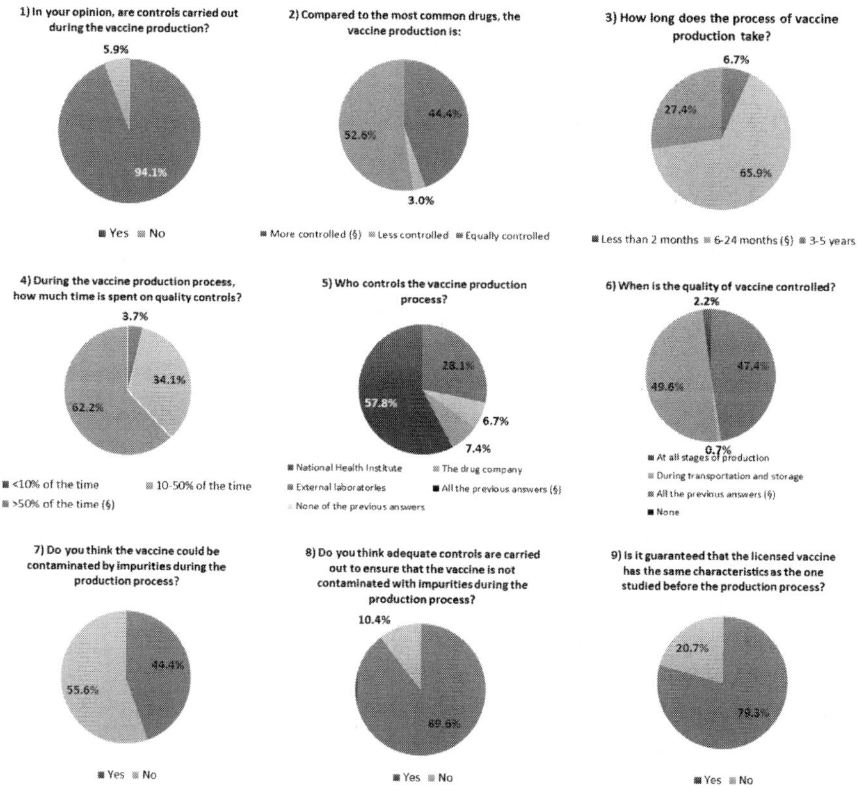

Below is the summary of the collected answers for the open question: "If one day you could visit a pharmaceutical industry, what would you like to ask or to see?" (N = 62): (Bechini A, Bonanni P, Zanella B, Di Pisa G, Moscadelli A, Paoli S, Ancillotti L, Bonito B, Boccalini S, 2021: © 2021 by the authors. Licensee MDPI, Basel, Switzerland. *This article is an open access article distributed under the terms and conditions of the Creative Commons Attribution (CC BY) license: https://creativecommons.org/licenses/by/4.0/):*

"If One Day You Could Visit a Pharmaceutical Industry, What Would You Like to Ask or to See?"	
	n (%)
The whole vaccine production process or specific phases	29 (46.8)
Quality control	15 (24.2)
Laboratories and Research & Development	5 (8.1)
Regulatory aspects	3 (4.8)
Research studies, efficacy studies	3 (4.8)
Production timing	2 (3.2)
Company structure (managers and workers)	2 (3.2)
Costs related to vaccine's production	2 (3.2)
Time spent in training	1 (1.6)

Described as "one of the most challenging industries", vaccine manufacture entails that all steps encompass efficacy and safety across its life cycle. A slew of factors dictate the net outcome ranging from the pathogen to the purification steps to the technician expertise and the environment. Risks arising from each of these steps in addition to improper analyses can result in product suspensions or failures. The production and testing processes are also issued a license by regulatory authorities necessitating care to circumvent small changes that can cause alterations in a product. The major cause of challenge in vaccine manufacturing as opposed to other pharma products emerges as this variability factor (Plotkin et al, 2017).

The major steps entailed in vaccine production are summarized below in table 1.3 (Plotkin S, Robinson JM, Cunningham G, Iqbal R, Larsen S: © 2017 The Authors:

Stage	Aim	Process
Exploratory & Pre-clinical	Assess safety and immunogenicity of a target antigen or cell in cell culture or animal disease models; assess a safe starting dose for human clinical studies	Small scale, often crude extracts or purchased antigens. The cost of manufacturing generally is not critical at this stage, although method of manufacturing is critical to the character of the ultimate product. Process development is often delayed until after some proof of concept in animal models is confirmed
Clinical Trial Authorization Application	Apply for approval to conduct human clinical studies	Outline all critical manufacturing steps and analytical methods used to produce and release the product and placebo, including all reagents, components, specifications, acceptable limits to manufacture and release the product ensuring the identify, strength, quality, and purity. Demonstrate stability of the drug product and placebo for at least the duration of the clinical studies
Phase I Vaccine Trials	Assess the safety of the candidate vaccine; determine the type and extent of immune response that the vaccine provokes	All human clinical materials are recommended to be made under cGMP. The state of the process development varies with strategy; complete process optimization is often deferred until after proof of concept in humans, but all process changes need to be qualified prior to advancing to the next clinical stage and deferring development can delay the next stages or risk the vaccine failure for unforeseen or unintended consequences of these changes
Phase 2 Vaccine Trials	Assess candidate vaccine safety, immunogenicity, dose response, schedule of immunizations, and method of delivery	Prior to initiating phase 2 studies it is recommended that all major process changes are incorporated and qualified. Significant changes after this step can risk repeating phase 1/2 studies. Projected cost of goods is confirmed to be appropriate for the intended use and markets

Phase 3 Vaccine Trials	Assess the candidate vaccine in the target populations for safety and rare adverse events Vaccine efficacy is estimated. Vaccine manufacturing consistency is confirmed. Concomitant testing with other prescribed vaccines may be required	Processes are finalized and validated. Analytical tests for manufacturing and release are completed and validated. Costs of goods are confirmed to be appropriate for the intended use (as changes to reduce costs would need to be re-validated and may require additional clinical testing)
Approval & Licensure	Submit and gain approval of the Biological Product Application	Full review and documentation of the manufacturing methods and analytical methods for licensed production; full shelf-life stability studies completed and in specification; completed process validation, facility validation, release testing validation; development of production and release protocols; launch lots prepared and released. Agency inspection of all manufacturing and release facilities and documentation of all manufacturing and quality systems
Post-Licensure Monitoring	Confirm filed use of vaccine is consistent with expectations from the clinical studies and finalized manufacturing and release process	Routine (annual, biennial) agency inspections of manufacturing and testing facilities. Annual product review and reporting demonstrating the process remains in control
License Amendments	Confirm any changes to the intended use of the vaccine in different populations or and changes to the manufacturing process (seeds, raw material sources, process steps, release steps, equipment, facilities, etc.) do not adversely affect the product purity, safety, or effectiveness	Process improvements after license approval are expected to keep the process optimized and to take advantage of advances in science and manufacturing methods, but can be expensive and risky (unintended consequence of a change). Significant changes should be carefully considered with respect to the risk/benefit of the change.

With a slew of production systems available for vaccines, the direct costs of vaccine manufacturing are inclusive of raw materials, staff, analyses, and documentation in lieu of Current Good Manufacturing Processes (cGMP) as necessitated by National Regulatory Authorities (NRAs). The indirect costs are inclusive of planning the production, quality system management, inventory, warehouse, distribution, marketing, sales and management.

Figure 1.3 below is illustrative of the costs (Plotkin S, Robinson JM, Cunningham G, Iqbal R, Larsen S: © 2017 The Authors: *This is an open access article under the CC BY license: http://creativecommons.org/licenses/by/4.0/*):

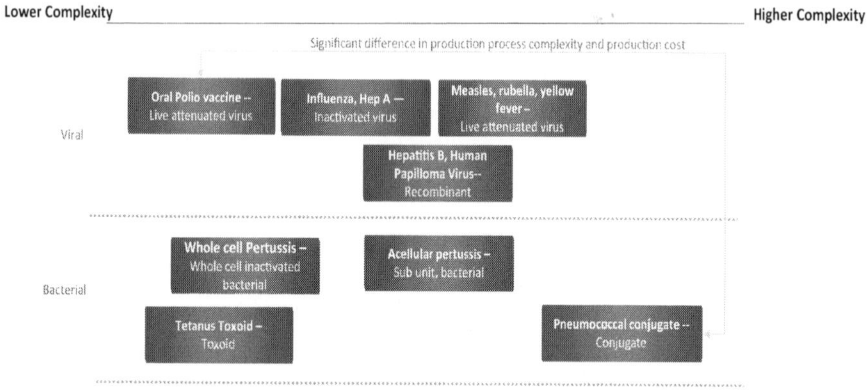

The predominant drivers and options to get an optimal Cost of Goods Sold (COGS) are listed in 1.4 below (Plotkin S, Robinson JM, Cunningham G, Iqbal R, Larsen S: © 2017 The Authors: *This is an open access article under the CC BY license: http://creativecommons.org/licenses/by/4.0/*):

Cost Driver	Major Cost Driver	Relative Impact of Cost Driver on overall costs	Cost range	Options to Reduce COGS	Potential Impact of COGS Reduction Strategy	Examples
Product Development	– R & D laboratories – R & D personnel	– High fixed costs/possible to be shared across antigens	>500 M USD Risk adjusted* cost of 135 to 350M USD *Risk adjusted cost incorporates the cost and probabilities of moving to the next phase of development [14]	– Copy originator process post patent expiration	– High	– MR vaccine copied from originator vaccine
				– Perform tech transfer with established product	– High	
				– Leverage correlates of protection to avoid large efficacy studies	– Medium	
				– Purchase antigens and execute form/fill as a means of gaining experience prior to full manufacturing end-to-end	– Medium	– OPV bulk can be sourced from an approved manufacture and formulated/filled
Facilities and Equipment	Capitalized costs that depreciate over time – Land – Buildings – Machinery Ongoing costs of upkeep – Repairs – Maintenance – Utilities	– High fixed costs/design for minimizing maintenance and utilities	50 to 700M USD Example: It took Pfizer five years and 600M USD to build a manufacturing site in the US [17]	– Design for very high facility utilization. Limit the number of production platforms; force fit new processes into established platforms to reduce need for new facilities; increase utilization of existing facilities. Use multi-dose vials.	– High	– Share filling lines across multiple vaccines, when possible

						– Shift production volumes to multi-dose vials to reduce filling costs (at the risk of higher vaccine wastage in field).
				– Use single-use disposable systems to reduce capital cost	– Medium	– Reduced capital offset by higher operating (consumable) costs
				– Minimize classified production space with closed systems and RABs	– Low/Med	
				– Limit automation and process/ equipment		
				– Leverage blow-fill-seal (BFS) filling technology to shrink clean room footprint and reduce final product component costs, and reduce labor		
				– Utilize Contract Manufacturing Organizations (CMO) for low volume products or until demand supports facility construction.	– Low/Med	– Seasonal influenza vaccines produced at a CMO. Reduced capital offset by CMO contract fees.

Direct Labor	Employee costs directly attributable to a specific vaccine – Wages – Benefits	– Low/typically less than 25% of total manufacturing costs	Costs can be significantly lower in China and India (25% lower for some manufacturers) of but manpower efficiency may be 120-130% of Western standards The difference is shrinking due to increasing labor costs as the requirements of cGMP practices increase	– Increase automation and single-use production technologies (balancing with potential increase in equipment or consumables costs)	– Medium	– Single-use, or disposable, bioreactors reduce cleaning and sterilization requirements, and complexity of qualification and validation
						– Pneumococcal conjugate vaccine assays are streamlined across multiple serotypes.
				– Standardize and streamline processes across as many steps and vaccines as possible.	– High	
				– Develop capacity progressively through reverse integration (packaging purchased products, filling and packing purchased products, form/fill/pack purchased products, then production of bulk drug substance for internal form/fill/pack)		

Overhead	Management, quality systems, IT systems	– High if company has few products – Low if overhead can be allocated across multiple products	Up to 45% of the cost of raw materials and labor combined [18]	– Invest in quality systems that can streamline quality practices and reduce costs over the long term	– Medium	– Introduce enterprise quality management software (EQMS)
				– Ensure management team has broad expertise to be leveraged across a portfolio of vaccines		
Licensing/ Regulatory and commercialization	Expenses paid for the right to use product-related IP (technology) Expenses to comply with regulatory requirements to produce either for domestic market or export	– Low if experienced teams are engaged early to prepare facilities and processes for regulatory review – High if review process requires considerable rework or if delays result in lost revenue	In addition to staff and consulting costs, the new WHO process assesses the following fees: A site audit fee of 30K USD and for: Simple/Traditional vaccines: – Evaluation fee of 25 to 100K, and Annual fees of 4.8K to 140K USD Combo or Novel Vaccines: – Evaluation fee of 66.5 to 232.8K USD, and Annual fees of 8.4 to 250K USD [19].	– Pursue WHO PQ as required by UNICEF/PAHO only when intending to access markets for which they procure (e.g., Gavi)		
				– Request royalty reductions or waivers for vaccine sold in low income countries (LICs)	– Low	– Royalty for HPV antigens waived for volumes sold in Gavi

				– Produce reagents in-house or seek viable alternatives rather than license.	– Medium	– CRM produced in-house to avoid licensing cost.
				– Differentiate originator production processes sufficiently to be considered a novel process		
				– Accelerate approval by seeking NRA or WHO priority review for vaccines for neglected diseases or emergency use.	– High	
				– Utilize FDA priority review vouchers for another product and allocate savings to the vaccine that secured the voucher.	– High	– Apply priority review voucher to a product intended for high income markets to maximize the value of accelerated approval

A 2017-published article in *Health policy and planning* by Geng and team scrutinized the cost structure of routine infant immunization services in low- and middle-income scenarios based on 316 immunization sites sourced from the EPIC studies (a multi-country study on routine immunization costing and financing) in Benin, Ghana, Honduras, Moldova, Uganda, Zambia. 77–93% of the costs were accounted by site-level costs with 14–69% amounting to labor and 13–69% of site-level costs accounted to vaccine costs. 48–78% of site-level costs for these countries were accounted for by outreach and servicing. The sites with more efficiency documented lower labor costs. Further, the cost structure of immunization services demonstrated a conspicuous variation between countries and across sites in a country based on site characteristics. Such studies can facilitate the probing of these differences to augment vaccine immunization efficiencies.

Table 1.5 below illustrates the categorization of costs by the researchers (Geng F, Suharlim C, Brenzel L, Resch SC, Menzies NA © The Author 2017. Published by Oxford University Press in association with The London School of Hygiene and Tropical Medicine. *This is an Open Access article distributed under the terms of the Creative Commons Attribution License (http://creativecommons.org/licenses/by/4.0/), which permits unrestricted reuse, distribution, and reproduction in any medium, provided the original work is properly cited*):

Categorization	Category	Details
Budget category	Labour	Shared and immunization-specific personnel salary and volunteer labour estimated as the market value.
	Vaccine	Vaccines, including wastage and supplies, including syringes, diluent, safety boxes and other supplies used for administration of vaccines.
	Transport	Value of all the vehicles and modes of transport, maintaining vehicles and other transport for immunization-related activities and other immunization-related transport, including both facility-based and outreach services.
	Cold chain	All cold chain equipment used to store and transport vaccines, related energy cost and the cost of ice.
	Infrastructure	Building areas, utilities and communication, costs related to building overheads, other equipment, such as computers, printers, furniture, other medical equipment used for immunization-related activities and printing costs,related to immunization-related materials.
	Per diem	Any allowances paid to paid or volunteer workers for immunization-related activities.

Programmatic activity	Facility-based services	Time and resources spent on the act of administering the vaccine to children within the facility/compound and costs of vaccines delivered through facilities.
	Surveillance	Disease surveillance, following-up post-vaccination events and active cases of diseases that were prevented by vaccination, record keeping, HMIS, monitoring and evaluation.
	Programme management	Programme management, training and supervision.
	Outreach services	Time and resources spent for outreach and costs of vaccines delivered through outreach.
	Social mobilization	Time and resources spent mobilizing the community and households, and advocating for vaccination.
	Supply chain	Cold chain equipment used to store and transport vaccines, cold chain energy cost, the cost of ice, and time and resources spent on vaccine collection, distribution and storage.

The routine cost of immunization is depicted below for the countries assayed in terms of the budget category (Panel A) and programmatic activity (Panel B) (Geng F, Suharlim C, Brenzel L, Resch SC, Menzies NA © The Author 2017. Published by Oxford University Press in association with The London School of Hygiene and Tropical Medicine. *This is an Open Access article distributed under the terms of the Creative Commons Attribution License (http://creativecommons.org/licenses/by/4.0/), which permits unrestricted reuse, distribution, and reproduction in any medium, provided the original work is properly cited*):

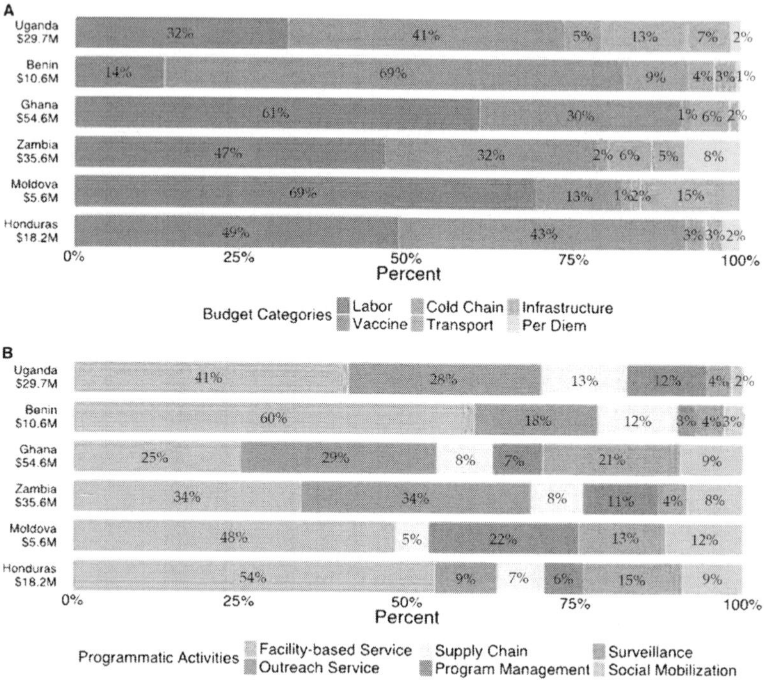

The costs of delivering vaccines in low- and middle-income countries (LMICs) were probed by Vaughan and team (2019) based on the following definition of immunization delivery costs (IDCs or operational costs) inclusive of "*(1) paid human resources, (2) volunteer human resources, (3) per diem and travel allowances, (4) cold chain equipment and their overheads (e.g. energy, maintenance, repairs), (5) vehicles, transport and fuel, (6) program management, (7) training and capacity building, (8) social mobilization and advocacy, (9) disease surveillance and activities related to adverse events following immunization (AEFI), (10) buildings, utilities, other overheads and shared costs, (11) vaccine supplies (e.g. safety boxes, diluents, reconstitution syringes), (12) waste management, (13) other supplies and recurrent costs, and (14) other non-vaccine costs*".

The geographic spread of IDCs are illustrated below in figure 1.4 with the EPI Costing and Financing Project (EPIC) amounting to 45% (Vaughan K, Ozaltin A, Mallow M, Moi F, Wilkason C, Stone J, Brenzel L © 2019 The

Author(s): *This is an open access article under the CC BY license: http://creativecommons.org/licenses/by/4.0/*):

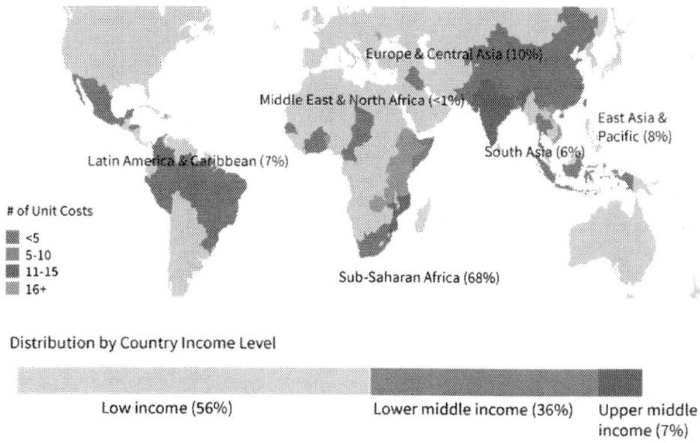

Country	Unit Costs (n)	Country	Unit Costs (n)
Uganda	89	China	6
Benin	42	Brazil	5
Moldova	42	Bhutan	5
Tanzania	35	Kenya	5
Rwanda	24	Cameroon	4
Zambia	23	Colombia	4
Vietnam	19	The Gambia	4
Ethiopia	14	Peru	4
India	14	Burkina Faso	2
Mozambique	11	Chad	2
Ghana	8	Indonesia	2
Cote d'Ivoire	7	Iraq	2
Honduras	7	Mexico	2
South Africa	6	Togo	2
Bangladesh	6	Pakistan	1
Haiti	6	Senegal	1
Thailand	6		

Analyses of 410 immunization delivery unit costs from more than 15,000 published and unpublished resources of 2005 to 2018 revealed higher ranges than what is pitched: all-inclusive incremental costs per dose of \$0.16–\$2 for single, newly introduced vaccines and economic costs of \$0.75–\$9.45 per dose for four to eight vaccine-schedules administered to children below one year of age.

Figure 1.5 below illustrates IDCs for ***TOP***: Single vaccines (Vaughan K, Ozaltin A, Mallow M, Moi F, Wilkason C, Stone J, Brenzel L © 2019 The Author(s): *This is an open access article under the CC BY license: http://creativecommons.org/licenses/by/4.0/*):

Key: BCG = Bacillus Calmette-Guérin; DTP = Diphtheria and tetanus toxoids and whole-cell pertussis vaccine, pediatric formulation; HepB = Hepatitis B; Hib = Haemophilus influenzae type b; HPV = Human Papillomavirus; JE = Japanese Encephalitis; OCV = Oral Cholera Vaccine; OPV = Oral Polio Vaccine; PCV = Pneumococcal Conjugate Vaccine (7-, 10-, or 13-valent); TT = Tetanus Toxoid

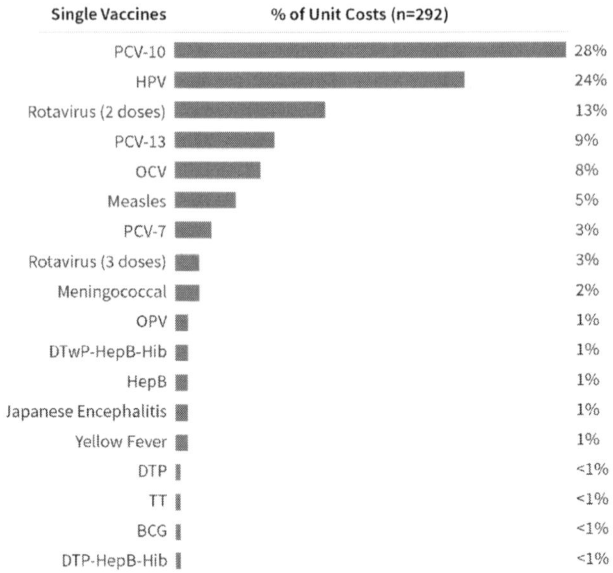

BOTTOM: Delivery costs (Vaughan K, Ozaltin A, Mallow M, Moi F, Wilkason C, Stone J, Brenzel L © 2019 The Author(s): *This is an open access article under the CC BY license: http://creativecommons.org/licenses/by/4.0/*):

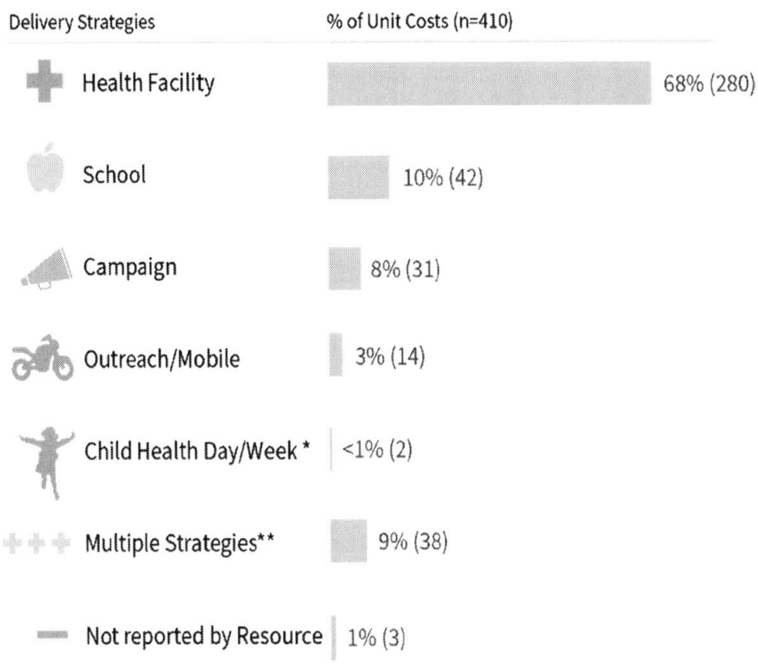

1.4 IMPACT

The impact of vaccines across three realms is depicted in figure 1.6 below (© 2020 Rodrigues and Plotkin. *This is an open-access article distributed under the terms of the Creative Commons Attribution License (CC BY). The use, distribution or reproduction in other forums is permitted, provided the original author(s) and the copyright owner(s) are credited and that the original publication in this journal is cited, in accordance with accepted academic practice. No use, distribution or reproduction is permitted which does not comply with these terms*):

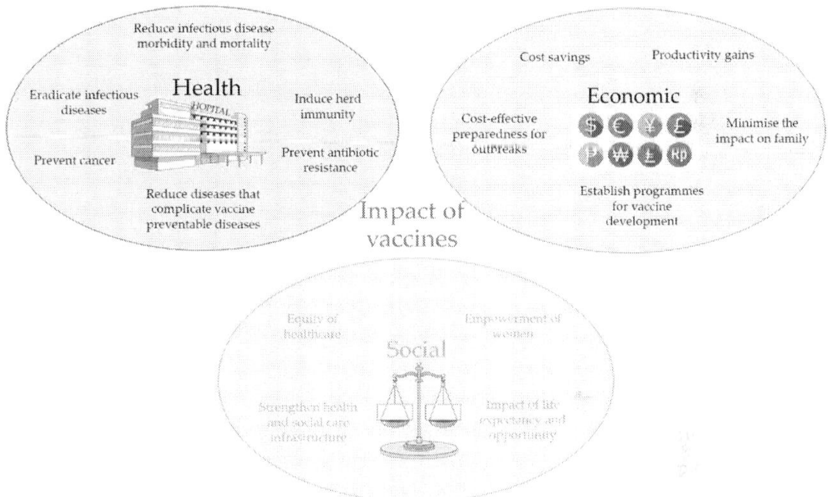

"While the vaccine discovery was progressive, the joy I felt at the prospect before me of being the instrument destined to take away from the world one of its greatest calamities [smallpox], blended with the fond hope of enjoying independence and domestic peace and happiness, was often so excessive that, in pursuing my favourite subject among the meadows, I have sometimes found myself in a kind of reverie"
— *Edward Jenner*
John Baron, The Life of Dr. Jenner (1827), 140.

1.5 REFERENCES

1. Lahariya C. (2014). A brief history of vaccines & vaccination in India. *The Indian journal of medical research*, *139*(4), 491–511.
2. Hajj Hussein I, Chams N, Chams S, El Sayegh S, Badran R, Raad M, Gerges-Geagea A, Leone A and Jurjus A (2015) Vaccines Through Centuries: Major Cornerstones of Global Health. *Front. Public Health* 3:269. doi: 10.3389/fpubh.2015.00269
3. Rodrigues CMC and Plotkin SA (2020) Impact of Vaccines; Health, Economic and Social Perspectives. *Front. Microbiol.* 11:1526. doi: 10.3389/fmicb.2020.01526
4. Bechini, A., Bonanni, P., Zanella, B., Di Pisa, G., Moscadelli, A., Paoli, S., Ancillotti, L., Bonito, B., & Boccalini, S. (2021). Vaccine Production Process: How Much Does the General Population Know about This Topic? A Web-Based Survey. *Vaccines*, *9*(6), 564. https://doi.org/10.3390/vaccines9060564
5. Plotkin, S., Robinson, J. M., Cunningham, G., Iqbal, R., & Larsen, S. (2017). The complexity and cost of vaccine manufacturing - An overview. *Vaccine*, *35*(33), 4064–4071. https://doi.org/10.1016/j.vaccine.2017.06.003
6. Geng, F., Suharlim, C., Brenzel, L., Resch, S. C., & Menzies, N. A. (2017). The cost structure of routine infant immunization services: a systematic analysis of six countries. *Health policy and planning*, *32*(8), 1174–1184. https://doi.org/10.1093/heapol/czx067
7. Vaughan, K., Ozaltin, A., Mallow, M., Moi, F., Wilkason, C., Stone, J., & Brenzel, L. (2019). The costs of delivering vaccines in low- and middle-income countries: Findings from a systematic review. *Vaccine: X*, *2*, 100034. https://doi.org/10.1016/j.jvacx.2019.100034

CHAPTER 2

Vaccines in Pandemics

"We do not need magic to change the world, we carry all the power we need inside ourselves already: we have the power to imagine better."

- J.K. Rowling

Contents

2.1 COVID-19 AND ROUTINE IMMUNIZATION

Among the various aspects of life and economy being hit by the COVID-19 pandemic, the vital framework of routine immunization (RI) was not spared too. This becomes pertinent given that outbreaks of vaccine-preventable diseases (VPDs) can further target the health system that is already reeling from COVID-19. For instance, post-Ebola, the year 2015 saw an outbreak of measles in Guinea while the Democratic Republic of Congo (DRC) saw a measles outbreak in 2019 causing the mortality of twice of those who succumbed to Ebola. Both these were attributed to impaired measles vaccinations due to Ebola (Mansour et al, 2021).

2020-published research in *Frontiers in pediatrics* by Buonsenso and team scrutinized the number of children (<5 years old) vaccinated for the most ubiquitous diseases at the Kent Community Health Post between March 1, 2020, to April 26, 2020, and compared these results to the same period of the earlier year (March 1, 2019, to April 26, 2019). Preliminary analyses revealed that the vaccination dipped by 50–80% with a variation on the vaccine type in Sierra Leone, West Africa.

Table 2.1 below documents the dip in vaccination in children <5 years of age in Sierra Leone, West Africa in the COVID-19 lockdown (© 2020 Buonsenso, Cinicola, Kallon and Iodice. *This is an open-access article distributed under the terms of the Creative Commons Attribution License (CC BY). The use, distribution or reproduction in other forums is permitted, provided the original author(s) and the copyright owner(s) are credited and that the original publication in this journal is cited, in accordance with accepted academic practice. No use, distribution or reproduction is permitted which does not comply with these terms*):

Key: BCG, Bacille Calmette Guerin; OPV, oral polio vaccine; PENTA, diphtheria, pertussis, tetanus, hepatitis B, and hemophilus; PCV, pneumococcal vaccine; ROTA, rotavirus; TT, tetanus toxoid vaccine; IPTi, intermittent preventive treatment in infants. The number associated with the abbreviation denotes initial and booster vaccinations. Data are collected as absolute numbers.

Vaccination	Children under 5 years of age			
	01/03/2019 to 26/04/2019	01/03/2020 to 26/04/2020	Change (%)	*P*-value
BCG	36	17	−52.7	$p < 0.0005$
OPV0	36	17	−52.7	$p < 0.0005$
OPV1	58	17	−70.7	$p < 0.0005$
PENTA1	58	17	−70.7	$p < 0.0005$
PCV1	58	17	−70.7	$p < 0.0005$
ROTA1	58	17	−70.7	$p < 0.0005$
OPV2	71	15	−78.9	$p < 0.0005$
PENTA2	71	15	−78.9	$p < 0.0005$
PCV2	71	15	−78.9	$p < 0.0005$
ROTA2	71	15	−78.9	$p < 0.0005$
IPTI1	49	15	−69.4	$p < 0.0005$
OPV3	67	15	−77.6	$p < 0.0005$
PENTA3	67	15	−77.6	$p < 0.0005$
PCV3	67	15	−77.6	$p < 0.0005$
IPTI2	44	15	−65.9	$p < 0.0005$
IPV	67	15	−77.6	$p < 0.0005$
IPTI3	45	22	−51.1	$p < 0.0005$
Measles	64	22	−65.6	$p < 0.0005$
Yellow fever	64	22	−65.6	$p < 0.0005$
Measles 2nd	49	8	−83.7	$p < 0.0005$

Another study by Mansour et al (2021) probed the changes in RI services in private and public sectors between October 2019 and April 2020 in Lebanon as opposed to the previous year. Nationally, a dip of 31% emerged for the utilization of vaccination services with a fall in half for private immunization.

Figure 2.1 below illustrates the dip in RI across ***TOP***: for measles, oral poliovirus 3 (OPV), pentavalent vaccine, inactivated poliovirus vaccine (IPV), diphtheria, tetanus, pertussis 3 (DTP), measles, mumps, rubella (MMR) and pneumococcal conjugate vaccine (PCV13) among private practitioners (© 2021 Mansour et al. *This is an open access article distributed under the terms of the Creative Commons Attribution License, which permits unrestricted use, distribution, and reproduction in any medium, provided the original author and source are credited*):

	Number	Percentage
Physicians witnessing a decrease in the utilization of routine immunization services between October 2019 and April 2020	267	77.4
Physicians witnessing a decrease in the utilization of the following vaccines between October 2019 and April 2020 (N = 267)		
OPV	132	49.4
IPV	186	69.7
Measles	189	70.8
DPT	195	73.0
Hepatitis B	171	64.0
MMR	189	70.8
PCV	219	82.0
Hepatitis A	213	79.8
Period with the most significant decrease rate in immunization services provision in the private sector (N = 267)		
October- November 2019	6	2.2
December 2019- January 2020	21	7.9
February-April 2020	234	87.6
Do not know	6	2.2

https://doi.org/10.1371/journal.pone.0246951.t001

and ***BOTTOM***: dip in RI in public systems (© 2021 Mansour et al. *This is an open access article distributed under the terms of the Creative Commons Attribution License, which permits unrestricted use, distribution, and reproduction in any medium, provided the original author and source are credited*):

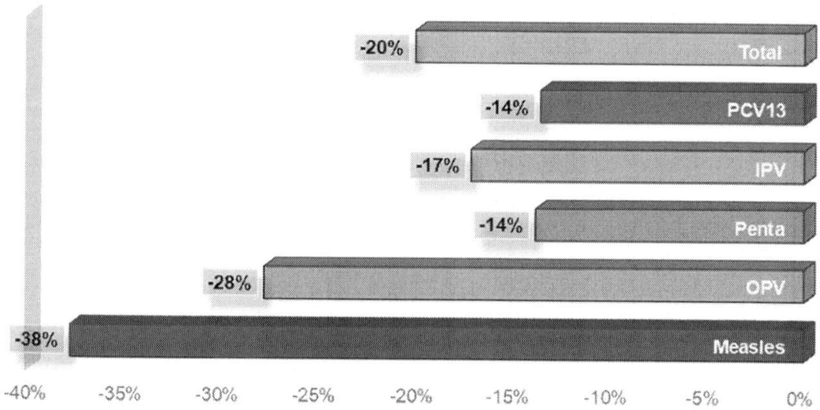

The impact of the COVID-19 pandemic curfew in Saudi Arabia on delayed vaccination in children below two years of age was scored by Baghdadi and team (2021). A cross-sectional study between March 4–July 6, 2020, of 577 caregivers (90.8% were mothers) revealed that the prevalence of intentional vaccination delay was 37%. The major cause emerged as fear of the epidemic. Interestingly, awareness of the home vaccination service prevailed in 82% of parents with 17.5% of them availing of the service during the lockdown. The reasons attributed to not looking at this option were no awareness of this system (17.9%), non-requirement (16.3%), no provision from their hospital (13.8%), clinic visit preference (12%), cost (7.4%), home visit vaccination doubts (7.1%), and fear of COVID-19 infection (7.1%).

The fear level of COVID-19 to delay child vaccination as computed by the researchers is presented below in figure 2.2 (© 2021 Baghdadi, Younis, Al Suwaidan, Hassounah and Al Khalifah. *This is an open-access article distributed under the terms of the Creative Commons Attribution License (CC BY). The use, distribution or reproduction in other forums is permitted, provided the original author(s) and the copyright owner(s) are credited and that the original publication in this journal is cited, in accordance with accepted academic practice. No use, distribution or reproduction is permitted which does not comply with these terms*):

Responses	*N* (%)	Mean	SD	*P*-value
Never thought of delaying vaccination	235 (40.7)	22.21	5.674	<0.001*
I thought about delaying the vaccination, but still vaccinated at the scheduled time	173 (30)	22.33	4.869	
I decided to delay the vaccination, but I could not find a reason to do so	14 (2.4)	18.43	4.052	
I delayed the vaccination	155 (26.9)	21.55	5.899	

*SD, standard deviation; *p-value is significant at 0.05.*

Reasons	Unadjusted model			**Adjusted model		
	#OR	*P*-value	95% CI	#OR	*P*-value	95% CI
Afraid that my child will get infected with COVID-19	3.56	>0.001*	2.44–5.19	4.14	>0.001*	2.74–6.27
Afraid of getting infected with COVID-19	2.86	>0.001*	1.97–4.16	3.34	>0.001*	2.19–5.09
Concerns of vaccine safety and effectiveness	4.19	>0.001*	2.09–8.41	4.68	>0.001*	2.14–10.27
I was traveling within Saudi Arabia	4.99	0.009*	1.48–16.8	4.52	0.02*	1.26–16.22
No transportation or access to a health-care facility	3.31	0.029*	1.13–9.7	3.04	0.059	0.96–9.61
Curfew timings were tight	3.43	>0.001*	2.15–5.47	4.57	>0.001*	2.73–7.66
Other	1.15	0.61	0.68–1.95	1.23	0.48	0.70–2.16

*CI, confidence interval; OR, odds ratio; #Reference group= did not delay the vaccine; *p-value is significant at 0.05; **Model adjusted for caregivers' and children's ages, chronic diseases, history of receiving scheduled vaccination for the older siblings, and COVID-19 factors: exposure to COVID-19, admission to the intensive care unit, and fear level of COVID-19.*

These studies outline the necessity for augmented RI and boosted awareness across healthcare systems and the public for sustaining immunization despite pandemics following the necessary precautions to circumvent vaccine-preventable disease outbreaks and add to the already strained nerves and healthcare systems.

2.2 THE PHASES: REGULAR VS. PANDEMIC

Usually, the process of vaccine development entails the initial **exploratory stage** where antigens are identified employing modeling and research. Assessing the immunogenicity and safety employing tissue or cell cultures ensues that constitute **pre-clinical studies**. Subsequent to the corroboration of safety in animals, three phases of human clinical trials are then conducted. These encompass the first stage of testing the vaccine in immunocompetent and healthy people to probe the safety, dosing and immune response (**Phase 1 – Safety**) followed by scrutinizing these same factors in albeit more people categorized by demographics (**Phase 2 - Expanded Safety**).

The efficacy is then assessed in **Phase 3 – Efficacy** with vaccine efficacy (VE) defined as the percentage of diminishing of the rate of disease incidence in vaccinated groups as opposed to the placebo. Subsequent to this is reviewing the results by regulatory bodies like the European Medicines Agency in EU or the Food and Drug Administration (FDA) of USA. Post review and approval, the vaccine is manufactured and marketed for use in the society where its large-scale effects are documented.

(Sharma et al, 2020)

Figure 2.3 below depicts the stages of vaccine development in a traditional scenario with the symbol representative of the number of human subjects in trials (© 2020 Sharma, Sultan, Ding and Triggle. *This is an open-access article distributed under the terms of the* ***Creative Commons Attribution License (CC BY)****. The use, distribution or reproduction in other forums is permitted, provided the original author(s) and the copyright owner(s) are credited and that the original publication in this journal is cited, in accordance with accepted academic practice. No use, distribution or reproduction is permitted which does not comply with these terms*):

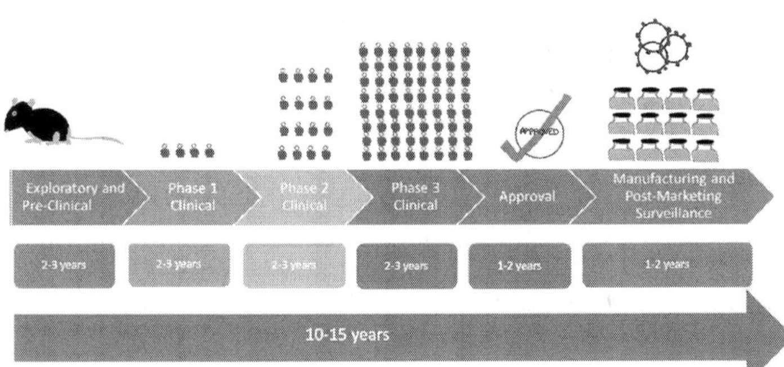

In the event of pandemics such as the COVID-19 pandemic, phases are combined to expedite this process. As an illustration, phases 1 and 2 may be combined or the entire approval system may not be followed. Figure 2.4 below depicts the scheme of expedited vaccine development for pandemics with the symbol representative of the number of human subjects in trials (© 2020 Sharma, Sultan, Ding and Triggle. *This is an open-access article distributed under the terms of the* ***Creative Commons Attribution License (CC BY)****. The use, distribution or reproduction in other forums is permitted, provided the original author(s) and the copyright owner(s) are credited and that the original publication in this journal is cited, in accordance with accepted academic practice. No use, distribution or reproduction is permitted which does not comply with these terms*):

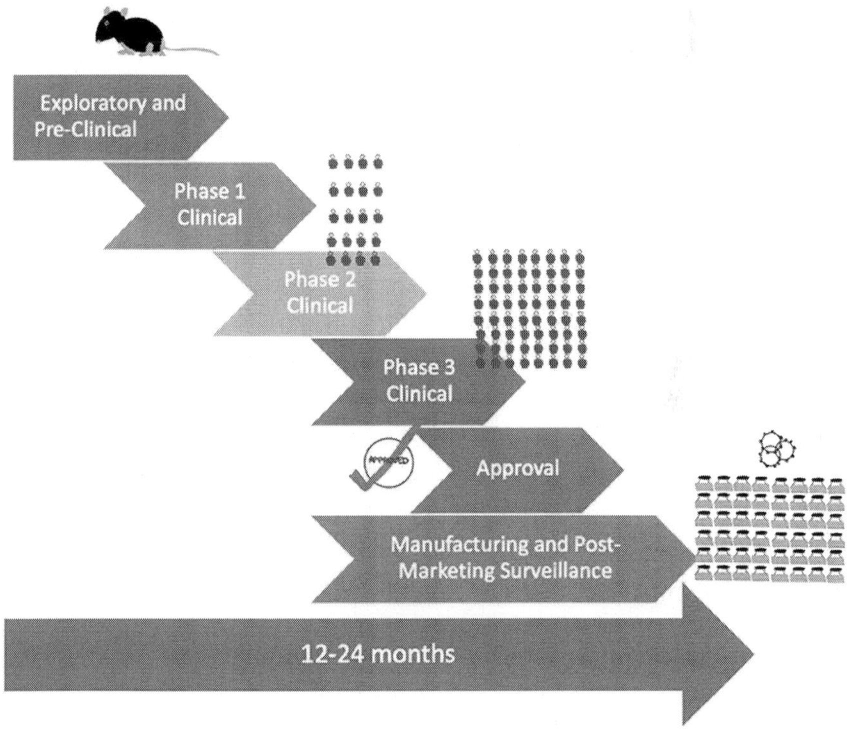

2.3 AN EXAMPLE OF MODELING

The initial spread of a pandemic may not immediately see a vaccine but the mitigation of the disease may be achieved once a vaccine is available. A second dose of vaccines becomes pertinent given the inadequate immune response in some people and the emergence of genetic variants of a pathogen to augment protection. Policymakers are expected to consider a slew of factors for efficient vaccination coverage ranging from supply to logistics to timelines. As an example, a model was presented in *Mathematical Biosciences and Engineering* by Bowman, Arino and Moghadas (2011) to explore the administration regimen of vaccines based on pandemic influenza transmission and its factors viz. distribution rate of vaccines, vaccine number, efficacy of the first/second dose and disease severity. The modeling system of single-dose and two-dose vaccination programs highlighted the start time of vaccination and the second dose efficiency as vital factors. In the event of limited vaccine supply, the epidemic can be addressed by a two-dose strategy provided the vaccination is commenced early say within 30 days post-onset. This point necessitates protection efficacy of the second dose. The benefits are diminished if there is a delay. A single-dose confers advantages if there is an issue with supply with the biggest advantage being the vaccination of twice the number of people. A reserve is warranted between dosages to facilitate the second dose. Another factor of the two-dose vaccination is dropping out on account of whims or infection.

Figure 2.5 below illustrates a model developed looking at vaccination regimens based on modeling (Bowman, Arino and Moghadas © 2011 the Author(s), licensee AIMS Press. *This is an open access article distributed under the terms of the Creative Commons Attribution License: http://creativecommons.org/licenses/by/4.0*):

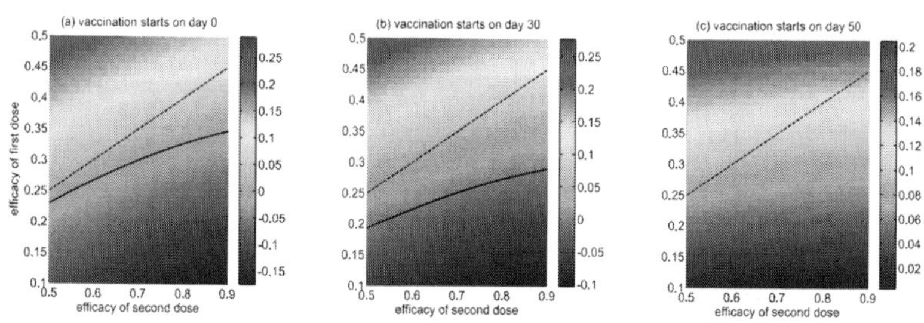

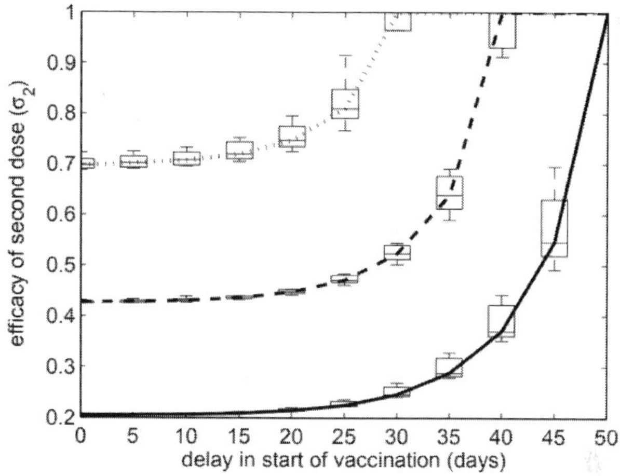

parameter	description	value (range)
β	baseline transmission rate	variable
σ_1	efficacy of first dose of vaccine	0.2 (0.1 – 0.5)
σ_2	efficacy of second dose of vaccine	0.8 (0.5 – 0.9)
γ	fraction of population vaccinated per day	0.01 (0.01 – 0.04)
δ_1	level of susceptibility after one dose of vaccine	$1 - \sigma_1$
δ_2	level of susceptibility after two doses of vaccine	$1 - \sigma_2$
p	probability of developing symptoms without vaccination	0.67 (0.5 – 0.75)
p_1	probability of developing symptoms after one dose of vaccine	$p\delta_1$
p_2	probability of developing symptoms after two doses of vaccine	$p\delta_2$
δ_A	relative transmissibility of asymptomatic infection	0.142
μ_A	recovery rate of asymptomatic infection	1/4.1 day^{-1}
μ	recovery rate of symptomatic infection	1/4.1 day^{-1}
a^*	time between first and second dose of vaccine	30 days
$\mathcal{R}_0$	basic Reproduction number	1.6

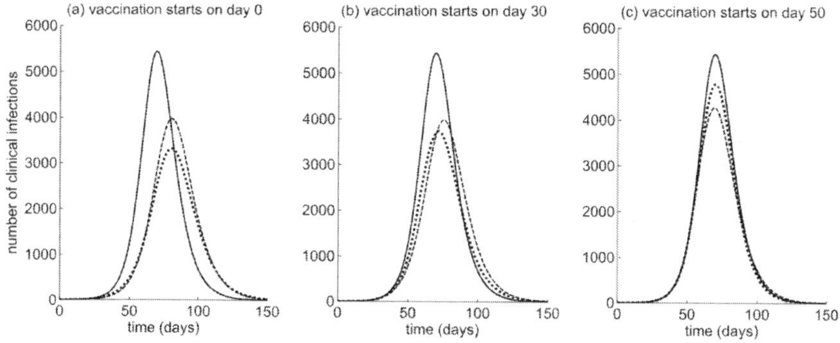

2.4 CASE STUDY OF COVID-19: PLATFORMS AND ETHICS

The quest for COVID-19 vaccines has entailed a slew of platforms ranging from inactivated vaccines to non-replicating vectors and RNA/DNA vaccines. Table 2.2 below gives a snapshot of the various systems and attributes (© 2020 Sharma, Sultan, Ding and Triggle. *This is an open-access article distributed under the terms of the* ***Creative Commons Attribution License (CC BY)****. The use, distribution or reproduction in other forums is permitted, provided the original author(s) and the copyright owner(s) are credited and that the original publication in this journal is cited, in accordance with accepted academic practice. No use, distribution or reproduction is permitted which does not comply with these terms*):

Platform	Candidates in clinical trials and phase[a]	Type of candidate vaccine	Target antigen	Single/ multiple dose	Speed[b]	Immune response	Advantages	Disadvantages
DNA	Inovio Pharmaceuticals - phase 1/2	DNA plasmid vaccine with electroporation	Spike protein	Multiple	Fast	Both humoral and cellular	-Electroporation generates a robust immune response -Made using genetic sequence and does not need to be cultured	-Although deemed to be safe, electroporation can be complicated and potentially problematic. -No DNA based vaccine has been previously produced
RNA	Moderna/NIAID - phase 3	Lipid nanoparticle [LNP]-encapsulated mRNA	Spike protein	Multiple	Fast	Both humoral and cellular	-Made using genetic sequence and does not need to be cultured	-LNP is temperature sensitive -Ability to manufacture large scale unknown -No RNA based vaccine has been produced before
	BioNTech/Fosun Pharma/Pfizer - phase 3	3 LNP-mRNAs	Spike protein					
Non-replicating viral vector	AstraZeneca/ University of Oxford - phase 3	AZD1222	Spike protein	Single	Medium	Both humoral and cellular	-Can be manufactured large scale -Safe and effective immunologically as shown with Ebola	-Pre-existing immunity could hamper clinical use and reduce immune response
	CanSino Biological Inc./Beijing Institute of Biotechnology - phase 2	Adenovirus type 5 vector	Spike protein					
Inactivated	Wuhan Institute of Biological Products/Sinopharm - phase 3	Inactivated	Whole virus	Multiple	Medium	Mostly humoral	-Pathogen is killed and hence, no risk of reversion	-Risk of vaccine-enhanced disease -Usually produce a weak immune response
	Beijing Institute of Biological Products/Sinopharm - phase 3		Whole virus					
	Sinovac - phase 3	Inactivated + aluminum adjuvant	Whole virus			Mostly humoral - aluminum adjuvant enhances response more robust		

[a]*Phase for vaccine development is as of August 20th, 2020.* [b]*Speed refers to how quickly a vaccine candidate is able to progress through different stages of vaccine development considering the nature of the platform being utilized. Adapted from the WHO COVID-19 Vaccine R&D Landscape (28) and others (29–31). Candidates listed in red are part of Operation Warp Speed to accelerate vaccine development and production.*

Figure 2.6 below summarizes the scheme of a few platforms for COVID-19 vaccine development research (Li *and* Li, 2021: *Open Access: This article is licensed under a Creative Commons Attribution 4.0 International License, which permits use, sharing, adaptation, distribution and reproduction in any medium or format, as long as you give appropriate credit to the original author(s) and the source, provide a link to the Creative Commons licence, and indicate if changes were made. The images or other third party material in this article are included in the article's Creative Commons licence, unless indicated otherwise in a credit line to the material.*

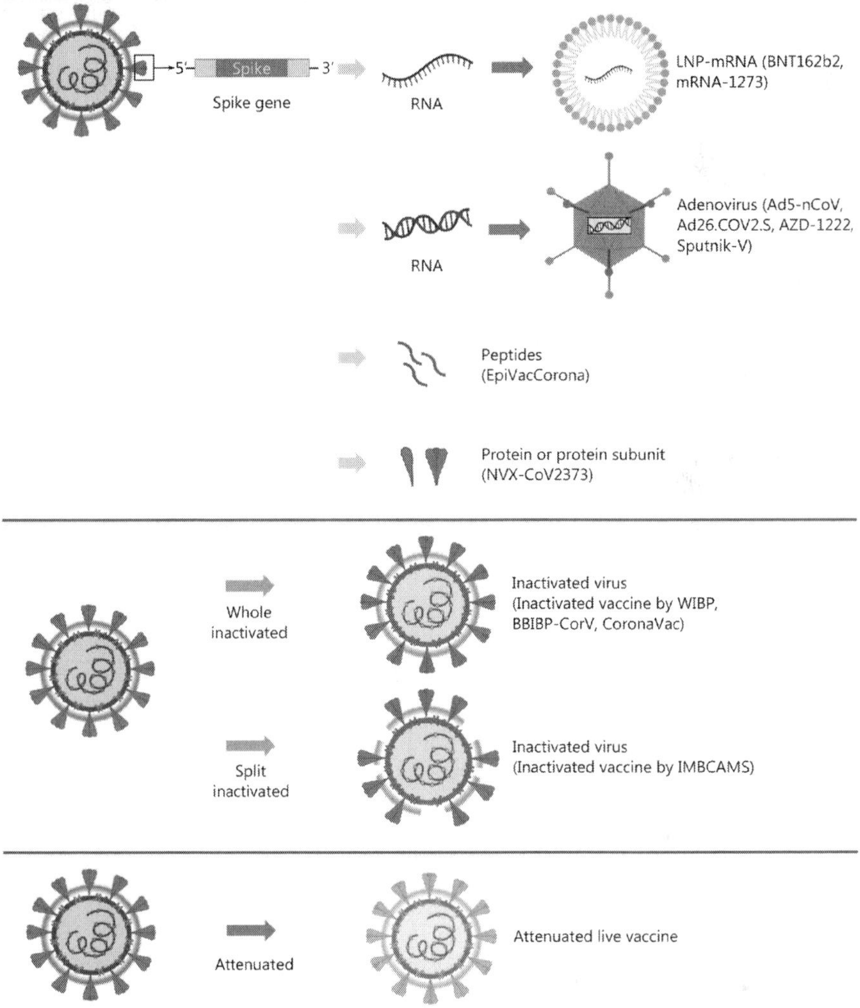

The urgency of a vaccine for the pandemic has resulted in expediting the process with the *Operation Warp Speed* program of the White House as an instance of hastening the development. A point emerges here as to the possibility of a sub-optimal vaccine in the event of such fast-tracking. It is warranted that complete transparency in all stages of vaccine development is available to the public to maintain trust. The post-marketing surveillance emerges as critical here given that the fact remains to be checked whether clinical trials were inclusive of age, comorbidity and ethnicity encompassing vital demographic factors.

Notwithstanding these aspects, another question emerges as to the ethics of fair allocation given that man-made boundaries and borders divide humanity. The maintenance of adequate doses to the elderly, health workers, immunocompromised and all socioeconomic strata is necessitated.

"We all should know that diversity makes for a rich tapestry, and we must understand that all the threads of the tapestry are equal in value no matter what their color."

-Maya Angelou

2.5 REFERENCES

1. Mansour Z, Arab J, Said R, Rady A, Hamadeh R, Gerbaka B, et al. (2021) Impact of COVID-19 pandemic on the utilization of routine immunization services in Lebanon. PLoS ONE 16(2): e0246951. https://doi.org/10.1371/journal.pone.0246951
2. Buonsenso, D., Cinicola, B., Kallon, M. N., & Iodice, F. (2020). Child Healthcare and Immunizations in Sub-Saharan Africa During the COVID-19 Pandemic. *Frontiers in pediatrics*, *8*, 517. https://doi.org/10.3389/fped.2020.00517
3. Baghdadi LR, Younis A, Al Suwaidan HI, Hassounah MM and Al Khalifah R (2021) Impact of the COVID-19 Pandemic Lockdown on Routine Childhood Immunization: A Saudi Nationwide Cross-Sectional Study. *Front. Pediatr.* 9:692877. doi: 10.3389/fped.2021.692877
4. Sharma O, Sultan AA, Ding H and Triggle CR (2020) A Review of the Progress and Challenges of Developing a Vaccine for COVID-19. *Front. Immunol.* 11:585354. doi: 10.3389/fimmu.2020.585354
5. Christopher S. Bowman, Julien Arino, S.M. Moghadas. Evaluation of vaccination strategies during pandemic outbreaks[J]. Mathematical Biosciences and Engineering, 2011, 8(1): 113-122. doi: 10.3934/mbe.2011.8.113
6. Li, DD., Li, QH. SARS-CoV-2: vaccines in the pandemic era. *Military Med Res* 8, 1 (2021). https://doi.org/10.1186/s40779-020-00296-y

CHAPTER

3 Vaccine Platforms: Old and New

Contents

3.1 PLATFORMS

A slew of vaccine technologies are being currently employed and under study ranging from viral vectors to attenuated pathogens to recombinant vaccines and nucleic acid-based vaccines. Several vaccines developed addressing a few major pandemics are depicted below in figure 3.1 to comprehend the time scale (© 2018 Rauch, Jasny, Schmidt and Petsch. *This is an open-access article distributed under the terms of the Creative Commons Attribution License (CC BY). The use, distribution or reproduction in other forums is permitted, provided the original author(s) and the copyright owner(s) are credited and that the original publication in this journal is cited, in accordance with accepted academic practice. No use, distribution or reproduction is permitted which does not comply with these terms*):

*1983 is indicated as the year of discovering HIV was discovered; **2003 is shown as increasing infections by H5N1(the first human infection was registered in 1997); Ad4, 5, 26, human adenovirus type 4, 5 or 26; ChAd, chimpanzee adenovirus; H1N1pdm09, influenza H1N1 2009 "swine flu"; H10N8, influenza H10N8; DNA, deoxyribonucleic acid-based vaccine, MVA, modified vaccinia Ankara; RNA, ribonucleic acid-based vaccine; VSV, vesicular stomatitis virus; HPIV3, human parainfluenza virus type 3; MV, measles virus.

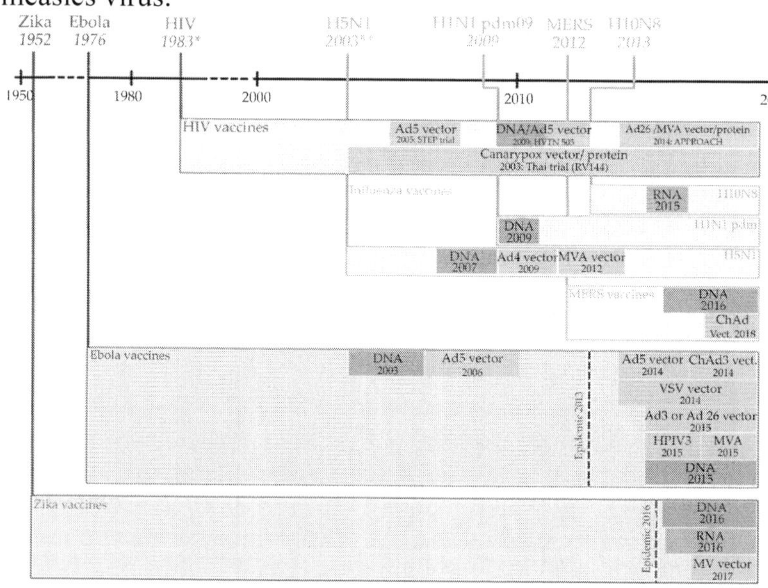

Table 3.1 below documents a snapshot ofvarious vaccine platforms (Li et al: 2020: *Open Access: This article is licensed under a Creative Commons Attribution 4.0 International License, which permits use, sharing,*

Vaccine platform	**Advantages**	**Disadvantages**	**Clinically approved examples**
Whole inactivated virus vaccine	Stronger immune response; Safer than the live attenuated virus	Potential epitope alteration by inactivation process	Typhoid, Cholera, Hepatitis A virus, Plague, Rabies, Influenza, Polio (Salk)
Live attenuated virus vaccine	Stronger immune response; Preservation of native antigen; Mimicking natural infection	Risk of residual virulence, especially for immunocompromised people	Measles, Mumps, Polio (Sabin), Rota virus, Yellow Fever, Bacillus Calmette–Guérin (BCG), Rubella, Varicella
Viral vector vaccine	Stronger immune response; Preservation of native antigen; Mimicking natural infection	More complicated manufacturing process; Risk of genomic integration; Response dampened by pre-existing immunity against the vector	Ebola virus
Subunit vaccine	Safe and well-tolerated	Lower immunogenicity; Requirement of adjuvant or conjugate to increase immunogenicity	Pertussis, Influenza, *Streptococcus pneumoniae, Haemophilusinfluenzae type b*
Viral-like particle vaccine	Safe and well-tolerated; mimicking native virus conformation	Lower immunogenicity; More complicated manufacturing process	Hepatitis B virus, Human Papillomavirus

DNA vaccine	Safe and well-tolerated; Stable under room temperature; Highly adaptable to new pathogens; Native antigen expression	Lower immunogenicity; Difficult administration route; Risk of genomic integration	NA
RNA vaccine	Safe and well-tolerated; Highly adaptable to new pathogens; Native antigen expression	Lower immunogenicity; Requirement of low temperature storage and transportation; Potential risk of RNA-induced interferon response	NA

3.2 NUCLEIC ACID VACCINE RESEARCH IN PANDEMICS

DNA vaccines entail the insertion of the desired antigen in a relevant plasmid vector that facilitates its replication and propagation. RNA vaccines encompass***non-replicating*** mRNA where 5′ and 3′ untranslated regions (UTRs) flank the desired antigen circumventing any additional expression product and *self-amplifying mRNA vaccines* with the target antigen sequencing replacing the structural protein of the alphavirus genome.

In pandemic scenarios, table 3.2 below summarizes trials probing ***TOP***: DNA vaccines (© 2018 Rauch, Jasny, Schmidt and Petsch. *This is an open-access article distributed under the terms of the Creative Commons Attribution License (CC BY). The use, distribution or reproduction in other forums is permitted, provided the original author(s) and the copyright owner(s) are credited and that the original publication in this journal is cited, in accordance with accepted academic practice. No use, distribution or reproduction is permitted which does not comply with these terms*):

Study start	N	Vaccine and delivery	Outcome
NCT00072605 Oct 2003	27	**EBOLA** **DNA,** trivalent; NF inj.dev. IM 2–8 mg in week 0, 4, and 8 Antigens: - GPΔTM EBOV - GPΔTM SUDV - NP	**Phase I** **Safety:** Acceptable safety profile **Immunogenicity:** - Specific antibody response to at least 1/3 antigens in all subjects - Specific $CD8^+$ T cell responses in 30% (6/20) subjects. - No detectable virus neutralizing responses
NCT00605514 Jan 2008	20	**EBOLA** **DNA,** mono or bivalent; NF inj.dev. IM 4 mg in week 0, 4, 8; (32) Antigens: - GP MARV - GP EBOV + GP SUDV	**Phase I** **Safety:** Acceptable safety profile **Immunogenicity:** - Specific antibody responses against one of the GPs at week 12 in 80% of subjects - $CD8^+$ T cell responses in some of the subjects
NCT00997607 Feb 2010	108	**EBOLA** **DNA,** mono or bivalent; NF inj.dev. IM 4 mg in week 0, 4, 8 Antigens: - GP MARV - GP EBOV + GP SUDV	**Phase Ib** **Safety:** Acceptable safety profile **Immunogenicity:** Specific antibody responses in 30% (MARV) and 50% (EBOV or SUDV) of subjects Antibody titers to near baseline levels by w 44 post vaccination
NCT02464670 May 2015	240	**EBOLA** **DNA,** mono-, bi- or trivalent; IM or ID + EP in 2 or 3 doses 0.8–4 mg GP; 0.2–1 mg IL12 Antigen: - GP EBOV pre 2013 - and/or GP EBOV 2014 - and IL-12 in trivalent vaccine	**Phase I** **Safety:** Acceptable safety profile **Immunogenicity:** Specific antibody responses in 88% (50/57) (IM) and 95% (119/122) (ID) of participants
NCT00709800 and NCT00694213 Aug 2007	103	**INFLUENZA H5N1** **DNA,** mono- or trivalent; needle or NF inj.dev. IM 0.1–1 mg in week 0, 3 Antigen: - HA of A/Vietnam/1203/04 - HA + NP + M2	**Phase I** **Safety:** Acceptable safety profile **Immunogenicity:** - HI titers ≥40, in 47- 67% (HA only) and 0- 20% (HA + NP + M2) of participants, peak at d56 - H5-specific T cell responses in 75–100% (HA only) and 50–57% % (HA + NP + M2) of subjects - Responses against HA unaffected by injection method
NCT00973895 Aug 2009	20	**INFLUENZA H1N1** **DNA,** monovalent; NF inj.dev. IM 4 mg in week 0, 4, 8 Antigen: HA of A/California/04/2009	**Phase I** **Safety:** Acceptable safety profile **Immunogenicity:** - HI titers ≥40 in 30% (6/20) of DNA vaccinated subjects - DNA + licensed vaccine HI titers ≥40 in 72% (13/18) - T cell responses in 25% (5/20) of subjects
NCT02809443 (NCT02887482) July 2016 (Aug 2016)	40 (160)	**ZIKA** **DNA,** monovalent; ID + EP 1 or 2 mg in week 0, 4, 12 Antigen: Consensus prM-E; IgE SP; removed glycosylation site	**Phase I** **Safety:** Acceptable safety profile (NCT02809443) **Immunogenicity (preliminary results NCT02809443):** - VNTs in 62% of the participants (Vero cell assay) - Protection of 92% (103/112) of mice by passive serum transfer in challenge model (IFN α/β receptor knockout)
NCT02840487; NCT02996461 Aug 2016 Dec 2016	125	**ZIKA** **DNA,** monovalent; needle or NF inj.dev. IM 4 mg in 2 or 3 doses Antigen: - prM-E; JEV SP (VRC5283) - prM-E; JEV SP and S/TM (VRC5288)	**Phase I/Ib** **Safety:** Acceptable safety profile **Immunogenicity:** - Humoral and T cell responses induced - VNTs in 60%–100% of subjects 4 w after the final vaccination - Best responses in VRC5283: Antibody responses in 100% (14/14) of participants in NF inj. in split doses group; best VNT and T cell responses
NCT03110770 Mar 2017	2500	**ZIKA** **DNA,** monovalent; NF inj.dev. IM in 3 doses 4 mg or 8 mg in 2 or 4 inj. Antigen: - prM-E; JEV SP (VRC5283)	**Phase II** Results pending, estimated study completion date Jan 2020

This table exclusively lists clinical trials discussed in the text. EBOV, Ebola virus; EP, electroporation; GP, glycoprotein; GPΔTM, glycoprotein delta transmembrane domain; HA, hemagglutinin influenza; HI, hemagglutination inhibition; ID, intradermal; IL-12, interleukin 12; IM, intramuscular; JEV, Japanese encephalitis virus; M2, ion channel protein influenza; MARV, Marburg virus; N, number of study participants; NF inj.dev, needle free injection device; NP, nucleoprotein influenza; prM-E, preMembrane-Envelope; SUDV, Sudan virus; VNT, virus neutralization titer; SP, signal peptide; S/TM, stem and transmembrane regions.

BOTTOM: RNA vaccines (© 2018 Rauch, Jasny, Schmidt and Petsch. *This is an open-access article distributed under the terms of the Creative Commons Attribution License (CC BY). The use, distribution or reproduction in other forums is permitted, provided the original author(s) and the* *copyright owner(s) are credited and that the original publication in this journal is cited, in accordance with accepted academic practice. No use, distribution or reproduction is permitted which does not comply with these terms*):

Study start	*N*	Vaccine and delivery	Outcome
NCT03014089 Dec 2016	90	**ZIKA** **mRNA 1325,** modified nucleotides; LNP-formulated, Antigen: prM-E polyprotein	**Phase I/II** Results pending; estimated primary completion date in Sept 2018
NCT03076385 Dec 2015	201	**INFLUENZA H10N8** **mRNA 1851,** modified nucleotides; LNP-formulated, Antigen: HA of H10N8 A/Jiangxi-Donghu/346/2013	**Phase I** Interim results published for 100 μg IM (*N* = 23) vs. placebo (*N* = 8) **Safety:** acceptable safety profile **Immunogenicity:** - HI titers ≥40 in 100% (23/23) of subjects at day 43 - MN ≥20 in 87% (20/23) at day 43
NCT03345043 May 2016	156	**INFLUENZA H7N9** **mRNA 1440,** modified nucleotides; LNP-formulated, Antigen: HA of H7N9 A/Anhui/1/2013	**Phase I** Results pending; estimated primary completion date in Sept 2018
NCT03325075 Aug 2017	60	**CHIKUNGUNYA** **mRNA 1388,** modified nucleotides; LNP-formulated Antigen: structural polyprotein	**Phase I** Results pending; estimated primary completion date in Sept 2019

This table exclusively lists clinical trials discussed in the text; prM-E, preMembrane-Envelope; HA, Hemagglutinin; HI, hemagglutination inhibition; MN, microneutralization titers; N, number of study participants; IM, intramuscular; ID, intradermal; LNP, lipid nanoparticle.

Despite their documented efficacy, the quest is on to augment their effects such as intradermal gene gun administration, *in vivo* electroporation, adjuvants, epigenetics, promoter design, codon optimization, nanocarriers and lowered immunogenicity by short hairpin RNA (shRNA). When mRNA vaccines are taken into account the use of thermostable mRNA, using pure RNA, encapsulation approaches, the addition of caps, poly-A tails and modified nucleosides to augment stability.

Figure 3.2 below summarizes the various approaches being probed to augment the functioning of nucleic acid vaccines (© 2020 Brisse, Vrba, Kirk, Liang and Ly. *This is an open-access article distributed under the terms of the Creative Commons Attribution License (CC BY). The use, distribution or reproduction in other forums is permitted, provided the original author(s) and the copyright owner(s) are credited and that the original publication in this journal is cited, in accordance with accepted academic practice. No use, distribution or reproduction is permitted which does not comply with these terms*):

DNA vaccines

Method Name	Method Diagram	Description of Method	Results
Gene Gun	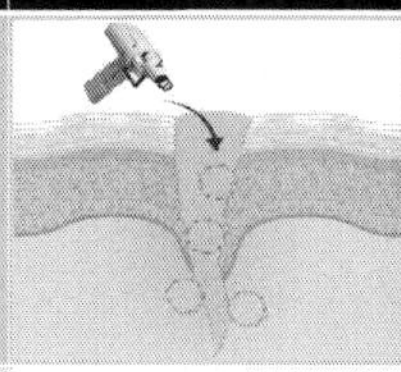	DNA plasmid is coated with gold particles and is taken up by cells via a blast of pressurized helium gas from a gene gun.	May increase humoral responses but is limited in allowable dosage
Electroporation	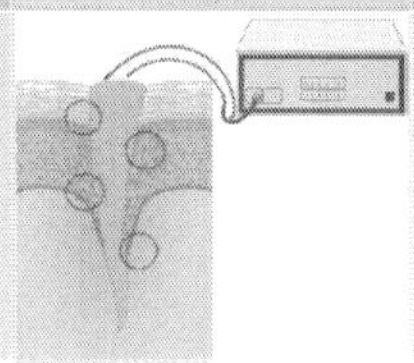	Plasmid is taken up by permeabilized skin followed by an electric current.	Maximizes DNA vaccine delivery efficiency. Currently in clinical trials

RNA Vaccines

Method Name	Method Diagram	Description of Method	Results
RNA Engineering	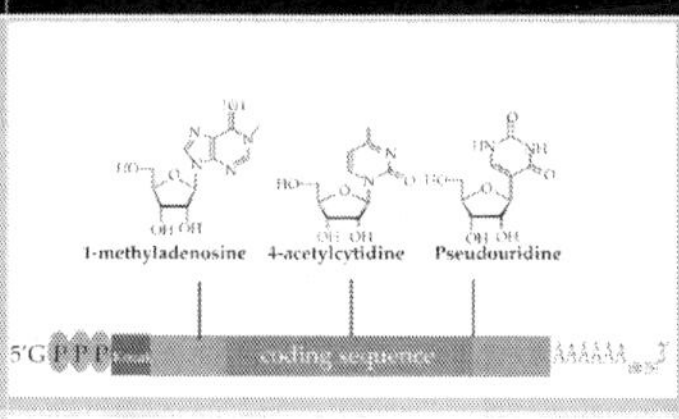	Adding 5' cap and Kozak sequences and 3' poly-A sequence to mimic structure of endogenous mRNA to avoid detection by innate immune sensors, and adding modified nucleosides to increase RNA stability	Widely used in vaccine design
Thermostable RNA	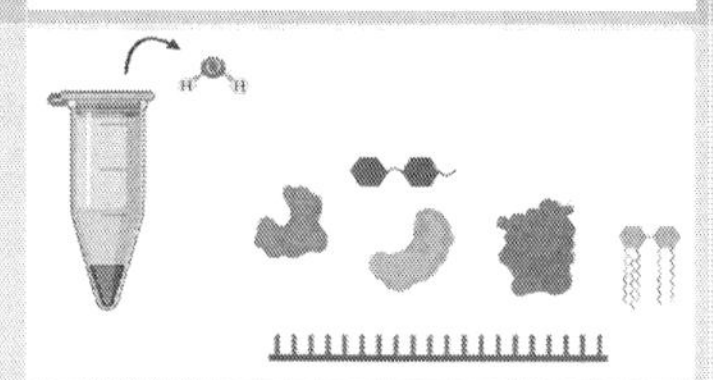	Freeze-drying RNA and incubation with various biomolecules increase RNA stability even at high temperature	Widely used in vaccine design
Dendritic Cell Vaccines	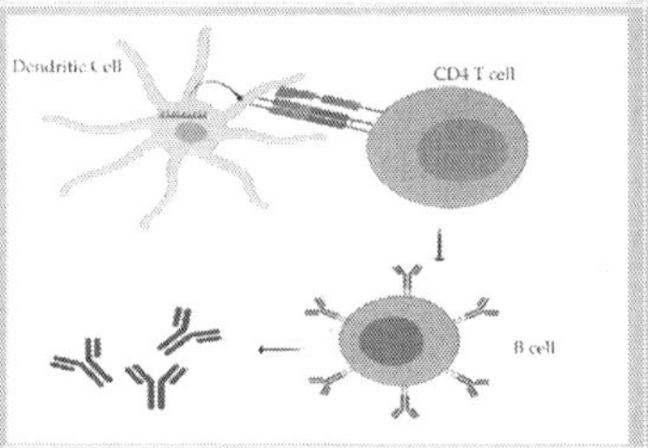	RNA is transfected into dendritic cells in-vitro, which can then activate CD4 T cells and stimulates antibody production in vivo.	Most explored for use as cancer vaccines- several in clinical trials.
Self Replicating RNA	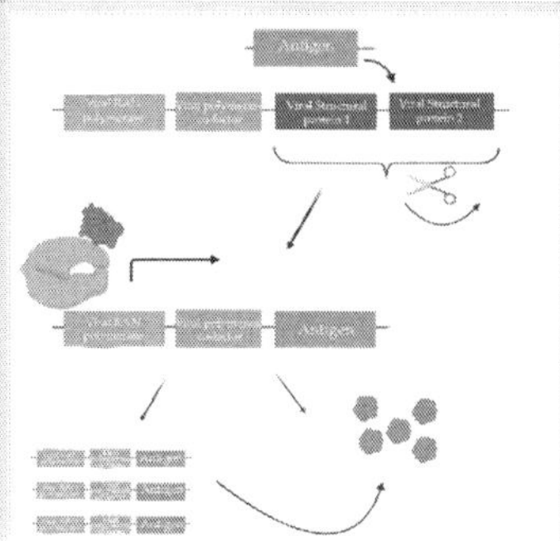	The structural proteins in the genome of an RNA virus are replaced with the coding sequence for the antigen- the viral polymerase can keep producing RNA which amplifies and maintains the amount of the antigenic protein.	Pre-clinical testing has shown effectiveness in inducing protection.

DNA and RNA Vaccines

Method Name	Method Diagram	Description of Method	Results
Codon Optimization	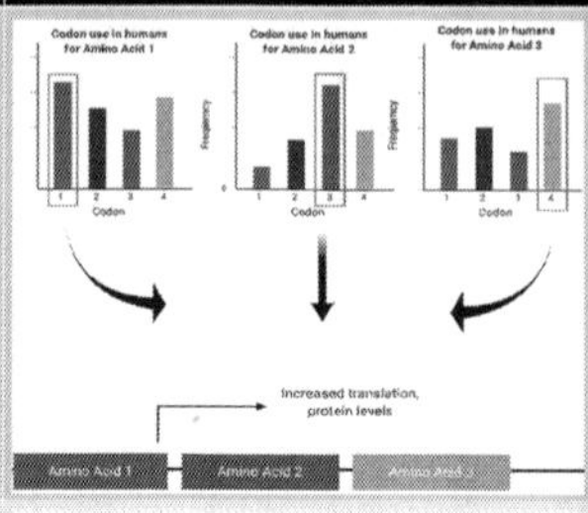	The codons with the most usage in the species of the vaccine target are used in the sequence of the antigen.	Widely used in vaccine design
Nanoparticle Vaccines	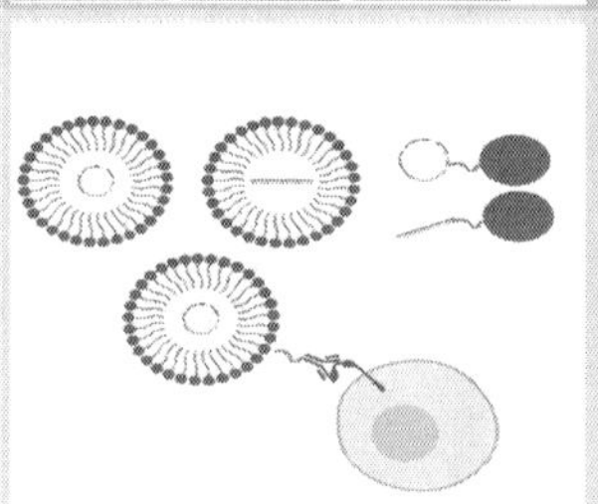	Nucleic acids are attached to or encapsulated by a carrier, which can also be attached to an antibody for specific cell targeting.	Wide variety of techniques undergoing preclinical and clinical testing

3.3 AUGMENTING THE EFFICACY

The augmented scalability and safety of subunit vaccines over whole-pathogen vaccines emerge on account of the lack of expressing the whole genome. Yet, their diminished immunogenicity warrants multiple dosages and adjuvant addition. In order to augment their efficacy, the following approaches are being probed.

Nanoparticle (NP) vaccines entail the cross-linking of carriers (organic lipids or inorganic metals or polymers) to augment immunogenicity. Combining an antigen with a linker and a carrier via chemical conjugation results in NP with optimal and adjustments of the shape or charge or size or hydrophobicity. The use of self-assembling monomers and carriers targeting immune cells has also been scrutinized.

Figure 3.3 below depicts the scheme of nanoparticle (NP) vaccines (© 2020 Brisse, Vrba, Kirk, Liang and Ly. *This is an open-access article distributed under the terms of the Creative Commons Attribution License (CC BY). The use, distribution or reproduction in other forums is permitted, provided the original author(s) and the copyright owner(s) are credited and that the original publication in this journal is cited, in accordance with accepted academic practice. No use, distribution or reproduction is permitted which does not comply with these terms*):

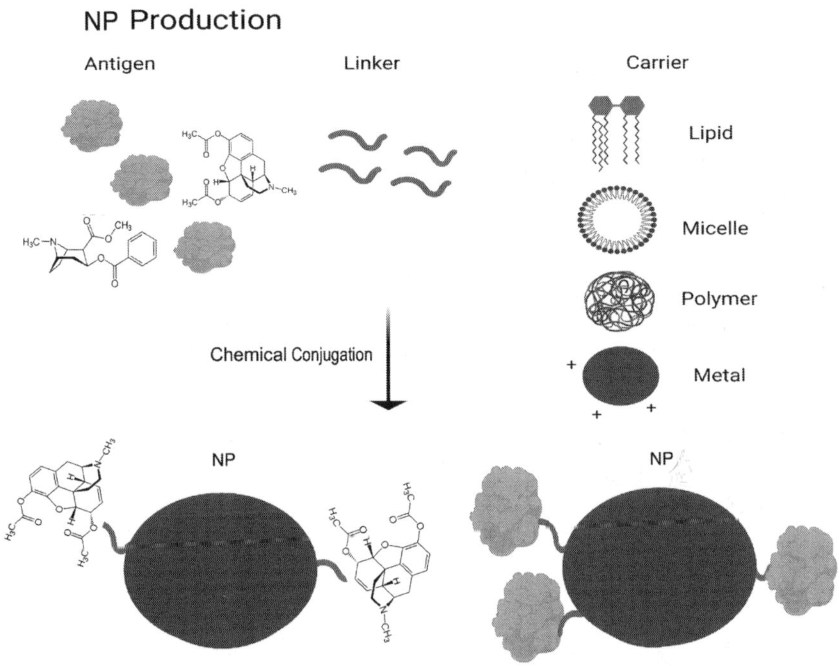

Virus-like particle (VLP) vaccines entail the self-assembly of antigenic particles via their expression in an appropriate system to mimic viruses. Chemically linking preformed particles and antigen peptides is another way. Their uptake by dendritic cells results in immune cell activation via MHC class I and MHC II molecules. Their property of possessing repetitive units results in a powerful immune response to find use in vaccination. An advantage here is the safety over live vaccines due to the lack of replication on account of an absent genome. Examples are inclusive of Gardasil® for human papillomavirus (HPV) and Engerix as a hepatitis B vaccine (HBV).

Figure 3.4 below depicts the scheme of VLPs augmenting the use of subunit vaccines (© 2020 Brisse, Vrba, Kirk, Liang and Ly. *This is an open-access article distributed under the terms of the Creative Commons Attribution License (CC BY). The use, distribution or reproduction in other forums is permitted, provided the original author(s) and the copyright owner(s) are credited and that the original publication in this journal is cited, in accordance with accepted academic practice. No use, distribution or reproduction is permitted which does not comply with these terms*):

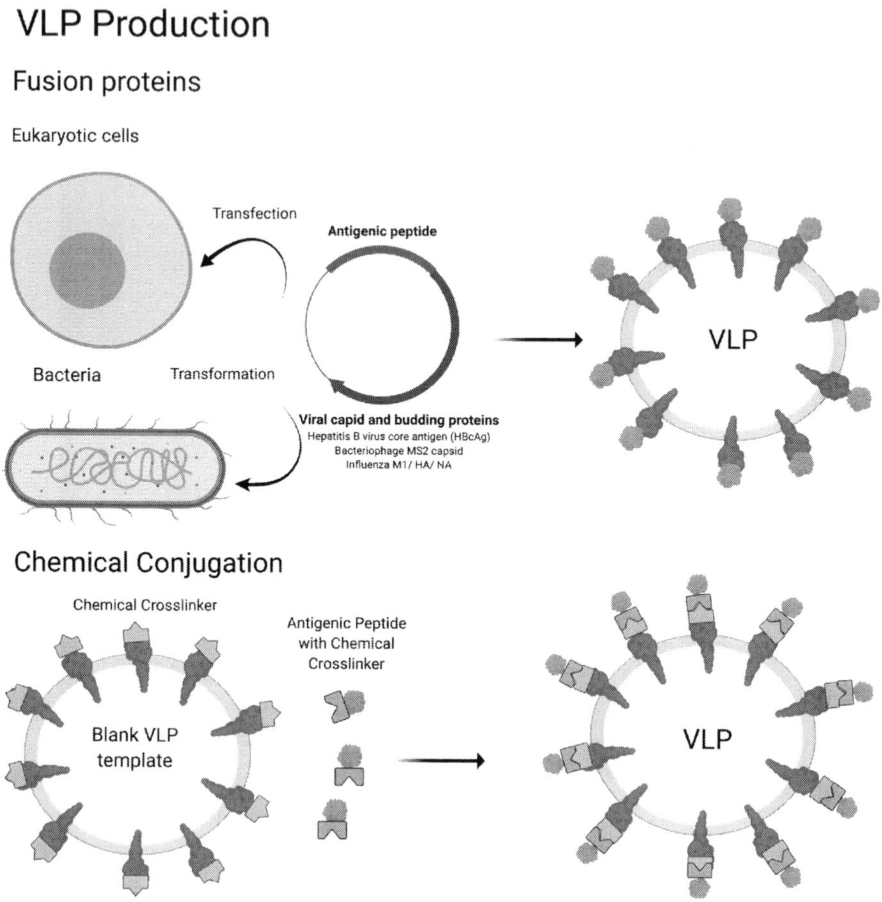

3.4 VACCINE PLATFORMS: A CASE STUDY OF COVID-19

Four million lives have been lost and 181 million confirmed cases have been reported for the severe acute respiratory syndrome coronavirus 2 (SARS-CoV-2) from its first detection in late 2019 to 29 June 2021 (WHO. WHO Coronavirus Disease (COVID-19) Dashboard. Available online: https://covid19.who.int/).

The quest for a vaccine was initiated since its sequence was released. A slew of candidates are being probed in trials while several have received emergency use authorization. The approaches employed range from subunit vaccines to nucleic acid vaccines to live attenuated virus vaccines, to name a few (Al-Jighefee et al, 2021). Figure 3.5 below illustrates ***TOP***: the virus and the vaccine platforms being scored

(a) SARS-CoV-2 structure encompassing the single-stranded RNA (ssRNA) genome and the four structural proteins: spike protein (S), envelop protein (E), membrane protein (M), and nucleocaspid protein (N). The platforms scrutinized are inclusive of (b) inactivated vaccine (c) live attenuated vaccine (d) viral vector vaccine (e) DNA vaccine (f) RNA vaccine (g) recombinant subunit vaccine (h) virus-like particles vaccine (Al-JighefeeHT, Najjar H, Ahmed MN, Qush A, Awwad S, Kamareddine© 2021 by the authors. Licensee MDPI, Basel, Switzerland.*This article is an open access article distributed under the terms and conditions of the Creative Commons Attribution (CC BY) license: https://creativecommons.org/licenses/by/4.0/*): *mRNA: messenger RNA, RBD: receptor-binding domain*

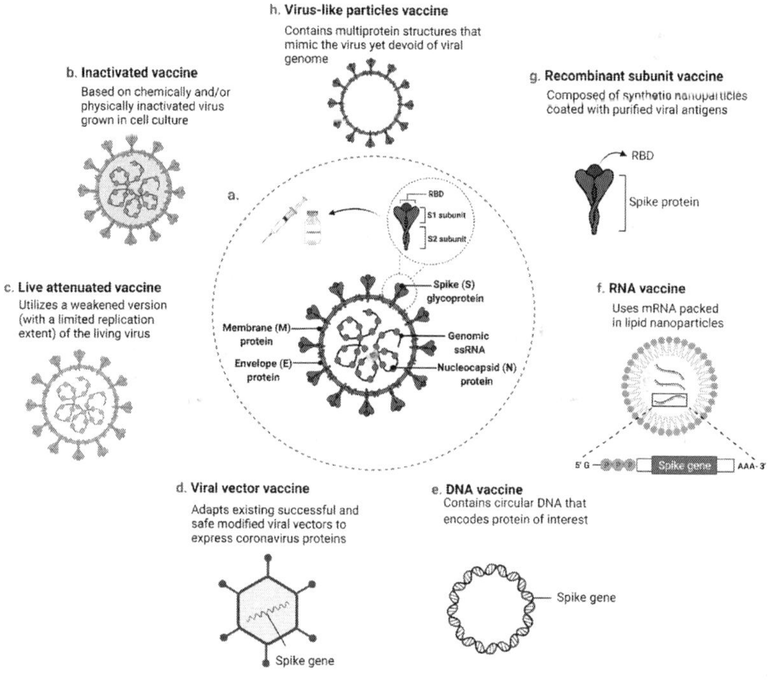

BOTTOM: (a) COVID-19 vaccine candidates in the pre-clinical and clinical phases of development (b) Candidates as of 29 June 2021from the World Health Organization (WHO) vaccine tracker and landscape website (Al-JighefeeHT, Najjar H, Ahmed MN, Qush A, Awwad S, Kamareddine © 2021 by the authors. Licensee MDPI, Basel, Switzerland.*This article is an open access article distributed under the terms and conditions of the Creative Commons Attribution (CC BY) license: https://creativecommons.org/licenses/by/4.0/*):

EU: emergency use, FDA: food and drug administration, EUA: emergency use authorization

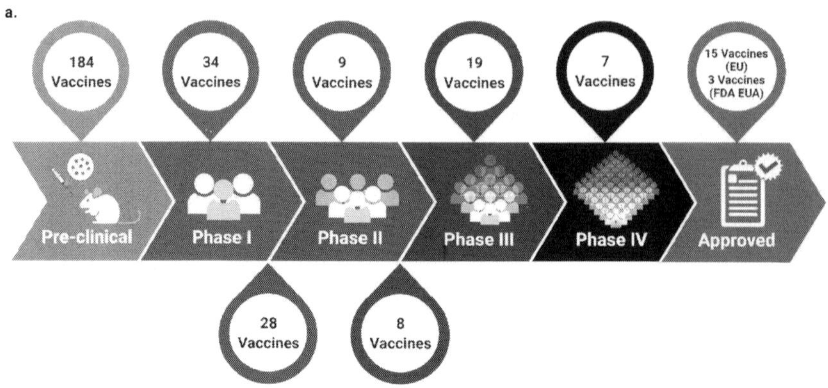

3.5 ADJUVANTS FOR VACCINES

While live attenuated pathogens or dead cells were employed for early work, the side effects led to the use of portions of the pathogen that led to a diminishing of their immunogenicity. In order to address this, adjuvants entered the platform. Apart from augmenting the immunogenicity, their impact also is inclusive of augmented stability that facilitates their storage under conditions of logistic challenges. Table 3.3 below summarizes the adjuvants that have received approval for humans (PogostinBH and McHugh, 2021: © 2021 by the authors. Licensee MDPI, Basel, Switzerland. *This article is an open access article distributed under the terms and conditions of the Creative Commons Attribution (CC BY) license: https:// creativecommons.org/licenses/by/4.0/*):

Adjuvant (Year)	Composition	Licensed Vaccine Targets	Immunological Function
Alum (1926)	Suspension of aluminum hydroxide or aluminum phosphate salts	Anthrax, hepatitis A, hepatitis B, human papillomavirus, diphtheria-pertussis-tetanus (DPT and TdaP), *Haemophilusinfluenzae* type b, Japanese encephalitis, pneumococcal conjugate vaccines, and COVID-19	Releases DAMPs at the injection site by causing cell death, resulting in the recruitment and activation of dendritic cells and neutrophils
Virosomes (1993)	Unilamellar liposomes composed of viral proteins and phospholipids of vaccine target virus	Seasonal flu and hepatitis A	PAMPs on the surface of virosomes stimulate and activate antigen-presenting cells while also facilitating antigen delivery
MF59 (1997)	Emulsion of Squalene, Tween (polysorbate) 80, and Span 85	Seasonal flu and pandemic flu (H1N1)	Known to recruit and activate macrophages and dendritic cells and cause chemokine secretion
AS03 (2009)	Emulsion of Squalene, α-tocopherol, and Tween (polysorbate) 80	H1N1	Activates human monocytes and macrophages and induces NF-κB activity and chemokine production
AS04 (2009)	MPL adsorbed onto alum	Human papillomavirus and hepatitis B	MPL activates TLR4 and NF-κB to stimulate antigen presenting cells and innate immune system while alum causes the release of DAMPs and local inflammation
AS01 (2015)	Liposome co-delivery of saponin QS-21 and MPL	Malaria and herpes zoster	Activates TLR4 in innate immune cells and caspase 1 in subcapsular sinus macrophages, induces differentiation of monocytes to DC, and activates NF-κB and production of IFNγ

CpG 1018 (2018)	22 nucleotide single-stranded DNA containing unmethylated cytosine phospho-guanosine dinucleotide	Hepatitis B	Activates TLR9 resulting in a type I interferon response
Viral Vectors (2020)	Adenoviruses carrying mRNA encoding for protein antigen	COVID-19	PAMPs on the surface of adenoviral carriers activate the innate immune system while also facilitating transfection for mRNA vaccines

The latest research is exploring new options given the pandemic scenarios we are facing. An example is self-assembling mesoporous silica rods (MPS) modified with granulocyte-macrophage colony-stimulating factor (GM-CSF) and CpG developed at Harvard University that have shown the induction of robust responses to cancer antigens and nicotine. Another example from the same center is ciVAX constituted by opsonin Fc-mannose-binding lectin- (FcMBL) coated magnetic particles that bind bacterial PAMPs(pathogen-associated molecular patterns) to which the MPS-GM-CSF and CpG mix is added. These beads conferred protection against methicillin-resistant *Staphylococcus aureus* in a murine system and *E. coli* in a swine system. A further promising system entails the use of self-assembling peptides facilitating tailor-made sequences and augmented safety.

3.6 FINAL POINTS

In addition to the ongoing COVID-19 pandemic, when the next one will strike is not known. The use of vaccines emerges as a crucial option to enable what we call "normal life". Apart from research on vaccine platforms and studies looking at augmenting their efficacies while diminishing their side effects, added impetus from policymakers can help fill the gaps in the system that serve as impediments to vaccine development. Latest technologies can also help fuel the development of safe and robust vaccines against pathogens- new and old.

"Vaccines are the tugboats of preventive health."

-William Foege

3.7 REFERENCES

1. Rauch S, Jasny E, Schmidt KE and Petsch B (2018) New Vaccine Technologies to Combat Outbreak Situations. *Front. Immunol.* 9:1963. doi: 10.3389/fimmu.2018.01963.
2. Li, YD., Chi, WY., Su, JH. *et al.* Coronavirus vaccine development: from SARS and MERS to COVID-19. *J Biomed Sci* 27, 104 (2020). https://doi.org/10.1186/s12929-020-00695-2.
3. Brisse M, Vrba SM, Kirk N, Liang Y and Ly H (2020) Emerging Concepts and Technologies in Vaccine Development. *Front. Immunol.* 11:583077. doi: 10.3389/fimmu.2020.583077
4. Al-JighefeeHT, Najjar H, Ahmed MN, Qush A, Awwad S, Kamareddine L. COVID-19 Vaccine Platforms: Challenges and Safety Contemplations. *Vaccines*. 2021; 9(10):1196. https://doi.org/10.3390/vaccines9101196.
5. PogostinBH, McHugh KJ. Novel Vaccine Adjuvants as Key Tools for Improving Pandemic Preparedness. *Bioengineering*. 2021; 8(11):155. https://doi.org/10.3390/bioengineering8110155.

CHAPTER

4 Vaccines for Neglected Tropical Diseases

Contents

"Only 0.6% of overseas development assistance for health is allocated to neglected tropical diseases, despite such diseases affecting at least 1 billion people".

- Bernhard Liese; sourced from Molyneux D. H. (2012). The 'Neglected Tropical Diseases': now a brand identity; responsibilities, context and promise. Parasites & vectors, 5, 23. https://doi.org/10.1186/1756-3305-5-23

4.1 BACKGROUND

Back in 2005, neglected tropical diseases (NTDs) encompassed the list below of the following African Neglected Tropical Diseases as published in PLOS Medicine in 2005 (Hotezet al, 2020: *This is an open access article, free of all copyright, and may be freely reproduced, distributed, transmitted, modified, built upon, or otherwise used by anyone for any lawful purpose. The work is made available under the Creative Commons CC0* public domain dedication):

Protozoan Infections

African trypanosomiasis	*Trypanosoma gambiense* and *T. rhodesiense*
Visceral leishmaniasis (Kala-azar)	*Leishmania donovani*

Helminth Infections

Soil-transmitted helminth infections	
Ascariasis	*Ascaris lumbricoides*
Trichuriasis	*Trichuris trichiura*
Hookworm infection	*Necator americanus*
Schistosomiasis	
Urinary schistosomiasis	*Schistosoma haematobium*
Hepatobiliary schistosomiasis	*Schistosoma mansoni*
Lymphatic filariasis	*Wuchereria bancrofti*
Onchocerciasis	*Onchocerca volvulus*
Dracunculiasis	*Dracunculus medinensis*

Bacterial Infections

Trachoma	*Chlamydia trachomitis*
Leprosy	*Mycobacterium leprae*
Buruli ulcer	*Mycobacterium ulcerans*

Subsequent expansion of the list has resulted in the following list as depicted in table 4.1 below (Hotez et al, 2020: *This is an open access article, free of all copyright, and may be freely reproduced, distributed, transmitted, modified, built upon, or otherwise used by anyone for any lawful purpose. The work is made available under the Creative Commons CC0* public domain dedication):

Classes of NTDs	NTDs recognized by WHO [2]	NTDs recognized by *PLOS Neglected Tropical Diseases* [3]	NTDs on the cusp
Helminth infections (and their vectors)	- Dracunculiasis - Echinoccocosis - Foodborne Trematodiases - Lymphatic filariasis - Onchocerciasis - Schistosomiasis - Soil-transmitted helminthiases (Ascariasis, Hookworm Diseases, Trichuriasis, Strongyloidiasis) - Taeniaisis/Cysticercosis	All human helminth infections including but not restricted to: - Dracunculiasis - Echinococcosis - Foodborne Trematodiases - Loiasis - Lymphatic Filariasis - Onchocerciasis - Other food-borne helminthiases, including Trichinosis, Anisakiasis, Gnathostomiasis, etc. - Schistosomiasis - Soil-transmitted helminthiases (Ascariasis, Hookworm Diseases, Trichuriasis, Strongyloidiasis) - Taeniasis-Cysticercosis - Toxocariasis and other larva migrans, e.g. *Baylisascaris*	- None (all human helminth infections are currently considered)
Protozoan infections (and their vectors)	- Chagas disease - Human African trypanosomiasis - Leishmaniasis	- Amebiasis including *Naegleria* - Babesiosis - Balantidiasis - Chagas Disease - Giardiasis - Human African Trypanosomiasis - Leishmaniasis - *Plasmodium vivax* and other non-*P. falciparum* malarias	- Blastocystiasis* - Cryptosporidiosis and cyclosporiasis - Toxoplasmosis
Bacterial infections (and their vectors)	- Buruli ulcer - Leprosy (Hansen's disease) - Trachoma - Yaws (Endemic treponematoses)	- Bartonella - Bovine Tuberculosis in Humans - Buruli Ulcer - Cholera - Enteric pathogens (Shigella, Salmonella, *E. coli*) - Leprosy - Leptospirosis - Melioidosis - Relapsing Fever - Trachoma - Yaws and other tropical treponematoses (Bejel, Pinta) - Q fever	- Group A Streptococcal disease
Fungal infections	- Mycetoma, chromoblastomycosis, and other deep mycoses	- Mycetoma, chromoblastomycosis, and other deep mycoses - Paracoccidiomycosis	- Cryptococcosis - Histoplasmosis
Viral infections (and their vectors)	-Dengue and Chikungunya -Rabies	- Arboviral infections including Dengue, Chikungunya, Zika, Japanese encephalitis, Jungle yellow fever and others - Enterovirus 71 and related viruses - HTLV-1, HTLV-2 and other non-HIV retrovirus infections - Rabies - Rift Valley fever - Viral hemorrhagic fevers	- *Henipavirus* species e.g. Nipah virus
Ectoparasitic infestations	- Scabies and other ectoparasites	- Scabies, Myiasis, and other ectoparasites	- None
Non-infectious diseases or conditions	- Snakebite envenoming	- Podoconiosis - Snakebite envenoming	- Sickle-cell Anemia

* The taxonomic position of the stramenopile-like, unicellular protist *Blastocystis* remains uncertain.

https://doi.org/10.1371/journal.pntd.0008001.t001

Table 4.2 below lists NTDs under the Center for Disease Control and Prevention (CDC) and World Health Organization (WHO) (Mitra and Mawson, 2017: © 2017 by the authors. Licensee MDPI, Basel, Switzerland. *This article is an open access article distributed under the terms and conditions of the Creative Commons Attribution (CC BY) license (http://creativecommons.org/licenses/by/4.0/*):

[a] Not under CDC list of neglected tropical diseases (NTDs); [b] Diseases that can be addressed via mass drug administration (MDA), or other interventions

Disease	**CDC**	**WHO**
Buruli ulcer (*Mycobacterium ulcerans* infection)	+	+
Chikungunya [a]	–	+
Chagas disease	+	+
Cysticercosis	+	+
Dengue fever	+	+
Dracunculiosis (or guinea worm disease) [b]	+	+
Echinococcosis	+	+
Fascioliasis	+	+
Foodborne trematodiasis [a]	–	+
Human African trypanosomiasis (or sleeping sickness)	+	+
Leishmaniasis (or kala-azar)	+	+
Leprosy	+	+
Lymphatic filariasis [b]	+	+
Mycetoma	+	+
Onchocerciasis (or river blindness) [b]	+	+
Rabies	+	+
Schistosomiasis [b]	+	+
Soil-transmitted helminthiasis [b]	+	+
Trachoma [b]	+	+
Yaws	+	+

The impact of these diseases cannot be pushed under the carpet. For instance, 1.3 billion people approximately have been pitched to have had a soil-transmitted helminth (STH) infection in 2010. Figure 4.1 below illustrates NTD distribution across our planet (Mitra and Mawson, 2017: © 2017 by the authors. Licensee MDPI, Basel, Switzerland. *This article is an open access article distributed under the terms and conditions of the Creative Commons Attribution (CC BY) license (http://creativecommons.org/licenses/by/4.0/*):

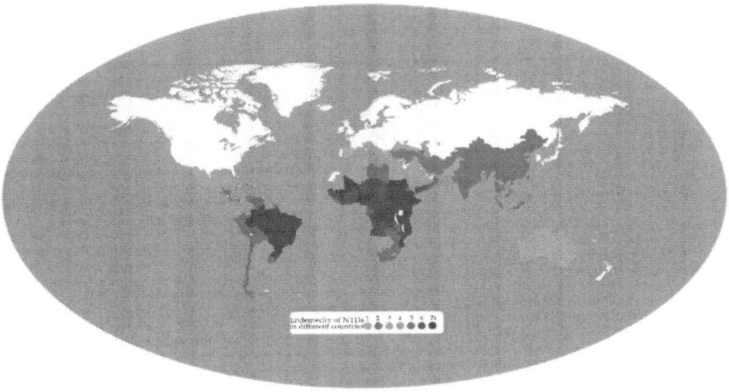

The disease burden is shown below in table 4.3 (Mitra and Mawson, 2017: © 2017 by the authors.

Licensee MDPI, Basel, Switzerland. *This article is an open access article distributed under the terms and conditions of the Creative Commons Attribution (CC BY) license (http://creativecommons.org/licenses/by/4.0/)*:

Key: Disability-Adjusted Life Years (DALYs), Years of Life Lost (YLL), and Years Lost Due to Disability (YLD) in WHO Member States; [a] Data source: WHO Health Statistics and Information Systems, 2015. Data on DALY are not available for *Buruli ulcer disease;* [b] *Bhutta et al. (2014);* [c] No data available for YLL

Disease	**2015 Data** [a]			**2010 Data** [b]
	YLL (thousand)	**YLD (thousand)**	**DALY = YLL + YDL (thousand)**	**DALY (thousand)** [b]
Soil-transmitted helminthiasis	449.50	3993.97	4443.47	5043
Ascariasis	225.30	869.37	1094.67	1254
Trichuriasis [c]	-	542.80	542.80	630
Hookworm [c]	-	1739.58	1739.58	3159
Schistosomiasis	1042.20	2471.65	3513.85	3971
Dengue fever	1848.79	761.29	2610.08	1243
Lymphatic filariasis [c]	-	2070.85	2070.85	2740

Cysticercosis	1258.27	598.09	1856.36	503
Rabies	1672.03	0.14	1672.17	2297
Leishmaniasis	1310.74	45.72	1356.46	3754
Onchocerciasis [c]	-	1135.57	1135.57	564
Foodborne trematodiasis	224.12	842.22	1066.34	665
Echinococcosis	568.20	73.23	641.43	600
Leprosy	457.67	30.97	488.64	215
Human African trypanosomiasis	368.68	2.97	371.65	1346
Trachoma [c]	-	278.97	278.97	308
Chagas disease	189.65	63.05	252.70	499

4.2 RESEARCH SCENARIO

The stigma associated with tuberculosis referred to as consumption or the white plague is well-known. Given the duration of treatment involved, the availability of only the BCG vaccine and a new vaccine aimed at by WHO to diminish the disease, researchers Theolis Costa Barbosa Bessa and team (2017) scrutinized the financing for research and development of preventive vaccines between 2007 and 2014 employing the Global Funding of Innovation for Neglected Diseases (G-FINDER) database. Vaccine candidates undergoing trial as reported by the researchers are listed below in table 4.4 as of 2016 (Theolis Costa Barbosa Bessa et al, 2017: Copyright © 2017 Theolis Costa Barbosa Bessa et al. *This is an open access article distributed under the Creative Commons Attribution License, which permits unrestricted use, distribution, and reproduction in any medium, provided the original work is properly cited*):

Candidate	Type	Sponsors/partners	Phase
Ad5 Ag85A	Viral vector	McMaster University (Canada); CanSino (China)	I
ID93 + GLA-SE	Adjuvanted subunit	*Infectious Disease Research Institute—IDRI* (USA); *Aeras* (USA)	I
Crucell Ad35/ MVA85A	Viral vector	Crucell Holland B.V (Netherlands); University of Oxford (UK); *Aeras* (USA)	I
Dar-901	*M.obuense* whole cell	Dartmouth-Hitchcock Medical Center (USA); *Aeras* (USA)	I
TB/FLU-04L	Viral vector	Research Institute for Biological Safety Problems—RIBSP (Kazakhstan)	I
MVA85A	Viral vector	Oxford University (UK)	I
ChAdOx1-85A/ MVA85A	Viral vectors	Oxford University (UK)	I
MVA85A-IMX313	Viral vector	Oxford University (UK); Imaxio (France)	I
MTBVAC	Live genetically attenuated MTB	*Tuberculosis Vaccine Institute—TBVI* (Netherlands); University of Zaragoza (Spain); Biofabri (Spain)	IIa
VPM 1002	Live rBCG	Max Planck Institute; Vakzine Projekt Management GmbH (Germany); *TBVI* (Netherlands); Serum Institute of India (India)	IIa
H1 (Ag85B/ESAT) + IC31	Adjuvanted subunit	Statens Serum Institute—SSI (Denmark); *TBVI* (Netherlands); Valneva SE (France)	IIa
RUTI (Tb lysate)	Fragmented MTB	Archivel Farma (Spain)	II a
H4 (Ag85B/ TB10.4) + IC31	Adjuvanted subunit	Statens Serum Institute—SSI (Denmark); Sanofi-Pasteur (France); *Aeras (*USA); Valneva SE (France)	II a
H56 (Ag85B/ESAT-6/Rv2660) + IC31	Adjuvanted subunit	Statens Serum Institute—SSI (Denmark); *Aeras (*USA); Valneva SE (France)	II a
M72 + AS01E	Adjuvanted subunit	GSK (UK); *Aeras (*USA)	II b
M. vaccae	*M. vaccae* whole cell	Anhui Zhifei Longcom Biologic Pharmacy Co. Ltd. (China)	III

Looking at the funding, the public sector and philanthropic institutions account for this research as outlined in table 4.5 below (Theolis Costa Barbosa Bessa et al, 2017: Copyright © 2017 Theolis Costa Barbosa Bessa et al. *This is an open access article distributed under the Creative Commons*

Attribution License, which permits unrestricted use, distribution, and reproduction in any medium, provided the original work is properly cited):

Type of funder	Funding to R&D	Funding to PDPs	Funding to intermediaries	Self-funding	Number of records
Academic and other research institutes	2				2
Philanthropic	51	20			71
Private sector		5	2	20	27
Public sector, governments	300	39	3	38	380
Unspecified	1				1
Total	354	64	5	58	481

Looking at the research conducted on NTDs in children, Rees and team (2019) probed the International Clinical Trials Registry Platform (ICTRP) of the WHO between January 1, 2006 and December 31, 2016. Of the 63 pediatric trials selected, vaccine trials constituted one-third (n=21) with the predominant ones among these accounted to by dengue vaccine trials. Overall, research has outlines the necessity of augmented research on NTDs given their incidence and current treatment options.

The details of vaccines along with other interventions for NTDs as depicted by the authors are presented below in table 4.6 (© 2019 Rees et al. *This is an open access article distributed under the terms of the Creative Commons Attribution License, which permits unrestricted use, distribution, and reproduction in any medium, provided the original author and source are credited*):

Condition	Adult Trials, n	Pediatric Trials, n	Characteristics of Pediatric Trials					
			Drug Trials, n (%)	Vaccine Trials, n (%)	Randomized Trials, n (%)	Industry Funded, n (%)	Enrollment, median	Trial Duration in Years, mean*
Dengue and chikungunya	42	21	5 (23.8)	16 (76.2)	20 (95.2)	17 (80.9)	504	3.3
Schistosomiasis	6	14	12 (85.7.0)	2 (14.3)	12 (85.7)	2 (14.3)	490	1.6
Soil Transmitted Helminthiases	12	9	9 (100.0)	0	7 (77.8)	2 (22.2)	350	1.3
Leishmaniasis	87	5	5 (100.0)	0	3 (60.0)	0	80	2.9
Rabies	33	4	0	4 (100.0)	2 (50.0)	3 (75.0)	246	0.8
Chagas	19	2	2 (100.0)	0	1 (50.0)	1 (50.0)	205	2.0
Scabies and Other Ectoparasites	18	2	1 (50.0)	1 (50.0)	1 (50.0)	0	266	1.4
Trachoma	11	2	2 (100.0)	0	2 (100.0)	0	1,657	6.1
Yaws	3	2	2 (100.0)	0	2 (100.0)	0	419	1.3
Human African Trypanosomiasis	5	1	1 (100.0)	0	0	0	125	3.2
Taeniasis/Cysticercosis	5	1	1 (100.0)	0	1 (100.0)	0	97	1.6
Dracunculiasis	0	0	-	-	-	-	-	-
Echinococcosis	1	0	-	-	-	-	-	-
Foodborne Trematodiases	1	0	-	-	-	-	-	-
Leprosy	18	0	-	-	-	-	-	-
Lymphatic filariasis	16	0	-	-	-	-	-	-
Mycetoma, chromoblastomycosis, and other deep mycoses	6	0	-	-	-	-	-	-
Onchocerciasis	8	0	-	-	-	-	-	-
Buruli Ulcer	3	0	-	-	-	-	-	-
Snake Bite Envenomation	12	0	-	-	-	-	-	-
Total	**306**	**63**	**42 (66.6)**	**21 (33.3)**	**52 (85.3)**	**25 (39.7)**	**315**	**2.3**

*Completion dates for the calculation of trial duration were missing for 4 trials on dengue, 2 trials on schistosomiasis, and 1 trial each on leishmaniasis, rabies, trachoma, and soil transmitted helminthiases.

https://doi.org/10.1371/journal.pntd.0007111.t001

Condition	WHO-Recommended Therapies	Drugs Studied	Vaccines Studied
Chagas	• Benznidazole* • Nifurtimox	• Benznidazole • Nifurtimox	
Dengue and Chikungunya	• Supportive care*	• Hypertonic sodium lactate and Ringer's lactate • Anti-CHIKV hyperimmune immunoglobulins • Zinc • Early corticosteroid therapy Vitamin E	• Tetravalent Dengue Vaccine Chimeric tetravalent dengue-serotype 1, 2, 3, 4 • Live, attenuated dengue serotype 1, 2, 3, and 4 • Takeda's Tetravalent Dengue Vaccine Candidate • CYD Dengue Vaccine • CYD Dengue Vaccine-5 dose formulation • Tetravalent Dengue Vaccine
Human African Trypanosomiasis	• Pentamidine* • Suramin • Eflornithine* and Nifurtimox • Melarsoprol	• Fexinidazole	
Leishmaniasis	• Meglumine antimoniate • Sodium stibogluconate • Amphotericin B deoxycholate* • Pentamidine isethionate* • Paromomycin ointment • Fluconazole* • Topical methylbenzethonium • Topical ketoconazole* • Miltefosine* • Pentavalent antimonials • Pentoxifylline • Liposomal amphotericin B*	• Miltefosine • Meglumine antimoniate • Azithromycin • Amphotericin B-deoxycholate	
Rabies	• Immediate vaccination • Rabies immune globulin	• Human rabies immunoglobulin	• Purified Vero Rabies Vaccine Serum Free and Imovax Human Diploid Cell Vaccine • Rabipur • CL184
Scabies and Other Ectoparasites	• Permethrin cream* • Malathion cream* • Benzyl benzoate emulsion cream • Sulphur ointment • Ivermectin*	• KD-357 I• vermectin • Permethrin	
Schistosomiasis	• Praziquantel* • Oxamniquine	• Praziquantel • Biltricide (racemate praziquantel) oral tablets • Racemate Praziquantel ODT • Levo Praziquantel ODT • Artemether-lumefantrine combination plus albendazole • Artemether-lumefantrine plus Praziquantel plus Albendazole • Albendazole plus Praziquantel • Moxidectin • Synriam • Artesunate+Sulfamethoxy-pyrazine/ pyrimethamine • Albendazole + praziquantel • Mebendazole + praziquantel • Sulfadoxine-pyrimethamine • Piperaquine • Albendazole • Praziquantel • Praziquantel and Arachidonic acid	• Bilhvax vaccine-Sh28GST • MVA85A
Soil Transmitted Helminthiases	• Pyrantel • Mebendazole* • Levamisole • Piperazine • Albendazole* • Flubendazole • Ivermectin*	• Albendazole • Mebendazole • Albendazole and Vitamin C • Oxantel pamoate • Pyrantel-oxantel + mebendazole • Ivermectin • Albendazole + Ivermectin • Mebendazole + Ivermectin	
Taeniasis/Cysticercosis	• Praziquantel* • Niclosamide	• Albendazole	
Trachoma	• Azithromycin*	• Azithromycin	
Yaws	• Azithromycin* • Benzathine penicillin*	• Azithromycin • Penicillin-G-benzathine + Azithromycin	

* Adequately labeled for use in children.

https://doi.org/10.1371/journal.pntd.0007111.t003

Insecticides to control vectors and drug regimens in patients have been employed for Chagas disease: the complications of prolonged treatment and contraindications have propelled research for vaccines (Dumonteil and Herrera, 2021). Studies probing the development of vaccines for this condition are depicted below in table 4.7 (Dumonteil and Herrera, 2021:

Key: PRF—perforin; GZN—granzyme; DTH—delated type hypersensitivity; TLR—Toll-like receptor; IL—interleukin; IFN—interferon; TNF—tumor necrosis factor; EKG—electrocardiogram; N/A—not applicable

	Antigens and Use	**Adjuvants and Delivery Systems**	**Immune Response**	**Efficacy Against Parasite**	**Efficacy Against Cardiac Damage and Dysfunction**	**Ref**
Therapeutic vaccines						
DNA	TcG2+TcG4, therapeutic during acute phase	Plasmids or nano plasmids	$CD4^+$ and $CD8^+$ producing IFNg, PRF, and GRZ	Decreased parasite burden in cardiac and skeletal muscle	Decrease in fibrosis in heart and skeletal muscle, decrease in oxidative stress	[40,41]
DNA	Cruzipain and Chagasin plasmids with *Salmonella* carrier, therapeutic during acute phase	GM-CSF expression plasmid	Increased IFNγ and antibodies	Decreased parasite burden in blood and heart, and increased survival	Decreased cardiac inflammation	[42]
Peptides	10 peptide epitopes mixture, therapeutic during acute phase	TLR4 agonist (MPLA)	Increased IFNγ	Decreased cardiac parasite burden, increased survival	N/A	[43]
Recombinant proteins	Tc24, TSA-1 and their optimized variants, therapeutic during acute phase	TLR4 agonists (E6020, MPLA, GLA), TLR9 agonist (CpG), nanoparticles	Antibodies, IFNγ, and $CD4^+$ and $CD8^+$ activation	Decreased cardiac parasite burden, increased survival	Decrease in cardiac inflammation and fibrosis	[44,45, 46,47]

Recombinant proteins	Tc24, therapeutic during chronic phase	TLR4 agonists (E6020)	High IFNγ and low IL4, and antibodies	Decreased parasitemia	Decrease in cardiac inflammation and fibrosis	[48]
Recombinant proteins	Tc24-C4, therapeutic combined with low dose Benznidazole	TLR4 agonist (E6020)	Increased IFNγ, IL12, TNFa, IL2, IL4 and IL10, and $CD4^+$ and $CD8^+$ T cell activation	Decreased parasitemia, increased survival	N/A	[49]
Viral vectors	Recombinant Adenovirus expression ASP2, therapeutic during acute phase		TNFa, iNOS, TLR4, and IL-10 expression in the liver	Increased survival, decreased parasite burden in liver	N/A	[50]
Viral vectors	Recombinant adenovirus expressing ASP2 and TS, therapeutic during chronic phase		IFNγ and $CD8^+$ T cells	Increased survival	Decreased cardiac fibrosis and dysfunction	[51]
Preventative vaccines						
Live parasites	Drug-cured primary infection, preventative	N/A	N/A	Sustained decrease in parasite burden in all body	N/A	[52]
Live parasites	Live attenuated parasite (inducible expression of alpha-toxin, and cecropin A), preventative	N/A	IFNγ, TNFa, $CD4^+$ and $CD8^+$ T cell activation, antibodies, and NK cells	No detectable parasites	Decrease in cardiac inflammation	[53,54]
Live parasites	Live attenuated parasite (TCC attenuated strain), preventative short term	IFNγ expressing plasmid	Antibodies and mixed Th1/Th2 response	Decreased parasitemia and increased survival	N/A	[55]
DNA	TcG2+TcG4, preventative short term	Plasmids alone or with *Trypanosoma rangeli* and/or Quil A as adjuvants	Antibodies, $CD4^+$ and $CD8^+$ producing IFNγ, TNFa, and PRF	Decreased parasite burden in cardiac and skeletal muscle		[56]

DNA	Cruzipain plasmid, preventative short term	GM-CSF plasmid	Antibodies and DTH	Decreased parasitemia, increased survival	Decreased cardiac tissue damage	[57]
DNA	Cruzipain, Tc52, Tc24 plasmids, preventative short term	*Salmonella enterica* carrier	Trypanolytic antibodies, DTH, IFNγ, IL12, and IL10	Decreased parasitemia, increased survival	Decreased cardiac tissue inflammation, necrosis	[58]
Recombinant proteins	Cruzipain fused with staphylococcal superantigen, preventative short term	TLR9 agonist (CpG)	Neutralizing antibodies and DTH	Decreased parasitemia and increased survival	N/A	[59]
Recombinant proteins	Recombinant Traspain, Cruzipain and ASP-2 fusion protein, preventative short term	c-di-AMP adjuvant (STING agonist)	Neutralizing antibodies, $CD4^+$ and $CD8^+$ T cell activation, IFNγ, TNFa, IL2, and IL17	Decreased parasitemia and increased survival	Decreased cardiac damage (CK, CK-MB), decreased necrosis and inflammation in the heart and skeletal muscle	[60]
Recombinant proteins	Recombinant Tc52 fragment, preventative short term	c-di-AMP adjuvant (STING agonist)	Antibodies, $CD4^+$ and $CD8^+$ T cell activation, IFNγ, and IL17	Decreased parasitemia and increased survival	N/A	[61]
Recombinant proteins	TcTASV, preventative short term	Unlipidated Outer Membrane Protein 19 of *Brucella abortus* (U-Omp19) as adjuvant	Trypanolytic antibodies, IFNγ, and IL17	Decreased parasitemia and increased survival	N/A	[59]
Recombinant proteins	Enolase, preventative short term	Freund complete/incomplete adjuvant	Antibodies, IFNγ, and IL2	Decreased parasitemia and increased survival	Decreased cardiac and skeletal muscle inflammation	[62]
Recombinant proteins	Trans-Sialidase fragment, Preventative short term	ISPA lipidic cages, ISCOMATRIX, or Freund adjuvant	Trypanolytic antibodies, IFNγ, $CD4^+$ and $CD8^+$ T cell activation, Treg activation	Decreased cardiac parasite burden, increased survival	Decreased cardiac inflammation	[63,64]

Glycotope	αGal glycotope, preventative short term	TLR4 agonist (Liposomal-monophosphoryl lipid A)	Trypanolytic antibodies, $CD4^+$ and $CD8^+$ T cell activation	Decreased parasite burden in multiple tissues, increased survival	Decreased cardiac inflammation and necrosis	[39]
Viral vectors	Recombinant Adenovirus and modified Vaccinia Ankara virus expressing ASP-2 and Trans-sialidase, preventative vaccination	PBS		Decreased parasite burden during the acute phase in all body, but no impact on long-term burden during chronic phase	N/A	[52]
Bacterial vectors	Recombinant *Mycobacterium bovis* (BCG) expressing trans-sialidase and cruzipain fragments, preventative short term		Trypanolytic antibodies, and DTH, $CD4^+$ expressing IFNγ, IL17, IL10, and $CD8^+$	Increased survival	Decreased cardiac inflammation and fibrosis	[65]
Heterologous prime-boost combination	*Salmonella enterica* expressing Traspain and ASP-2	TLR9 agonist (CpG)	Increased IFNγ, IL17, low IL4, $CD4^+$, and $CD8^+$ T cell activation	Decreased parasite burden in blood, heart and skeletal muscle, and increased survival	Decreased inflammation and improved EKG	[66]
Heterologous prime-boost combination	Recombinant 80 kDa prolyl oligopeptidase (Tc80) and plasmid DNA	TLR9 agonist (CpG)	Increased IFNγ, IL2, TNFa, $CD4^+$, and $CD8^+$ T cell activation	Decreased parasitemia, and increased survival	Decreased cardiac inflammation, damage (CK and CK-MB), improved EKG	[67]
Heterologous prime-boost combination	Recombinant Tc52 and plasmid DNA with Salmonella carrier, preventative short term	TLR9 agonist (CpG)	Trypanoplytic antibodies, Increased IFNγ, IL10, $CD4^+$, and $CD8^+$ T cell activation	Decreased parasitemia, and increased survival	Decreased cardiac inflammation	[68]
Heterologous prime-boost combination	TcG1, TcG2, and TcG4 expression plasmids and recombinant proteins, preventative long term	IL2 and GM-CSF plasmids	Increased IFNγ, TNFa, $CD4^+$ and $CD8^+$ T cell activation	N/A	N/A	[69]

Overall, the potential of vaccines in addressing NTDs is being documented in studies. Further investment and political push are warranted for propelling more scrutiny in this field. Given the costs and impact, vaccines for preventing NTDs warrant a deeper thrust.

"As for the future, your task is not to foresee it, but to enable it."

- *Antoine de Saint Exupery*

4.3 REFERENCES

1. HotezPJ, Aksoy S, BrindleyPJ, Kamhawi S (2020) What constitutes a neglected tropical disease? PLoS Negl Trop Dis 14(1): e0008001. https://doi.org/10.1371/journal.pntd.0008001
2. Mitra AK, Mawson AR. Neglected Tropical Diseases: Epidemiology and Global Burden. *Trop Med Infect Dis*. 2017;2(3):36. Published 2017 Aug 5. doi:10.3390/tropicalmed2030036
3. Theolis Costa Barbosa Bessa, Erika Santos de Aragão, Jane Mary Medeiros Guimarães, Bethânia de Araújo Almeida, "R&D in Vaccines Targeting Neglected Diseases: An Exploratory Case Study Considering Funding for Preventive Tuberculosis Vaccine Development from 2007 to 2014", BioMed Research International, vol. 2017, Article ID 4765719, 12 pages, 2017. https://doi.org/10.1155/2017/4765719
4. Rees CA, Hotez PJ, Monuteaux MC, Niescierenko M, Bourgeois FT (2019) Neglected tropical diseases in children: An assessment of gaps in research prioritization. PLoS Negl Trop Dis 13(1): e0007111. https://doi.org/10.1371/journal.pntd.0007111
5. Dumonteil E, Herrera C. The Case for the Development of a Chagas Disease Vaccine: Why? How? When? *Tropical Medicine and Infectious Disease*. 2021; 6(1):16. https://doi.org/10.3390/tropicalmed6010016

CHAPTER 5

Vaccination Against Tuberculosis: Revamping BCG by Molecular Genetics Guided by Immunology

Stefan H. E. Kaufmann[1,2]

[1]Max Planck Institute for Infection Biology, Berlin, Germany

[2]Hagler Institute for Advanced Study, Texas A&M University, College Station, TX, United States

ABSTRACT

Tuberculosis (TB) remains a major health threat. Although a vaccine has been available for almost 100 years termed Bacille Calmette-Guérin (BCG), it is insufficient and better vaccines are urgently needed. This treatise describes first the basic immunology and pathology of TB with an emphasis on the role of T lymphocytes. Better understanding of the immune response to *Mycobacterium tuberculosis (Mtb)* serves as blueprint for rational design

---

Citation: Kaufmann S. (2020). Vaccination Against Tuberculosis: Revamping BCG by Molecular Genetics Guided by Immunology. Frontiers in immunology, 11, 316. https://doi.org/10.3389/fimmu.2020.00316.

of TB vaccines. Then, disease epidemiology and the benefits and failures of BCG vaccination will be presented. Next, types of novel vaccine candidates are being discussed. These include: (i) antigen/adjuvant subunit vaccines; (ii) viral vectored vaccines; and (III) whole cell mycobacterial vaccines which come as live recombinant vaccines or as dead whole cell or multi-component vaccines. Subsequently, the major endpoints of clinical trials as well as administration schemes are being described. Major endpoints for clinical trials are prevention of infection (PoI), prevention of disease (PoD), and prevention of recurrence (PoR). Vaccines can be administered either pre-exposure or post-exposure with *Mtb*. A central part of this treatise is the description of the viable BCG-based vaccine, VPM1002, currently undergoing phase III clinical trial assessment. Finally, new approaches which could facilitate design of refined next generation TB vaccines will be discussed.

Keywords: tuberculosis, vaccine, Bacille Calmette-Guérin, subunit, bio-marker, macrophage, T lymphocyte, clinical trial

"Commit to advancing research for basic science, public health research and the development of innovative products and approaches, ..., without which ending the tuberculosis epidemic will be impossible, including towards delivering, as soon as possible, new, safe, effective, equitable, affordable, available vaccines, ..." Resolution adopted by General Assembly of the United Nations from the High Level Meeting on the fight against TB, 2018 (1).

INTRODUCTION

The only tuberculosis (TB) vaccine in use until today, Bacille Calmette Guérin (BCG), was introduced in 1921 after intensive research & development (R&D) for more than a decade (2). It was not the first tryout to immunize against TB. The very first attempt was made by Robert Koch who used a subunit-adjuvant formulation (3). Subsequently, several other approaches were tested including killed mycobacterial vaccines and live non-tuberculous mycobacterial strains. Yet, these all failed and the only vaccine with proven safety and efficacy until today remains BCG. In fact, today BCG is the most widely used vaccine, which has been given more than 4 billion times. BCG was developed to protect newborns at high risk of TB (2). This mission has been accomplished at least partially since BCG was proven to protect against severe extra pulmonary, but less against pulmonary TB in infants (4–6). Yet, even today infant TB takes a worrisome toll in TB endemic countries with high coverage of BCG immunization (7–9). Later, BCG was also tested as a vaccine against pulmonary TB in adolescents and adults, but this ambitious target was not reached and no vaccine has ever succeeded in reliably protecting against pulmonary TB, the most prevalent form of the disease, in any age group. A better vaccine is urgently needed since *Mycobacterium tuberculosis* (*Mtb*), the cause of TB, remains on top of the infamous list of deadly infectious agents (10). In 2018, 10 million individuals fell ill with this disease and 1.5 million died (11) (Figure 1). The early 21st century has witnessed increasing R&D efforts for novel TB vaccines (12–19). These include subunit-adjuvant formulations comprising fusion proteins of *Mtb*, viral vectored vaccines expressing one or more antigens of *Mtb*, killed mycobacterial vaccines and viable attenuated mycobacterial vaccines.

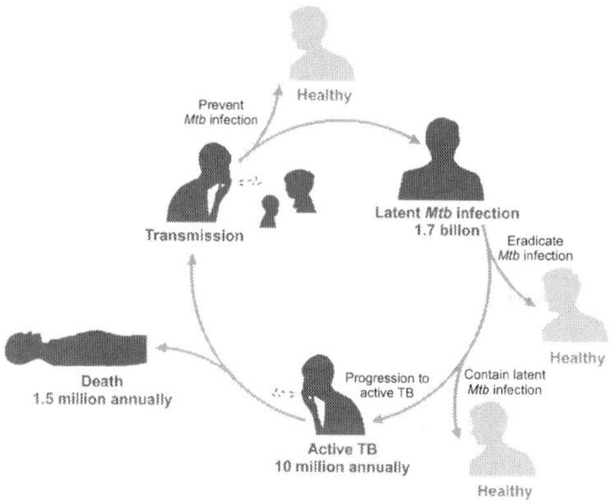

Figure 1: Epidemiologic data for tuberculosis (TB).

IMMUNOPATHOLOGY OF TUBERCULOSIS

Tuberculosis is a chronic infectious disease caused by the intracellular pathogen *Mtb* (20). This acid-fast bacillus is shielded by a unique lipoid-rich cell wall containing various wax-like substances and glycolipids which contribute to resistance against immune attack. *Mtb* is generally transmitted by aerosols in which it enters alveoli in lower lung lobes. Once the pathogen has been engulfed by alveolar phagocytes, it ends up in a phagosome, where it keeps the local pH neutral (21). Moreover, *Mtb* is capable of egressing into the cytosol (22). These and other mechanisms facilitate resistance of *Mtb* to professional phagocytes including polymorphonuclear neutrophilic granulocytes (in short neutrophils) and mononuclear phagocytes (tissue macrophages and monocytes) (23–27). Resting tissue macrophages generally fail to eliminate *Mtb* and serve as its retreat due to their long lifespan. Blood monocytes are slightly more aggressive but fail to achieve sterile elimination of *Mtb*. Neutrophils are highly aggressive phagocytes with the potential to harm *Mtb*. Due to their short lifespan, they generally will not succeed in completely eliminating *Mtb* and they do not serve as a harbor, in which *Mtb* can persist. Once activated by cytokines, notably Interferon-γ (IFN-γ), mononuclear phagocytes increase their anti-bacterial capacities and pose a more serious threat to *Mtb* although they generally fail to eradicate it completely. The innate immune response mediated by professional phagocytes serves as a first barrier for *Mtb*. Recent evidence suggests that epigenetic changes induced by *Mtb* in professional phagocytes leads to trained immunity. Such trained immunity could play a role in early defense against repeated *Mtb* infections (28, 29). However, thus far compelling evidence for this notion is still incomplete.

In addition, subtypes of dendritic cells (DC) can engulf *Mtb* (30, 31). They likely translocate *Mtb* into the lung parenchyma, where the formation of a granuloma is initiated.

Granuloma formation is strongly regulated by T lymphocytes originally stimulated in the draining lymph nodes to which DC harboring *Mtb* navigate (25, 32). T lymphocytes orchestrate formation of solid granulomas which are primarily composed of macrophages, DCs, and T and B lymphocytes. Within these granulomas *Mtb* is contained and the infected individual remains healthy and develops latent TB infection (LTBI) (24, 33, 34). CD4 T cells have been proven to be central to acquired resistance against and containment of *Mtb* (19, 25). According to the cytokines, these CD4 T cells secrete, they can be categorized into TH1, TH2, and TH17 cells. TH1 cells are preferentially stimulated during *Mtb* infection and are of major importance for defense. They produce cytokines such as IFN-γ, interleukin-2 (IL-2) and tumor necrosis-α (TNF-α). TH2 cells are only weakly induced. They are often considered harmful in TB since they induce inappropriate effector mechanisms. Their major cytokines are IL-4, IL-5, IL-10, and IL-13. However, evidence has been provided that TH2 cytokines, at least in part, can contribute to tissue healing. TH17 cells induce rapid proinflammatory responses by secreting IL-17. They are stimulated during *Mtb* infection and evidence has been published that they participate in protection against TB, notably at early stages of infection. The role of CD8 T cells in protection and containment – although less profound – is also widely accepted. CD8 T cells often produce cytokines of TH1 type and in addition express cytolytic activity (19, 25, 26). Contribution of cytolytic mechanisms to killing of *Mtb* has been demonstrated (35). The role of other lymphoid cells including innate lymphoid cells (iLC), NK T cells, mucosa associated immune T cells (MAIT), γδ T lymphocytes, and B lymphocytes is a matter of ongoing discussion (32, 36–45). B lymphocytes could participate in immunity against TB via two mechanisms: First, as regulatory B lymphocytes and second as antibody producing plasma cells. Evidence for regulatory B lymphocytes in immunity against TB is scarce (46, 47). A role for distinct antibody isotypes in defense against TB has been provided (36, 42, 45). Perhaps these antibodies modulate professional phagocytes through their binding to distinct Fc receptors. Convincing evidence has been generated that γδ T cells contribute to early immune defense by secreting IL-17 (38). The iLC can be categorized into iLC-1, iLC-2 and iLC-3 according to their cytokine secretion pattern (40). Cytokines produced by iLC-1 are of TH1 type, iLC-2 cytokines are of TH2 and iLC-3 cytokines are of TH17 type. The iLC-1 and iLC-3 probably contribute to resistance to *Mtb* and the iLC-2 to healing of lesions (37). During chronic infection, canonical CD4 and CD8 T lymphocytes develop into memory T cells which can be grouped into effector memory T cells (T_{EM}), central memory T cells (T_{CM}), and tissue resident memory T cells (T_{RM}) (48). Although the role of the different memory T cells in protection against *Mtb* is incompletely understood, evidence for a particular role of T_{RM} and T_{CM} in protection against *Mtb* has been provided (49, 50). It is likely that different types of memory T cells participate in protective immunity at different stages of infection.

During LTBI, *Mtb* is contained in solid granuolomas (24, 33, 51). LTBI transforms into active TB disease when granulomas become necrotic and then caseous. This happens in about 5% of individuals with LTBI within the first 2 years and in another 5% at later time points. Thus, only ca. 10% of the 1.7 billion individuals with LTBI develop active TB disease (52). Progression to active disease is due to weakening of the immune response via several incompletely understood mechanisms. It is likely that myeloid-derived suppressor cells and regulatory T lymphocytes participate in dampening of protective immunity (53, 54). These cells produce inhibitory cytokines including IL-4, IL-10, and transforming growth factor-β (TGF-β). Moreover, excessive checkpoint control through inhibitory surface molecules including PD-1/PDL-1 and CTLA-4/B7 co-receptor interactions is likely involved (55, 56).

Hybrid 1 (**H1**, phase I completed) comprising either **IC31** or **CAF01** as adjuvant and a fusion protein of Antigen 85B and ESAT-6 as antigen (110, 111).
H4 (phase II completed) and **H56** (phase IIb) formulated in **IC31** as adjuvant and fusion proteins of Antigen 85B and TB10.4 (H4) or Antigen 85B, ESAT-6 and Rv2660c (H56) (73, 112–114).
ID93 (phase IIa) composed of **GLA-SE** as adjuvant and a fusion protein of 4 antigens, namely Rv2608, Rv3619, Rv3620 and Rv1813 (115, 116).
M72 (phase IIb completed) composed of **$AS01_E$** as adjuvant and a fusion protein of 2 antigens, Rv1196, and Rv0125. M72 has completed a phase IIb trial revealing its partial protective efficacy (for further details see text) (65, 66, 117).

- Compositions of adjuvants:

IC31, cationic peptides plus TLR-9 agonist;
CAF01, cationic liposome vehicle plus immunomodulatory glycolipid;
GLA-SE, Squalen oil-in-water emulsion plus TLR-4 agonist;
$AS01_E$, liposomes with monophosphoryl lipid A plus saponin QS21.

- Viable vaccines undergoing clinical testing are:

MTBVAC (phase IIa completed), a genetically attenuated *Mtb* vaccine (118, 119).
VPM1002 (several phase III trials), a **rBCG** vaccine (for further details see text) (84, 85).

- Killed whole cell vaccines include:

DAR-901 (killed *M. obuense*) which had already completed a phase III trial under a different name (120–123) and is now under re-evaluation (phase I trial completed) (124).
MIP (phase III) based on killed *M. indicus pranii* organisms (125–127).
M. vaccae (phase III) based on killed *M. vaccae* (128–132).
RUTI (phase IIa) a purified killed vaccine of *Mtb* fragments (133–135).

- Therapeutic vaccines: The above vaccine trials assess outcome of preventive vaccination. Several candidates are also tested as therapeutic vaccines either for TB treatment in adjunct to canonical chemotherapy or for PoR of TB patients who were cured from TB by canonical chemotherapy but may undergo recurrence (136).
- Therapeutic vaccines in clinical trials include:

H56:IC31 (phase I), a subunit protein formulation;
ID93:GLA-SE (phase I), a subunit protein formulation;
RUTI (phase IIa), a purified killed vaccine of *Mtb* fragments;
TB-FLU-04L (phase IIa), a viral vectored vaccine;
MIP (phase III completed), a killed *M. indicus pranii* preparation;
M. vaccae (phase III completed), a killed *M. vaccae* preparation;
VPM1002 (phase III), a live rBCG vaccine.

- Prevention of Infection (PoI): This clinical endpoint can be applied for pre-exposure vaccination, i.e. vaccination of individuals who have not yet encountered *Mtb*. The most important target group for PoI are neonates. The WHO has prioritized a vaccine to lower the risk of *Mtb* infection (11).
- Prevention of Disease (PoD): It is obvious that PoI will result in PoD. The major target population for PoD, however, are individuals with LTBI. Cutting the risk of TB disease in individuals with LTBI has also been prioritized by the WHO (11).

- Prevention of Recurrence (PoR): In high endemic areas, ca. 10% of TB patients who had been cured by canonical drug treatment undergo recurrence, either due to reinfection or relapse (64).
- Therapeutic Vaccination in Adjunct to Canonical Drug Treatment: Such a vaccination regimen gains increasing importance for patients with multi or extensively drug-resistant TB (MDR / XDR-TB) (16). An estimated half million of active TB patients suffer from MDR-TB and 50,000 to 100,000 individuals from XDR-TB (1). Vaccines for PoR are sometimes considered as therapeutic vaccines, as well.

This review will focus on vaccines that prevent active TB either through PoI, PoD, or PoR.

PREVENTION OF DISEASE BY THE SUBUNIT VACCINE M72 IN A PHASE IIB CLINICAL TRIAL

The M72 vaccine candidate developed by GlaxoSmithKline has successfully completed a phase IIb clinical trial (65, 66). Participants of this study were HIV^{-} adults with LTBI who had been immunized with BCG as infants. Hence, the study was a post-exposure booster immunization with a subunit vaccine with PoD as clinical endpoint. The clinical endpoint was determined after 2 years of follow-up as pulmonary TB in absence of HIV infection (66). The study revealed ca. 50% protection over placebo control. The follow-up study confirmed the efficacy after the third year (65). This is the first vaccine trial to provide evidence for PoD in human TB. A positive control with BCG was not included in this study. It is hoped that global gene expression profiling and immunologic data will provide information about potential mechanisms underlying PoD induced by this vaccine. The vaccine comprises two TB antigens formulated in a potent adjuvant, AS_{01E} (see Box 1). This adjuvant had been developed as part of the adjuvant system (AS) series and is also used in the shingles vaccines, Shingrix, and the malaria vaccine, Mosquirix (67). Availability of AS_{01E} is limited and production cost is high. It has to be seen whether and how these limitations affect supply of this vaccine for broad-scale immunization programs. Satisfactory supply of vaccines for poverty-related diseases including TB and malaria strongly depends on an affordable price (68).

PROMISING PREVENTION OF DISEASE DATA IN NON-HUMAN PRIMATES (NHP) BY A VIRAL VECTORED TUBERCULOSIS VACCINE CANDIDATE

Cytomegalovirus (CMV) based vaccines have been studied in a number of infectious diseases (69). Notably, in a simian immunodeficiency virus (SIV) model of rhesus macaques, CMV vectored vaccines expressing SIV antigens have shown profound protection mediated by CD4 and CD8 T lymphocytes (69). These T cells have been characterized as effector T_{EM} cells and transitional effector memory T cells. Based on these findings, a TB vaccine candidate was designed which is based on a CMV vector expressing 6 or 9 *Mtb* antigens (70). This vector has been tested for PoD in rhesus macaques and was shown to induce profound protection against TB disease (70). Importantly, in a proportion of animals, evidence for sterile eradication of *Mtb* by this CMV-vectored TB vaccine was obtained. As expected, the vaccine induced profound CD4 and CD8 T cell responses as well as marked IFN-γ and TNF secretion. In contrast, antibody responses were not induced significantly. The protective CD8 T cell population was not only restricted by MHC I, but also by MHC-E or MHC II. BCG administered intradermally also induced protection, albeit weaker. Intriguingly, prime with BCG and boost with the CMV-based TB vaccine reverted the strong protective effect of the CMV vaccine to levels of protection induced by BCG. Gene expression profiling of vaccinated animals indicated a role for neutrophils in protection induced by the CMV vectored TB vaccine. In conclusion, despite certain disadvantages of CMV-vectored vaccines in general, the CMV-based TB vaccine represents a promising candidate which deserves further investigation. Obviously, the nullifying effect of BCG prime on protective efficacy induced by the CMV TB vaccine boost needs particular attention. Neonatal BCG immunization is done routinely in high TB endemic areas as part of the expanded program of immunization (EPI) recommended by WHO. Hence a new vaccine that provides no added value for BCG-immunized individuals will face major issues before it can be further developed. Similarly, a recent study revealed that in NHP boosting BCG with M72 or H56 (see Box 1) vaccines failed to enhance protection induced by BCG (71).

RECENT FINDINGS WITH THE CANONICAL BACILLE CALMETTE-GUÉRIN VACCINE

Two recent studies on BCG immunization have revealed marked impact of the vaccination regimen (72, 73). In the first study, NHP were immunized with BCG intravenously (72). Earlier research in the 1970s had already provided compelling evidence that intravenous immunization with live BCG induces superior protection against TB as compared to other routes of administration in NHP with evidence for sterile eradication of *Mtb* (74, 75). Thus, in one study 3/3 animals were markedly protected against TB as measured by hematogenous spread, lymphadenopathy and lung involvement (74). On the other hand, profound splenomegaly was reported after intravenous administration of live BCG. Probably this significant adverse event was the major reason that such studies were not followed up. Only very recently this approach was investigated in greater depth. It was shown that intravenous immunization of NHP with BCG induced more profound protection than intradermal or aerogenic vaccination (72). Indeed, from a proportion of animals receiving BCG by the intravenous route no *Mtb* could be recovered. This study also included a series of highly sophisticated immunologic and pathologic analyses. It was found that antigen responses of CD4 and CD8 T lymphocytes were induced substantially by intravenous immunization prior to *Mtb* challenge, whereas γ/δ T cells and MAIT cells were, similarly, activated as in groups receiving other routes of immunization. The T cell response was mostly of TH1 type with some contribution of TH17 type. On the negative side, splenomegaly was observed after intravenous immunization with a ca. twofold enlargement of spleens compared to controls. However, splenomegaly was transient and 6 months after BCG immunization no differences were observed in spleen size across the different experimental groups including intravenous administration. Six months after immunization, animals were challenged with a low dose of *Mtb*. Positron emission tomography – computed tomography (PET/CT) scans revealed fewer granulomas in the intravenously immunized animals compared to controls. These findings provide proof of concept that BCG immunization can induce profound, in some cases sterile, protection in NHP. It needs to be seen how far the splenomegaly observed will be prohibitive for clinical studies in humans.

The second recent study tested the outcome of BCG booster vaccination in *Mtb* unexposed adults (73). Booster vaccination with BCG had been performed previously although generally it was not endorsed because of the potential risk of adverse events. This assumption was largely based on anecdotal reports describing occasional adverse events after repeated BCG immunization in individuals with LTBI and frequent severe events in TB patients. Principally, BCG revaccination in *Mtb* uninfected individuals does not cause major side effects and in the recent formal clinical trial, BCG revaccination of *Mtb* unexposed individuals demonstrated partial prevention of stable *Mtb* infection (73). More precisely, exposure was determined indirectly via an IFN-γ release assay (IGRA) which determines IFN-γ secretion by canonical T cells after *in vitro* restimulation with *Mtb* specific antigens (76–78). This assay is mostly based on CD4 T cell responses with some contribution of CD8 T cells. Whilst initial IGRA conversion did not differ between BCG immunized and untreated study participants, sustained IGRA conversion was significantly reduced by ca. 45% in BCG immunized study participants over controls (73). These findings can be interpreted to mean that stable *Mtb* infection is prevented by BCG revaccination although in fact it is based on reduced T cell responses as measured by IGRA. It remains to be established more precisely whether prevention of sustained IGRA conversion directly translates into long-term PoI and consequently PoD. Previous observational studies had evaluated PoD by BCG revaccination based on epidemiologic data. Generally, they did not find significant differences between controls and BCG revaccinated individuals (79–81).

These two studies provide strong evidence that the outcome of BCG vaccination is markedly influenced by the kind of administration, notably route of immunization (intravenous) and type of vaccine schedule (pre-exposure revaccination). In conclusion, the BCG vaccine still provides room for improvement.

VPM1002

One of the most advanced TB vaccines, VPM1002, was improved by genetic modification (82). VPM1002 is a recombinant BCG (rBCG) which expresses listeriolysin from *Listeria monocytogenes* and is devoid of urease C (83). Development of this vaccine had started in the 1990s with the aim to improve BCG by endowing it with the capacity to stimulate a broader, more efficacious T cell response.

VPM1002 has successfully completed phase I and phase IIa clinical trials proving its safety and immunogenicity in adults and neonates (84, 85). A phase II clinical trial in HIV exposed and unexposed neonates has been completed and awaits unblinding (NCT 02391415). A phase III clinical trial in HIV exposed and unexposed neonates is being prepared and expected to start in 2020. This trial has been designed as pre-exposure BCG replacement for infants with PoI as clinical endpoint. In this clinical trial, termed **priMe**, neonates will be immunized with VPM1002 or BCG as comparator at several sites in Sub-Saharan Africa. A phase III clinical trial with VPM1002 assessing PoR is currently ongoing in India (NCT 03152903). For this trial, patients with TB who had been cured by drug treatment are being recruited. An estimated 10 % of these individuals will develop active TB disease due to reinfection or relapse within 1 year after completion of drug treatment. The clinical trial therefore will reveal whether vaccination with VPM1002 given 3 months after completion of drug treatment can prevent recurrence. A phase III household contact trial has been launched in July 2019 by the Indian Council for Medical Research (ICMR), in which VPM1002 and another vaccine candidate (MIP, see Box 1) will be assessed for PoD in household contacts of patients with active pulmonary TB disease. In addition, VPM1002 is also being assessed as therapeutic agent against non-muscle invasive bladder cancer as substitute for BCG (NCT 02371447). The canonical TB vaccine BCG is the preferred immunomodulatory medicine for treatment of bladder cancer and the current clinical trial assesses whether VPM1002 is safer than and at least equally efficient as BCG against recurrence of bladder cancer. In conclusion, a century after the introduction of the original vaccine BCG, there is hope for a revival of an improved BCG-based TB vaccine. A rationally revamped BCG could contribute to the solution of the TB crisis.

HOW DOES THE INTRACELLULAR BEHAVIOR DIFFER BETWEEN VPM1002, BACILLE CALMETTE-GUÉRIN, AND *MTB*?

Both BCG and *Mtb* reside in phagosomes, which are arrested at an early stage by neutralization of the phagosomal pH to prevent its acidification (21). Consequently, phagolysosome fusion is diminished. Yet, BCG is degraded in the phagosome whereas *Mtb* survives in phagocytes for prolonged periods of time. Only recently, virulence mechanisms of *Mtb* absent from BCG have been elucidated. Although several sub-strains of BCG exist, it is now clear that the critical step which occurred during attenuation of the parental *Mycobacterium bovis* strain was the loss of the region of difference (RD) 1 which encodes a number of gene products mediated through the ESX/type VII secretion system and capable of perturbating the phagosomal membrane (86). Membrane perturbation by the RD-1 gene products of *Mtb* leads to inflammasome activation, apoptosis and autophagy (Figure 2). The signaling cascades involve nod-like receptor protein 3 (NLRP-3) and absent in melanoma 2 (AIM-2), responsible for IL-1 and IL-18 processing from their respective precursor molecules by the inflammasome as well as STING responsible for autophagy and type I IFN dependent responses (87). STING senses cyclic guanosine monophosphate-adenosine monophosphate (cGAMP) derived from double-stranded DNA of *Mtb* via the enzyme cyclic guanosine monophosphate-adenosine monophosphate synthase (cGAS). All these sequelae are caused by *Mtb* but not or less so by BCG.

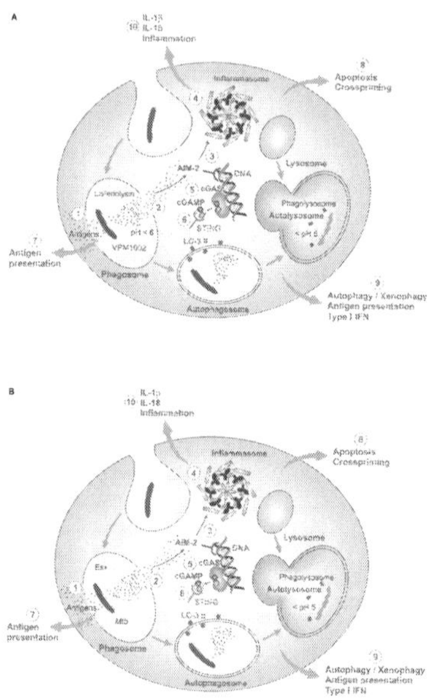

Figure 2: Major mechanisms underlying induction of the host immune response by VPM1002 and *M. tuberculosis* (*Mtb*) (for further details see text). (A) VPM1002. VPM1002 (rBCGΔureC::Hly) expresses listeriolysin and lacks urease C activity. Following phagocytosis, VPM1002 ends up in a phagosome. Principally phagosomes become acidic after uptake of particles, but BCG and *Mtb* actively keep the phagosomal pH neutral. Due to the absence of ureaseC in VPM1002, acidification takes place. This facilitates perturbation of the phagosomal membrane by biologically active listeriolysin. (1) Membrane perturbation allows egress of antigens into the cytosol for processing through the MHC class I pathway. (2) Perturbation can lead to apoptosis. (3) Double-strand DNA released into the cytosol is sensed by absence in melanoma 2 (AIM2). (4) AIM2 activates the inflammasome to generate IL-1β and IL-18. (5) Cyclic GMP-AMP synthase (cGAS) is formed which is then transformed into cyclic guanosine monophosphate-adenosine monophosphate (cGAMP).

(6) The latter molecule is sensed by stimulator of IFN genes (STING) which induces autophagy and type I IFN responses. (7) Antigen egress into the cytosol allows stimulation of CD8 T cells in addition to CD4 T cells. (8) Apoptosis promotes crosspriming. (9) Autophagy accelerates elimination of VPM1002 and improves antigen presentation and T cell stimulation. (10) IL-1β and IL-18 induce an inflammatory response. Through these mechanisms, VPM1002 induces an immune response with more depth and breadth than parental BCG (B) *Mtb*. The genome of *Mtb* comprises the region of difference 1 (RD-1) which encodes numerous virulence factors which are absent in BCG. Notably genes for Esx dependent mechanisms cause perturbation of phagosomal membranes, very similar to VPM1002. For further details see (A). Because the RD-1 encoded gene products are not degraded after their egress into the cytosol, pathologic consequences prevail. Moreover, RD-1 encoded gene products are not controlled by pH. Hence, inbuilt safety mechanisms of VPM1002 are absent from *Mtb* (see also Figure 3).

For the design of VPM1002, BCG was equipped with listeriolysin from *L. monocytogenes* which facilitates perturbation of the phagosomal membrane thereby inducing stronger T cell responses (83). Listeriolysin is a thiol-activated perforin, which perforates cholesterol containing membranes at an acidic pH (88–90). This pH restriction generally prevents listeriolysin activity in the extracellular milieu with neutral pH, e.g. blood and interstitial space. It is however, achieved during natural infection of phagocytes with *L. monocytogenes* which allows secretion of biologically active listeriolysin. Because BCG neutralizes the phagosomal compartment, acidification is not achieved. For this reason, the urease C encoding gene was deleted in VPM1002 (83). This enzyme is responsible for ammonia production and thereby participates in neutralization of the phagosome where BCG resides (21). Accordingly, VPM1002 lacking urease C favors phagosomal acidification and thereby secretion of biologically active listeriolysin (Figure 2). Once listeriolysin has reached the cytosol, it is rapidly degraded. This is due to the amino acid sequence proline-glutamate-serine-threonine (PEST) in the listeriolysin amino acid sequence which promotes its ubiquitination (Figure 3) (88–90). This represents an inbuilt safety mechanism, which restricts listeriolysin-activity to the perturbation of the membrane of the phagosome, where VPM1002 resides and prevents further potentially detrimental effects on cell membranes. The RD-1 encoded machinery of *Mtb* is not endowed with such a safety mechanism.

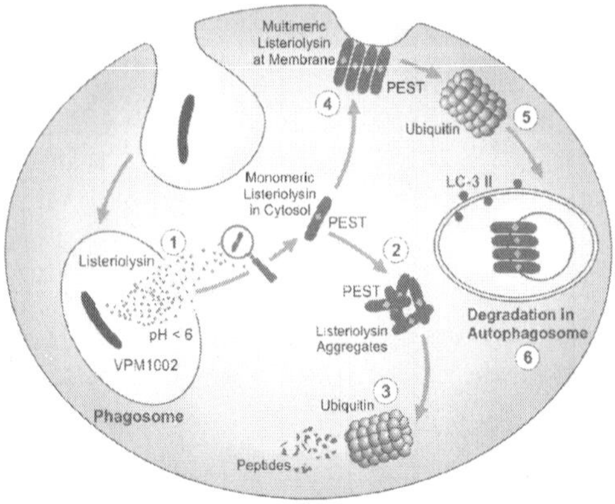

Figure 3: Safety mechanisms of listeriolysin render VPM1002 less virulent than parental BCG. Listeriolysin contains a PEST-like sequence which promotes its degradation. (1) Only at acidic pH, listeriolysin is biologically active and hence perturbates the phagosomal membrane. (2) In the cytosol, monomeric listeriolysin aggregates. (3) Aggregated listeriolysin is degraded by ubiquitin resulting in inactive peptides. (4) Multimeric listeriolysin complexes are formed at the plasma membrane. (5) These complexes are translocated into autophagosomes by ubiquitin. (6) These listeriolysin complexes are inactivated in the phagosome. PEST = Proline (P), Gutalate (E), Serine (S), and Threonine (T). Modified from (88–90).

Similar to the RD-1 machinery in *Mtb*, listeriolysin-mediated perturbation of the phagosomal membrane by VPM1002 results in inflammasome activation through AIM-2 (91). Hence, IL-1 and IL-18 are processed from their respective precursors. These proinflammatory cytokines create a milieu favorable for activation of TH1 and TH17 cells. Listeriolysin also facilitates autophagy via AIM-2 and STING by promoting sensing of double-strand mycobacterial DNA derived from VPM1002 via cGAS and cGAMP after its egress into the cytosol (91). In addition, membrane perturbation by listeriolysin causes apoptosis, which leads to cross priming of T cells (92). Together these mechanisms improve vaccine efficacy of VPM1002 as compared to canonical BCG (82, 93). Moreover, VPM1002 was shown to be safer than BCG in preclinical studies (83). In experimental models improved stimulation of both CD4 and CD8 T cells has been demonstrated (92) as well as more profound activation of TH17 cells in addition to TH1 cells (94). In addition, central memory T cells were more strongly stimulated by VPM1002 as compared to canonical BCG (50). Finally, VPM1002 was shown to stimulate higher serum levels of specific antibodies both in animal models and in human (50, 84, 85). In conclusion, VPM1002 stimulates an immune response of more depth and breadth and at the same time expresses lower virulence as compared to BCG (82).

LEARNING FROM INDIVIDUALS RESISTANT TO STABLE *MTB* INFECTION AND THOSE CAPABLE OF ERADICATING *MTB* AFTER STABLE INFECTION

Individuals with LTBI are generally identified by tuberculin skin test (TST) or IGRA (Figure 4) (76–78). Accordingly, identification of the 1.7 billion individuals on this globe with LTBI is based on measurements of T cell responses against *Mtb* antigens. These antigens are relatively undefined mixtures of *Mtb* proteins (purified protein derivative, PPD) in the case of TST and well defined *Mtb* proteins and/or peptides in the case of IGRA.

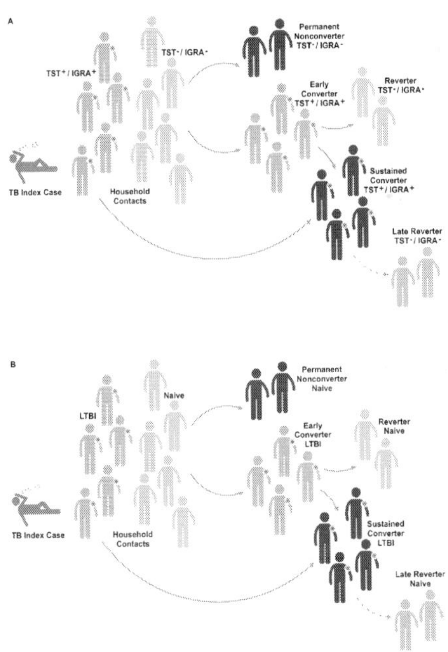

Figure 4: Fate of household contacts of a TB index case. Household contacts of a TB index case are either already latently TB infected (LTBI) or do not show evidence for immunity against *Mtb* infection. After sustained contact with a TB index case, the majority of naïve individuals will rapidly convert to LTBI because they mount an immune response against *Mtb* infection. Most of these early converters will remain LTBI and hence become sustained converters. A small proportion of early converters reverts to naïve, i.e. devoid

of a measurable immune response to *Mtb* infection. Some naïve individuals will remain permanent non-converters, i.e. they do not change their status of absent immunity indicating absence of *Mtb* infection. Finally, some individuals with LTBI will revert to naïve, i.e. they lose their detectable immune response to *Mtb* indicating elimination of *Mtb*. The mechanisms underlying these conversions/reversions remain elusive. (A) Indicates response in TST/IGRA and (B) depicts resulting conclusions on conversion/reversion (for further details see text).

Several lines of evidence suggest that a distinct population of individuals remain *Mtb* uninfected despite their close and prolonged contact with patients with active pulmonary TB who continuously expel *Mtb* (95–98). This notion is based on the finding that such individuals do not convert when tested by TST or IGRA. Assuming that the lack of the canonical immune response determined by TST and/or IGRA reflects absence of *Mtb* infection, the following scenario arises (Figure 3): Initially, household contacts of a patient with active pulmonary TB fall into two groups; those who are already LTBI because of previous contact and hence are TST^+/$IGRA^+$, and naïve individuals who are TST^-/$IGRA^-$. Due to the intensive contact, most of the naïve individuals will convert to TST^+/$IGRA^+$ and most of them remain TST^+/$IGRA^+$ over longer periods of time, if not lifelong. However, a small group may revert to TST^-/$IGRA^-$ indicating that they are capable of eradicating *Mtb* before they become permanently infected. The recent BCG revaccination trial on PoI (73) described above did not reveal significant differences between BCG-immunized and control groups in early conversion to $IGRA^+$. Yet, a 45% reduction in sustained $IGRA^+$ (determined at later time points) was observed in the BCG immunized group as compared to controls without BCG immunization. Moreover, observational studies have identified a distinct group of permanent non-converters (TST^-/$IGRA^-$) generally in the order of 20% (95–98).

Obviously, the described effects could also be due to technical reasons and the TST^-/$IGRA^-$ group could be infected with *Mtb* but missed by TST/IGRA because these individuals develop a protective immune response which is not detected by TST and IGRA. Underlying mechanisms could include antibodies, MAIT cells, γδ T cells, NK cells and NKT cells (32, 36–45). Furthermore, it remains unclear whether all TST^+/$IGRA^+$ individuals are indeed *Mtb* infected or whether at least a subgroup has succeeded in eliminating *Mtb* but remains TST^+/$IGRA^+$ because of a strong memory T cell response which persists in absence of *Mtb* antigens. Principally, immunology defines memory as a state of immunity in absence of nominal antigen(s).

Another interesting group may arise years after primary *Mtb* infection. Whilst many individuals with LTBI remain $TST^+/IGRA^+$ livelong, some individuals revert to $TST^-/IGRA^-$. It is likely that in these individuals, reversion from $TST^+/IGRA^+$ to $TST^-/IGRA^-$ reflects sterile eradication of *Mtb*. Yet, it cannot be excluded formally that these individuals remain *Mtb* infected and control infection by unknown immune mechanisms not detected by TST/IGRA such as antibodies and unconventional T cells. In any case, the permanent non-converters and the late reverters are highly interesting study groups which provide the opportunity to gain deeper insights into the mechanisms of protection against *Mtb*. TB vaccines which prevent stable infection with *Mtb* and thereby prevent LTBI and active TB disease would be highly desirable. The specific mechanisms underlying permanent non-conversion and late reversion could be elucidated by determining transcriptomic, metabolomic and immunologic markers and signatures which distinguish permanent non-converters and late reverters from sustained and livelong reverters, respectively (26, 99).

OUTLOOK AND FUTURE

Over the last decade, the TB vaccine pipeline has significantly progressed. First, a number of vaccines is ready for clinical efficacy testing for PoI, PoD, or PoR (see Box 1). This implies that several vaccine candidates have already proven their safety and immunogenicity. Second, several positive signals arose from clinical trials over the last years including proof of concept that a subunit vaccine empowered by a strong adjuvant can partially protect against active TB when given post-exposure to individuals with LTBI (65, 66). Third, BCG revaccination of *Mtb* unexposed individuals has provided indirect evidence for partial prevention of sustained *Mtb* infection (73). Obviously, major issues remain to be solved. These include: First, BCG revaccination outcome was determined by IGRA which measures canonical T-cell immune responses rather than *Mtb* infection *per se* (see above). This raises the question whether BCG indeed prevented *Mtb* infection or whether infection occurred but was controlled by alternative immune mechanisms such as antibodies and/or unconventional T cells. Second, the M72 clinical trial did not include BCG vaccination as a positive control. Third, studies in NHP revealed that different subunit vaccine constructs including M72, H56, and the CMV vectored TB vaccine failed to increase protection when given as booster on BCG prime (70, 71). In sum, there is well justified hope for better vaccines; but it remains difficult to predict when and to which degree TB can be controlled by improved vaccination strategies.

Nassim Nicholas Taleb is best known for describing the Black Swan Concept which basically includes the notions (100): (i) rare and improbable events do occur more frequently than we assume; (ii) these extreme events can have enormous consequences; (iii) experts generally provide explanations post-hoc which were not plausible ex-ante. This concept was aptly illustrated by the financial stock market crisis in 2008 when numerous stock owners who had gradually accumulated a financial depot went bankrupt through a single event. The most illustrative description for the Black Swan concept is the life of a Thanksgiving Day turkey which is taken care of very well over the first 1000 days by feeding it with most nourishing food. An observer (including the turkey if it could do so) could conclude that the quality of life of this animal increases constantly. Yet, on day 1001, the butcher kills the animal unexpectedly in preparation for Thanksgiving Day. Obviously, this is an extreme event with a major impact on the animal. This scenario can also be turned upside down into a positive direction, i.e. that an unexpected and improbable event turns into something markedly better (an event which would perhaps be better described by the term Pink Swan). With respect to TB vaccine design, continuous funding into R&D (from basic research to preclinical and clinical development) will increase our knowledge about the underlying mechanisms of protection against TB and how this information can be harnessed for TB vaccine design (Figure 5).

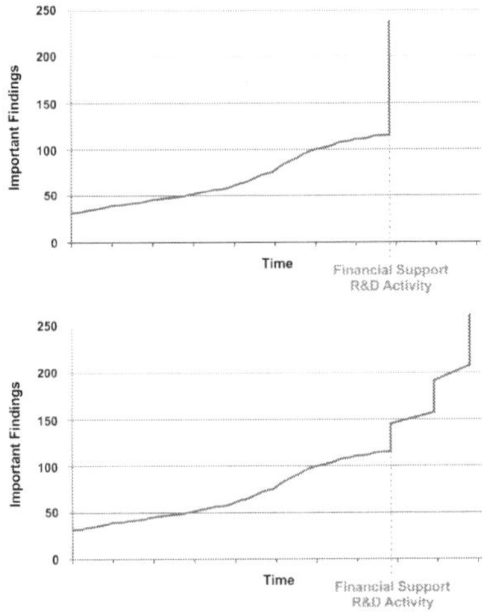

Figure 5: Possible scenarios of TB vaccine development given that adequate financial funding is provided for research & development (R&D). Upper, single step event; Lower, multistep event.

For long periods of time research crawls, but every now and then it jumps. By increasing funding, a fertile soil can be prepared for R&D on better vaccines. Maybe this leads to an extreme event (a single jump), resulting in a novel vaccine that fits all purposes. More likely a couple of smaller, yet significant events will occur which ultimately lead to TB vaccines for different purposes. The type of vaccine and the time when it will be ready for clinical licensing remain unclear as expected for a Black/Pink Swan. Yet, increased funding for R&D will favor chances of success. Undoubtedly, this cannot be accomplished free of cost; yet ultimately it will save cost by reducing the enormous expenses caused by the TB endemic. After all, annual cost for treating active TB disease globally has been estimated to be in the range of 2 billion US dollars and the total burden of TB on the global economy in the order of 100 billion US dollars (101).

It is worth to conclude with a citation from the resolution adopted by the General Assembly of the United Nations from the High Level Meeting on the fight against TB in 2018 (1):

"Commit to mobilize sufficient and sustainable financing, with the aim of increasing overall global investments to 2 billion dollars, in order to close the estimated 1.3 billion dollar gap in funding annually for tuberculosis research, ensuring that all countries contribute appropriately to research and development …."

AUTHOR CONTRIBUTIONS

SK conceived the idea and wrote the manuscript.

ACKNOWLEDGMENTS

The author thanks Diane Schad for superb graphics and Souraya Sibaei for excellent help in preparation of the manuscript.

REFERENCES

1. United Nations *UN General Assembly High-Level Meeting on the Fight Against Tuberculosis.* (2018). Available online at: www.who.int/tb/unhlmonTBDeclaration.pdf (accessed September 26, 2018).
2. Calmette A, Guérin C, Boquet A, Négre L.*La Vaccination Préventive Contre La Tuberculose Par Le "BCG".* Paris: Masson et Cie; (1927).
3. Kaufmann SH, Winau F. From bacteriology to immunology: the dualism of specificity. *Nat Immunol.* (2005) 6:1063–66.
4. Colditz GA, Berkey CS, Mosteller F, Brewer TF, Wilson ME, Burdick E, et al. The Efficacy of bacillus Calmette-Guerin vaccination of newborns and infants in the prevention of tuberculosis: meta-analyses of the published literature. *Pediatrics.* (1995) 96(1 Pt 1):29–35.
5. Colditz GA, Brewer TF, Berkey S, Wilson ME, Burdick E, Fineberg HV, et al. Efficacy of BCG vaccine in the prevention of tuberculosis. Meta-analysis of the published literature. *JAMA.* (1994) 271:698–702. 10.1001/jama.271.9.698
6. Roy A, Eisenhut M, Harris RJ, Rodrigues LC, Sridhar S, Habermann S, et al. Effect of BCG vaccination against *Mycobacterium tuberculosis* infection in children: systematic review and meta-analysis. *Br Med J.* (2014) 349:g4643. 10.1136/bmj.g4643 [PMC free article]
7. Chiang SS, Khan FA, Milstein MB, Tolman AW, Benedetti A, Starke JR, et al. Treatment outcomes of childhood tuberculous meningitis: a systematic review and meta-analysis. *Lancet Infect Dis.* (2014) 14:947–57. 10.1016/S1473-3099(14)70852-7
8. Hesseling AC, Marais BJ, Gie RP, Schaaf HS, Fine PE, Godfrey-Faussett P, et al. The risk of disseminated Bacille Calmette-Guerin (BCG) disease in HIV-infected children. *Vaccine.* (2007) 25:14–8.
9. Trunz BB, Fine P, Dye C. Effect of BCG vaccination on childhood tuberculous meningitis and miliary tuberculosis worldwide: a meta-analysis and assessment of cost-effectiveness. *Lancet.* (2006) 367:1173–80.
10. Abu-Raddad LJ, Sabatelli L, Achterberg JT, Sugimoto JD, Longini IM, Jr., Dye C, et al. Epidemiological benefits of more-effective tuberculosis vaccines, drugs, and diagnostics. *Proc Natl Acad Sci USA.* (2009) 106:13980–85. 10.1073/pnas.0901720106
11. World Health Organization. *WHO Global Tuberculosis Report 2019.* Geneva: World Health Organization; (2019).

12. Andersen P, Kaufmann SH. Novel vaccination strategies against tuberculosis. *Cold Spring Harb Perspect Med.* (2014) 4:a018523. 10.1101/cshperspect.a018523
13. Andersen P, Scriba TJ. Moving tuberculosis vaccines from theory to practice. *Nat Rev Immunol.* (2019) 19:550–62. 10.1038/s41577-019-0174-z
14. Ginsberg AM. Designing tuberculosis vaccine efficacy trials – lessons from recent studies. *Expert Rev Vaccines.* (2019) 18:423–32. 10.1080/14760584.2019.1593143
15. Kaufmann SH. Future vaccination strategies against tuberculosis: thinking outside the box. *Immunity.* (2010) 33:567–77. 10.1016/j.immuni.2010.09.015
16. Kaufmann SH, Weiner J, von Reyn CF. Novel approaches to tuberculosis vaccine development. *Int J Infect Dis.* (2017) 56:263–67. 10.1016/j.ijid.2016.10.018
17. Kaufmann SHE. Tuberculosis vaccines: time to think about the next generation. *Semin Immunol.* (2013) 25:172–81. 10.1016/j.smim.2013.04.006
18. Kaufmann SHE, Dockrell HM, Drager N, Ho MM, McShane H, Neyrolles O, et al. TBVAC2020: advancing tuberculosis vaccines from discovery to clinical development. *Front Immunol.* (2017) 8:1203. 10.3389/fimmu.2017.01203
19. Ottenhoff TH, Kaufmann SH. Vaccines against tuberculosis: where are we and where do we need to go? *PLoS Pathog.* (2012) 8:e1002607. 10.1371/journal.ppat.1002607
20. Pai M, Behr MA, Dowdy D, Dheda K, Divangahi M, Boehme CC, et al. Tuberculosis. *Nat Rev Dis Primers.* (2016) 2:16076. 10.1038/nrdp.2016.76
21. Gordon AH, Hart PD, Young MR. Ammonia inhibits phagosome-lysosome fusion in macrophages. *Nature.* (1980) 286:79–80. 10.1038/286079a0
22. van der Wel N, Hava D, Houben D, Fluitsma D, van Zon M, Pierson J, et al. *M. tuberculosis* and *M. leprae* translocate from the phagolysosome to the cytosol in myeloid cells. *Cell.* (2007) 129:1287–98.
23. Dorhoi A, Kaufmann SH. Versatile myeloid cell subsets contribute to tuberculosis-associated inflammation. *Eur J Immunol.* (2015) 45:2191–202. 10.1002/eji.201545493

24. Gengenbacher M, Kaufmann SHE. *Mycobacterium tuberculosis*: success through dormancy. *FEMS Microbiol Rev.* (2012) 36:514–32. 10.1111/j.1574-6976.2012.00331.x
25. O'Garra A, Redford PS, McNab FW, Bloom CI, Wilkinson RJ, Berry MP. The immune response in tuberculosis. *Annu Rev Immunol.* (2013) 31:475–527. 10.1146/annurev-immunol-032712-095939
26. Ottenhoff TH, Ellner JJ, Kaufmann SH. Ten challenges for TB biomarkers. *Tuberculosis.* (2012) 92 (Suppl. 1):S17–20. 10.1016/S1472-9792(12)70007-0
27. Schorey JS, Schlesinger LS. Innate immune responses to tuberculosis. *Microbiol Spectr.* (2016) 4:TBTB2-0010-2016.
28. Aaby P, Benn CS. Stopping live vaccines after disease eradication may increase mortality. *Vaccine.* (2019) 38:10–4. 10.1016/j.vaccine.2019.10.034
29. Netea MG, Schlitzer A, Placek K, Joosten LAB, Schultze JL. Innate and adaptive immune memory: an evolutionary continuum in the host's response to pathogens. *Cell Host Microbe.* (2019) 25:13–26. 10.1016/j.chom.2018.12.006
30. Lozza L, Farinacci M, Fae K, Bechtle M, Staber M, Dorhoi A, et al. Crosstalk between human DC subsets promotes antibacterial activity and CD8+ T-cell stimulation in response to Bacille Calmette-Guerin. *Eur J Immunol.* (2014) 44:80–92. 10.1002/eji.201343797
31. Randall TD. Pulmonary dendritic cells: thinking globally, acting locally. *J Exp Med.* (2010) 207:451–4. 10.1084/jem.20100059
32. Behar SM, Carpenter SM, Booty MG, Barber DL, Jayaraman P. Orchestration of pulmonary T cell immunity during *Mycobacterium tuberculosis* infection: immunity interruptus. *Semin Immunol.* (2014) 26:559–77. 10.1016/j.smim.2014.09.003
33. Barry CE, III, Boshoff HI, Dartois V, Dick T, Ehrt S, Flynn J, et al. The spectrum of latent tuberculosis: rethinking the biology and intervention strategies. *Nat Rev Microbiol.* (2009) 7:845–55. 10.1038/nrmicro2236
34. Ulrichs T, Kaufmann SH. New insights into the function of granulomas in human tuberculosis. *J Pathol.* (2006) 208:261–9.
35. Stenger S, Hanson DA, Teitelbaum R, Dewan P, Niazi KR, Froelich CJ, et al. An antimicrobial activity of cytolytic T cells mediated by granulysin. *Science.* (1998) 282:121–5.

36. Achkar JM, Casadevall A. Antibody-mediated immunity against tuberculosis: implications for vaccine development. *Cell Host Microbe.* (2013) 13:250–62. 10.1016/j.chom.2013.02.009
37. Ardain A, Domingo-Gonzalez R, Das S, Kazer SW, Howard NC, Singh A, et al. Group 3 innate lymphoid cells mediate early protective immunity against tuberculosis. *Nature.* (2019) 570:528–32. 10.1038/s41586-019-1276-2
38. Coulter F, Parrish A, Manning D, Kampmann B, Mendy J, Garand M, et al. IL-17 production from T helper 17, mucosal-associated invariant T, and γδ cells in tuberculosis infection and disease. *Front Immunol.* (2017) 8:1252.
39. Downey AM, Kaplonek P, Seeberger PHMAIT. MAIT cells as attractive vaccine targets. *FEBS Lett.* (2019) 593:1627–40. 10.1002/1873-3468.13488
40. Klose CS, Artis D. Innate lymphoid cells as regulators of immunity, inflammation and tissue homeostasis. *Nat Immunol.* (2016) 17:765–74. 10.1038/ni.3489
41. Li H, Javid B. Antibodies and tuberculosis: finally coming of age? *Nat Rev Immunol.* (2018) 18:591–6. 10.1038/s41577-018-0028-0
42. Lu LL, Chung AW, Rosebrock TR, Ghebremichael M, Yu WH, Grace PS, et al. A functional role for antibodies in tuberculosis. *Cell.* (2016) 167:433–43.e14. 10.1016/j.cell.2016.08.072
43. Lu LL, Smith MT, Yu KKQ, Luedemann C, Suscovich TJ, Grace PS, et al. IFN-gamma-independent immune markers of *Mycobacterium tuberculosis* exposure. *Nat Med.* (2019) 25:977–87. 10.1038/s41591-019-0441-3
44. Roy Chowdhury R, Vallania F, Yang Q, Lopez Angel CJ, Darboe F, Penn-Nicholson A, et al. A multi-cohort study of the immune factors associated with *M. tuberculosis* infection outcomes. *Nature.* (2018) 560:644–8.
45. Zimmermann N, Thormann V, Hu B, Kohler AB, Imai-Matsushima A, Locht C, et al. Human isotype-dependent inhibitory antibody responses against *Mycobacterium tuberculosis*. *EMBO Mol Med.* (2016) 8:1325–39. 10.15252/emmm.201606330
46. Benard A, Sakwa I, Schierloh P, Colom A, Mercier I, Tailleux L, et al. B cells producing Type I IFN modulate macrophage polarization in tuberculosis. *Am J Respir Crit Care Med.* (2018) 197:801–13. 10.1164/rccm.201707-1475OC

47. Cerqueira C, Manfroi B, Fillatreau S. IL-10-producing regulatory B cells and plasmocytes: molecular mechanisms and disease relevance. *Semin Immunol.* (2019) 44:101323. 10.1016/j.smim.2019.101323
48. Masopust D, Soerens AG. Tissue-resident T cells and other resident leukocytes. *Annu Rev Immunol.* (2019) 37:521–46. 10.1146/annurev-immunol-042617-053214
49. Perdomo C, Zedler U, Kuhl AA, Lozza L, Saikali P, Sander LE, et al. Mucosal BCG vaccination induces protective lung-resident memory T cell populations against tuberculosis. *mBio.* (2016) 7:e01686-16. 10.1128/mBio.01686-16
50. Vogelzang A, Perdomo C, Zedler U, Kuhlmann S, Hurwitz R, Gengenbacher M, et al. Central memory CD4+ T cells are responsible for the recombinant Bacillus Calmette-Guerin Δurec::hly vaccine's superior protection against tuberculosis. *J Infect Dis.* (2014) 210:1928–37.
51. Simmons JD, Stein CM, Seshadri C, Campo M, Alter G, Fortune S, et al. Immunological mechanisms of human resistance to persistent *Mycobacterium tuberculosis* infection. *Nat Rev Immunol.* (2018) 18:575–89. 10.1038/s41577-018-0025-3
52. Houben RM, Dodd PJ. The global burden of latent tuberculosis infection: a re-estimation using mathematical modelling. *PLoS Med.* (2016) 13:e1002152. 10.1371/journal.pmed.1002152
53. Brighenti S, Ordway DJ. Regulation of immunity to tuberculosis. *Microbiol Spectr.* (2016) 4:TBTB2-0006-2016. 10.1128/microbiolspec.TBTB2-0006-2016
54. Dorhoi A, Du Plessis N. Monocytic myeloid-derived suppressor cells in chronic infections. *Front Immunol.* (2017) 8:1895. 10.3389/fimmu.2017.01895
55. Wei SC, Duffy CR, Allison JP. Fundamental mechanisms of immune checkpoint blockade therapy. *Cancer Discov.* (2018) 8:1069–86. 10.1158/2159-8290.CD-18-0367
56. Barber DL, Sakai S, Kudchadkar RR, Fling SP, Day TA, Vergara JA, et al. Tuberculosis following PD-1 blockade for cancer immunotherapy. *Sci Transl Med.* (2019) 11:eaat2702. 10.1126/scitranslmed.aat2702
57. Dodd CE, Schlesinger LS. New concepts in understanding latent tuberculosis. *Curr Opin Infect Dis.* (2017) 30:316–21. 10.1097/QCO.0000000000000367

58. Lin PL, Ford CB, Coleman MT, Myers A, Gawande R, Ioerger T, et al. Sterilization of granulomas is common in active and latent tuberculosis despite within-host variability in bacterial killing. *Nat Med.* (2014) 20:75–9. 10.1038/nm.3412
59. Drain PK, Bajema KL, Dowdy D, Dheda K, Naidoo K, Schumacher SG, et al. Incipient and subclinical tuberculosis: a clinical review of early stages and progression of infection. *Clin Microbiol Rev.* (2018) 31:e00021-18. 10.1128/CMR.00021-18
60. Suliman S, Thompson E, Sutherland J, Weiner Rd J, Ota MOC, Shankar S, et al. Four-gene pan-African blood signature predicts progression to tuberculosis. *Am J Respir Crit Care Med.* (2018) 197:1198–208. 10.1371/journal.pmed.1002781
61. Weiner J, III, Maertzdorf J, Sutherland JS, Duffy FJ, Thompson E, Suliman S, et al. Metabolite changes in blood predict the onset of tuberculosis. *Nat Commun.* (2018) 9:5208. 10.1038/s41467-018-07635-7
62. Zak DE, Penn-Nicholson A, Scriba TJ, Thompson E, Suliman S, Amon LM, et al. A blood RNA signature for tuberculosis disease risk: a prospective cohort study. *Lancet.* (2016) 387:2312–22. 10.1016/S0140-6736(15)01316-1
63. Weiner J, III, Kaufmann SH. Recent advances towards tuberculosis control: vaccines and biomarkers. *J Intern Med.* (2014) 275:467–80. 10.1111/joim.12212
64. Rosser A, Marx FM, Pareek M. Recurrent tuberculosis in the pre-elimination era. *Int J Tuberc Lung Dis.* (2018) 22:139–50. 10.5588/ijtld.17.0590
65. Tait DR, Hatherill M, Van Der Meeren O, Ginsberg AM, Van Brakel E, Salaun B, et al. Final analysis of a trial of M72/AS01E vaccine to prevent tuberculosis. *N Engl J Med.* (2019) 381:2429–39. 10.1056/NEJMoa1909953
66. Van Der Meeren O, Hatherill M, Nduba V, Wilkinson R, Muyoyeta M, Van Brakel E, et al. Phase 2b controlled trial of M72/AS01E vaccine to prevent tuberculosis. *N Engl J Med.* (2018) 379:1621–34. 10.1056/NEJMoa1803484
67. Del Giudice G, Rappuoli R, Didierlaurent AM. Correlates of adjuvanticity: a review on adjuvants in licensed vaccines. *Semin Immunol.* (2018) 39:14–21. 10.1016/j.smim.2018.05.001

68. Kaufmann SHE. Highly affordable vaccines are critical for our continued efforts to reduce global childhood mortality. *Hum Vaccin Immunother.* (2019) 15:2660–5. 10.1080/21645515.2019.1605817

69. Picker LJ, Hansen SG, Lifson JD. New paradigms for HIV/AIDS vaccine development. *Annu Rev Med.* (2012) 63:95–111. 10.1146/annurev-med-042010-085643

70. Hansen SG, Zak DE, Xu G, Ford JC, Marshall EE, Malouli D, et al. Prevention of tuberculosis in rhesus macaques by a cytomegalovirus-based vaccine. *Nat Med.* (2018) 24:130–43. 10.1038/nm.4473

71. Darrah PA, DiFazio RM, Maiello P, Gideon HP, Myers AJ, Rodgers MA, et al. Boosting BCG with proteins or rAD5 does not enhance protection against tuberculosis in rhesus macaques. *NPJ Vaccines.* (2019) 4:21.

72. Darrah PA, Zeppa JJ, Hackney JA, Wadsworth IIMH, Hughes TK, Pokkali S, et al. Prevention of tuberculosis in macaques after intravenous BCG immunization. *Nature.* (2020) 577:95–102. 10.1038/s41586-019-1817-8

73. Nemes E, Geldenhuys H, Rozot V, Rutkowski KT, Ratangee F, Bilek N, et al. Prevention of *M. tuberculosis* infection with H4:IC31 vaccine or BCG revaccination. *N Engl J Med.* (2018) 379:138–49. 10.1056/NEJMoa1714021

74. Anacker RL, Brehmer W, Barclay WR, Leif WR, Ribi E, Simmons JH, et al. Superiority of intravenously administered BCG and BCG cell walls in protecting rhesus monkeys (*Macaca Mulatta*) against airborne tuberculosis. *Z Immunitatsforsch Exp Klin Immunol.* (1972) 143:363–76.

75. Barclay WR, Anacker RL, Brehmer W, Leif W, Ribi E. Aerosol-induced tuberculosis in subhuman primates and the course of the disease after intravenous BCG vaccination. *Infect Immun.* (1970) 2:574–82.

76. Andrews JR, Hatherill M, Mahomed H, Hanekom WA, Campo M, Hawn TR, et al. The dynamics of quantiferon-TB gold in-tube conversion and reversion in a cohort of South African adolescents. *Am J Respir Crit Care Med.* (2015) 191:584–91. 10.1164/rccm.201409-1704OC

77. Diel R, Goletti D, Ferrara G, Bothamley G, Cirillo D, Kampmann B, et al. Interferon-gamma release assays for the diagnosis of latent *Mycobacterium tuberculosis* infection: a systematic review and meta-analysis. *Eur Respir J.* (2011) 37:88–99. 10.1183/09031936.00115110

78. Pai M, Zwerling A, Menzies D. Systematic review: T-cell-based assays for the diagnosis of latent tuberculosis infection: an update. *Ann Intern Med.* (2008) 149:177–84.
79. Barreto ML, Pereira SM, Pilger D, Cruz AA, Cunha SS, Sant'Anna C, et al. Evidence of an effect of BCG revaccination on incidence of tuberculosis in school-aged children in Brazil: second report of the BCG-revac cluster-randomised trial. *Vaccine.* (2011) 29:4875–77. 10.1016/j.vaccine.2011.05.023
80. Karonga Prevention Trial Group. Randomised controlled trial of single BCG, repeated BCG, or combined BCG and killed *Mycobacterium leprae* vaccine for prevention of leprosy and tuberculosis in Malawi. *Lancet.* (1996) 348:17–24. 10.1016/s0140-6736(96)02166-6
81. Rodrigues LC, Pereira SM, Cunha SS, Genser B, Ichihara MY, de Brito SC, et al. Effect of BCG revaccination on incidence of tuberculosis in school-aged children in Brazil: the BCG-revac cluster-randomised trial. *Lancet.* (2005) 366:1290–5.
82. Nieuwenhuizen NE, Kulkarni PS, Shaligram U, Cotton MF, Rentsch CA, Eisele B, et al. The recombinant Bacille Calmette-Guerin vaccine VPM1002: ready for clinical efficacy testing. *Front Immunol.* (2017) 8:1147. 10.3389/fimmu.2017.01147
83. Grode L, Seiler P, Baumann S, Hess J, Brinkmann V, Nasser Eddine A, et al. Increased vaccine efficacy against tuberculosis of recombinant *Mycobacterium bovis* Bacille Calmette-Guérin mutants that secrete listeriolysin. *J Clin Invest.* (2005) 115:2472–9.
84. Grode L, Ganoza CA, Brohm C, Weiner IIIJ, Eisele B, Kaufmann SH. Safety and immunogenicity of the recombinant BCG vaccine VPM1002 in a phase 1 open-label randomized clinical trial. *Vaccine.* (2013) 31:1340–8. 10.1016/j.vaccine.2012.12.053
85. Loxton AG, Knaul JK, Grode L, Gutschmidt A, Meller C, Eisele B, et al. Safety and immunogenicity of the recombinant *Mycobacterium bovis* BCG vaccine VPM1002 in HIV-unexposed newborn infants in South Africa. *Clin Vaccine Immunol.* (2017) 24:e00439-16. 10.1128/CVI.00439-16
86. Vaziri F, Brosch R. ESX/Type VII secretion systems-an important way out for mycobacterial proteins. *Microbiol Spectr.* (2019) 7:PSIB-0029-2019. 10.1128/microbiolspec.PSIB-0029-2019

87. Wassermann R, Gulen MF, Sala C, Perin SG, Lou Y, Rybniker J, et al. *Mycobacterium tuberculosis* differentially activates cGAS- and inflammasome-dependent intracellular immune responses through ESX-1. *Cell Host Microbe.* (2015) 17:799–810. 10.1016/j.chom.2015.05.003
88. Chen C, Nguyen BN, Mitchell G, Margolis SR, Ma D, Portnoy DA. The listeriolysin O pest-like sequence Co-opts AP-2-mediated endocytosis to prevent plasma membrane damage during listeria infection. *Cell Host Microbe.* (2018) 23:786–95.e5. 10.1016/j.chom.2018.05.006
89. Decatur AL, Portnoy DAA. A pest-like sequence in listeriolysin O essential for *Listeria monocytogenes* pathogenicity. *Science.* (2000) 290:992–5.
90. Nguyen BN, Peterson BN, Portnoy DA. Listeriolysin O: a phagosome-specific cytolysin revisited. *Cell Microbiol.* (2019) 21:e12988. 10.1111/cmi.12988
91. Saiga H, Nieuwenhuizen N, Gengenbacher M, Koehler AB, Schuerer S, Moura-Alves P, et al. The recombinant BCG Δurec::hly vaccine targets the AIM2 inflammasome to induce autophagy and inflammation. *J Infect Dis.* (2015) 211:1831–41. 10.1093/infdis/jiu675
92. Farinacci M, Weber S, Kaufmann SH. The recombinant tuberculosis vaccine rBCG Δurec::hly(+) induces apoptotic vesicles for improved priming of CD4(+) and CD8(+) T cells. *Vaccine.* (2012) 30:7608–14.
93. Winau F, Weber S, Sad S, de Diego J, Hoops SL, Breiden B, et al. Apoptotic vesicles crossprime CD8 T cells and protect against tuberculosis. *Immunity.* (2006) 24:105–17.
94. Desel C, Dorhoi A, Bandermann S, Grode L, Eisele B, Kaufmann SH. Recombinant BCG Δurec hly+ induces superior protection over parental BCG by stimulating a balanced combination of Type 1 and Type 17 cytokine responses. *J Infect Dis.* (2011) 204:1573–84.
95. Badger TL, Spink WW. First-infection type of tuberculosis in adults. *N Engl J Med.* (1937) 217:424–31.
96. Dickie HA. Tuberculosis in student nurses and medical students at the University of Wisconsin. *Ann Intern Med.* (1950) 33:941–59.
97. Houk VN, Baker JH, Sorensen K, Kent DC. The epidemiology of tuberculosis infection in a closed environment. *Arch Environ Health.* (1968) 16:26–35.

98. Kaipilyawar V, Salgame P. Infection resisters: targets of new research for uncovering natural protective immunity against *Mycobacterium tuberculosis. F1000Research.* (2019) 8:F1000 Faculty Rev-1698. 10.12688/f1000research.19805.1
99. Kaufmann SH, Evans TG, Hanekom WA. Tuberculosis vaccines: time for a global strategy. *Sci Transl Med.* (2015) 7:276fs8. 10.1126/scitranslmed.aaa4730
100. Taleb NN. *The Black Swan. The Impact of the Highly Improbable.* 2nd ed New York, NY: Penguin Random House LLC; (2010).
101. APPG *The Price of a Pandemic. Counting the Cost of MDR-TB.* London: All-Party Parliamentary Group on Global Tuberculosis; (2015).
102. Ndiaye BP, Thienemann F, Ota M, Landry BS, Camara M, Dieye S, et al. Safety, immunogenicity, and efficacy of the candidate tuberculosis vaccine MVA85a in healthy adults infected with HIV-1: a randomised, placebo-controlled, Phase 2 trial. *Lancet Respir Med.* (2015) 3:190–200. 10.1016/S2213-2600(15)00037-5
103. Tameris MD, Hatherill M, Landry BS, Scriba TJ, Snowden MA, Lockhart S, et al. Safety and efficacy of MVA85A, a new tuberculosis vaccine, in infants previously vaccinated with BCG: a randomised, placebo-controlled Phase 2b trial. *Lancet.* (2013) 381:1021–8.
104. Manjaly Thomas Z-R, Satti I, Marshall JL, Harris SA, Lopez Ramon R, Hamidi A, et al. Alternate aerosol and systemic immunisation with a recombinant viral vector for tuberculosis, MVA85A: a Phase I randomised controlled trial. *PLoS Med.* (2019) 16:e1002790. 10.1371/journal.pmed.1002790
105. Satti I, Meyer J, Harris SA, Manjaly Thomas ZR, Griffiths K, Antrobus RD, et al. Safety and immunogenicity of a candidate tuberculosis vaccine MVA85a delivered by aerosol in BCG-vaccinated healthy adults: a Phase 1, double-blind, randomised controlled trial. *Lancet Infect Dis.* (2014) 14:939–46. 10.1016/S1473-3099(14)70845-X
106. Smaill F, Jeyanathan M, Smieja M, Medina MF, Thanthrige-Don N, Zganiacz A, et al. A human Type 5 adenovirus-based tuberculosis vaccine induces robust T cell responses in humans despite preexisting anti-adenovirus immunity. *Sci Transl Med.* (2013) 5:205ra134. 10.1126/scitranslmed.3006843

107. Smaill F, Xing Z. Human Type 5 adenovirus-based tuberculosis vaccine: is the respiratory route of delivery the future? *Expert Rev Vaccines.* (2014) 13:927–30. 10.1586/14760584.2014.929947

108. Stylianou E, Griffiths KL, Poyntz HC, Harrington-Kandt R, Dicks MD, Stockdale L, et al. Improvement of BCG protective efficacy with a novel chimpanzee adenovirus and a modified vaccinia ankara virus both expressing Ag85A. *Vaccine.* (2015) 33:6800–8. 10.1016/j.vaccine.2015.10.017

109. Sergeeva MV, Pulkina AA, Vasiliev KA, Romanovskaya-Romanko EA, Komissarov A, Kuchur OA, et al. Safety and immunogenicity of cold-adapted recombinant influenza vector expressing ESAT-6 and Ag85A antigens of *M. tuberculosis*. *Vopr Virusol.* (2017) 62:266–72.

110. van Dissel JT, Arend SM, Prins C, Bang P, Tingskov PN, Lingnau K, et al. Ag85B-ESAT-6 adjuvanted with IC31 promotes strong and long-lived *Mycobacterium tuberculosis* specific T cell responses in naive human volunteers. *Vaccine.* (2010) 28:3571–81. 10.1016/j.vaccine.2010.02.094

111. van Dissel JT, Joosten SA, Hoff ST, Soonawala D, Prins C, Hokey DA, et al. A novel liposomal adjuvant system, CAF01, promotes long-lived *Mycobacterium tuberculosis*-specific T-cell responses in human. *Vaccine.* (2014) 32:7098–107. 10.1016/j.vaccine.2014.10.036

112. Geldenhuys H, Mearns H, Miles DJ, Tameris M, Hokey D, Shi Z, et al. The tuberculosis vaccine H4:IC31 is safe and induces a persistent polyfunctional CD4 T cell response in South African adults: a randomized controlled trial. *Vaccine.* (2015) 33:3592–9. 10.1016/j.vaccine.2015.05.036

113. Luabeya AK, Kagina BM, Tameris MD, Geldenhuys H, Hoff ST, Shi Z, et al. First-in-human trial of the post-exposure tuberculosis vaccine H56:IC31 in *Mycobacterium tuberculosis* infected and non-infected healthy adults. *Vaccine.* (2015) 33:4130–40. 10.1016/j.vaccine.2015.06.051

114. Suliman S, Luabeya AKK, Geldenhuys H, Tameris M, Hoff ST, Shi Z, et al. Dose optimization of H56:IC31 vaccine for tuberculosis-endemic populations. a double-blind, placebo-controlled, dose-selection trial. *Am J Respir Crit Care Med.* (2019) 199:220–31. 10.1164/rccm.201802-0366OC

115. Coler RN, Day TA, Ellis R, Piazza FM, Beckmann AM, Vergara J, et al. The TLR-4 agonist adjuvant, GLA-SE, improves magnitude and

quality of immune responses elicited by the ID93 tuberculosis vaccine: first-in-human trial. *NPJ Vaccines.* (2018) 3:34. 10.1038/s41541-018-0057-5

116. Penn-Nicholson A, Tameris M, Smit E, Day TA, Musvosvi M, Jayashankar L, et al. Safety and immunogenicity of the novel tuberculosis vaccine ID93 + GLA-SE in BCG-vaccinated healthy adults in South Africa: a randomised, double-blind, placebo-controlled Phase 1 trial. *Lancet Respir Med.* (2018) 6:287–98. 10.1016/S2213-2600(18)30077-8
117. Leroux-Roels I, Forgus S, De Boever F, Clement F, Demoitie MA, Mettens P, et al. Improved CD4(+) T cell responses to *Mycobacterium tuberculosis* in PPD-negative adults by M72/AS01 as compared to the M72/AS02 and Mtb72F/AS02 tuberculosis candidate vaccine formulations: a randomized trial. *Vaccine.* (2013) 31:2196–206. 10.1016/j.vaccine.2012.05.035
118. Spertini F, Audran R, Chakour R, Karoui O, Steiner-Monard V, Thierry AC, et al. Safety of human immunisation with a live-attenuated *Mycobacterium tuberculosis* vaccine: a randomised, double-blind, controlled Phase I trial. *Lancet Respir Med.* (2015) 3:953–62. 10.1016/S2213-2600(15)00435-X
119. Tameris M, Mearns H, Penn-Nicholson A, Gregg Y, Bilek N, Mabwe S, et al. Live-attenuated *Mycobacterium tuberculosis* vaccine MTBVAC versus BCG in adults and neonates: a randomised controlled, double-blind dose-escalation trial. *Lancet Respir Med.* (2019) 7:757–70. 10.1016/S2213-2600(19)30251-6
120. Johnson JL, Kamya RM, Okwera A, Loughlin AM, Nyole S, Hom DL, et al. Randomized controlled trial of *Mycobacterium vaccae* immunotherapy in non-human immunodeficiency virus-infected Ugandan adults with newly diagnosed pulmonary tuberculosis. the Uganda-Case Western Reserve University research collaboration. *J Infect Dis* (2000) 181:1304–12.
121. Lahey T, Arbeit RD, Bakari M, Horsburgh CR, Matee M, Waddell R, et al. Immunogenicity of a protective whole cell mycobacterial vaccine in HIV-infected adults: a Phase III study in Tanzania. *Vaccine.* (2010) 28:7652–8. 10.1016/j.vaccine.2010.09.041
122. von Reyn CF, Mtei L, Arbeit RD, Waddell R, Cole B, Mackenzie T, et al. Prevention of tuberculosis in Bacille Calmette-Guerin-primed, HIV-infected adults boosted with an inactivated whole-cell mycobacterial

vaccine. *AIDS.* (2010) 24:675–85. 10.1097/QAD.0b013e3283350f1b

123. Vuola JM, Ristola MA, Cole B, Jarviluoma A, Tvaroha S, Ronkko T, et al. Immunogenicity of an inactivated mycobacterial vaccine for the prevention of HIV-associated tuberculosis: a randomized, controlled trial. *AIDS.* (2003) 17:2351–5.
124. von Reyn CF, Lahey T, Arbeit RD, Landry B, Kailani L, Adams LV, et al. Safety and immunogenicity of an inactivated whole cell tuberculosis vaccine booster in adults primed with BCG: a randomized, controlled trial of DAR-901. *PLoS One.* (2017) 12:e0175215. 10.1371/journal.pone.0175215
125. Gupta A, Ahmad FJ, Ahmad F, Gupta UD, Natarajan M, Katoch V, et al. Efficacy of *Mycobacterium indicus pranii* immunotherapy as an adjunct to chemotherapy for tuberculosis and underlying immune responses in the lung. *PLoS One.* (2012) 7:e39215. 10.1371/journal.pone.0039215
126. Mayosi BM, Ntsekhe M, Bosch J, Pandie S, Jung H, Gumedze F, et al. Prednisolone and *Mycobacterium indicus pranii* in tuberculous pericarditis. *N Engl J Med.* (2014) 371:1121–30. 10.1056/NEJMoa1407380
127. Sharma SK, Katoch K, Sarin R, Balambal R, Kumar Jain N, Patel N, et al. Efficacy and safety of *Mycobacterium indicus pranii* as an adjunct therapy in category II pulmonary tuberculosis in a randomized trial. *Sci Rep.* (2017) 7:3354. 10.1038/s41598-017-03514-1
128. Butov DA, Efremenko YV, Prihoda ND, Zaitzeva SI, Yurchenko LV, Sokolenko NI, et al. Randomized, placebo-controlled Phase II trial of heat-killed *Mycobacterium vaccae* (Immodulon Batch) formulated as an oral pill (V7). *Immunotherapy.* (2013) 5:1047–54. 10.2217/imt.13.110
129. de Bruyn G, Garner P. *Mycobacterium vaccae* immunotherapy for treating tuberculosis. *Cochrane Database Syst Rev.* (2003) 2003:CD001166. 10.1002/14651858.CD001166
130. Efremenko YV, Butov DA, Prihoda ND, Zaitzeva SI, Yurchenko LV, Sokolenko NI, et al. Randomized, placebo-controlled Phase II trial of heat-killed *Mycobacterium vaccae* (Longcom Batch) formulated as an oral pill (V7). *Hum Vaccin Immunother.* (2013) 9:1852–6. 10.4161/hv.25280

131. Weng H, Huang J-Y, Meng X-Y, Li S, Zhang G-Q. Adjunctive therapy of *Mycobacterium vaccae* vaccine in the treatment of multidrug-resistant tuberculosis: a systematic review and meta-analysis. *Biomed Rep.* (2016) 4:595–600. 10.3892/br.2016.624

132. Yang XY, Chen QF, Li YP, Wu SM. *Mycobacterium vaccae* as adjuvant therapy to anti-tuberculosis chemotherapy in never-treated tuberculosis patients: a meta-analysis. *PLoS One.* (2011) 6:e23826. 10.1371/journal.pone.0023826

133. Cardona PJ. RUTI: a new chance to shorten the treatment of latent tuberculosis infection. *Tuberculosis (Edinb).* (2006) 86:273–89.

134. Nell AS, D'Lom E, Bouic P, Sabate M, Bosser R, Picas J, et al. Safety, tolerability, and immunogenicity of the novel antituberculous vaccine ruti: randomized, placebo-controlled Phase II clinical trial in patients with latent tuberculosis infection. *PLoS One.* (2014) 9:e89612. 10.1371/journal.pone.0089612

135. Vilaplana C, Montane E, Pinto S, Barriocanal AM, Domenech G, Torres F, et al. Double-blind, randomized, placebo-controlled Phase I clinical trial of the therapeutical antituberculous vaccine RUTI. *Vaccine.* (2010) 28:1106–16. 10.1016/j.vaccine.2009.09.134

136. Groschel MI, Prabowo SA, Cardona PJ, Stanford JL, van der Werf TS. Therapeutic vaccines for tuberculosis – a systematic review. *Vaccine.* (2014) 32:3162–8. 10.1016/j.vaccine.2014.03.047

CHAPTER 6

From Bench to Field: A Guide to Formulating and Evaluating Anti-Tick Vaccines Delving beyond Efficacy to Effectiveness

Charles Ndawula, Jr.[1,2]

[1]National Agricultural Research Organization, Entebbe, Wakiso 256, Uganda

[2]National Livestock Resources Research Institute, Vaccinology Research Programme, Nakyesasa, Wakiso 256, Uganda

ABSTRACT

Ticks are ubiquitous blood-sucking ectoparasites capable of transmitting a wide range of pathogens such as bacteria, viruses, protozoa, and fungi to animals and humans. Although the use of chemicals (acaricides) is the predominant method of tick-control, there are increasing incidents of acaricide tick resistance. Furthermore, there are concerns over accumulation of acaricide residues in meat, milk and in the environment. Therefore, alternative methods of tick-control have been proposed, of which anti-tick cattle vaccination is regarded as sustainable and user-friendly. Over the years,

Citation: Ndawula, C., Jr. From Bench to Field: A Guide to Formulating and Evaluating Anti-Tick Vaccines Delving beyond Efficacy to Effectiveness. Vaccines 2021, 9, 1185. https://doi.org/ 10.3390/vaccines9101185

tremendous progress has been made in identifying and evaluating novel candidate tick vaccines, yet none of them have reached the global market. Until now, Bm86-based vaccines (Gavac™ in Cuba and TickGARD^{PLUS}™ Australia-ceased in 2010) are still the only globally commercialized anti-tick vaccines. In contrast to Bm86, often, the novel candidate anti-tick vaccines show a lower protection efficacy. Why is this so? In response, herein, the potential bottlenecks to formulating efficacious anti-tick vaccines are examined. Aside from Bm86, the effectiveness of other anti-tick vaccines is rarely assessed. So, how can the researchers assess anti-tick vaccine effectiveness before field application? The approaches that are currently used to determine anti-tick vaccine efficacy are re-examined in this review. In addition, a model is proposed to aid in assessing anti-tick vaccine effectiveness. Finally, based on the principles for the development of general veterinary vaccines, a pipeline is proposed to guide in the development of anti-tick vaccines.

Keywords: anti-tick vaccine, anti-tick vaccine efficacy, anti-tick vaccine effectiveness, anti-tick vaccine protection

INTRODUCTION

Ticks are ubiquitous blood-sucking ectoparasites capable of transmitting bacteria, virus, protozoa, and fungi to animals and humans [1,2]. Ticks are categorized into three families: Ixodidae (hard ticks), Argasidae (soft ticks), and Nuttalliellidae (monotypic ticks), of which the Ixodidae are the most significant with about 700 known species [3,4,5]. Of the three families, the argasid and ixodid ticks are of greater veterinary and medical importance [1,2]. However, more efforts have been devoted to the control of ixodid than to argasid ticks. This can mainly be attributed to the differences in feeding habits and, behaviors [6]. Additionally, in comparison to the two- or three-host ixodid ticks, the effectiveness of tick-control measures is likely to be higher against one-host ixodid ticks. Three-host ticks (e.g., *Rhipicephalus appendiculatus*, *Amblyomma variegatum*) are those that feed on a different vertebrate host at every stage of the life cycle, whereas one-host ixodid ticks (e.g., *Rhipicephalus microplus*, *Rhipicephalus decoloratus*, *Rhipicephalus annulatus*) feed on one host throughout their life cycle.

Until now, chemicals (acaricides) remain the frontline tool for livestock tick control [7]. However, the practical significance of acaricides is undermined by increasing acaricide tick resistance [8] and the concern over acaricide residue accumulation in meat, milk, and the environment [9]. For these reasons, researchers have proposed numerous alternative methods for livestock tick control [10]. Relative to the other alternative methods, the immunological tick control (anti-tick vaccination) is regarded as a user-friendly and sustainable treatment method [11]. Ideally, unlike with the two- or three-host ticks, three out of the four stages of the one-host tick life cycle (larvae, nymph, and adult), are pre-exposed to the mode of treatment, for instance, chemicals or anti-tick vaccine-induced antibodies. What this suggests is that, in comparison to the one-host ticks, the control of two- or three-host ticks would require a more expansive treatment approach in order to successfully interfere with the different stages of the tick life cycle. However, it remains unanswered whether, as with acaricides, a proportion of the tick population could acquire tolerance or resistance against anti-tick vaccines.

What Is an Anti-Tick Vaccine?

Anti-tick vaccines are tick proteins or peptides that induce host antibodies which confer protection against ticks. The anti-tick vaccine-constituting proteins (antigens) are broadly categorized into concealed and non-concealed antigens. Unlike concealed antigens, non-concealed antigens are derivatives of tick saliva proteins that are exposed to the vertebrate host during blood meal acquisition [12]. However, based on the above definition of anti-tick vaccines, one may unwittingly conclude that upon feeding on vaccinated cattle, the ticks will instantly be sterilized. On the contrary, unlike with the acaricides or insecticides, anti-tick vaccines do not instantly sterilize the ticks. In broad terms, anti-tick vaccines are defined as tick proteins which are administered to a vertebrate host to induce antibodies with potential to bind and interfere with or inhibit (A) the tick salivary proteins that are secreted at the host–tick feeding site to promote tick pathogen transmission and/or blood acquisition through modulation of the host hemostasis and immune responses, and (B) specific or related proteins that are expressed in the tick tissues to promote its biological parameters (tick feeding, egg laying, egg hatching, larvae, and/or nymph molting). Over time, vaccination against ticks leads to a reduction in the overall tick population and/or interferes with pathogen transmission, and therefore a reduction in the host–tick burden and tick-borne diseases. Specifically, the principle of anti-tick vaccination aims to mimic and exploit the immunological responses that transpire during routine tick feeding (Figure 1) toward affecting the tick biological parameters. Additionally, it should be noted that, potentially, some concealed (e.g., subolesin [13]) or non-concealed (e.g., 64TPR [14]) anti-tick vaccines interfere with pathogen transmission at the tick–pathogen or at the host–pathogen interface, respectively. The phenomenon can be explained through three possible mechanisms: the antibodies may (1) bind to tick proteins (e.g., midgut receptors proteins), which facilitate pathogen entry into the tick tissues for further development; (2) bind to tick salivary proteins that act a vehicle for pathogen entry at the host-tick feeding site; or (3) cross-bind to the pathogen proteins expressed at the host–pathogen and/or at the tick–pathogen interface.

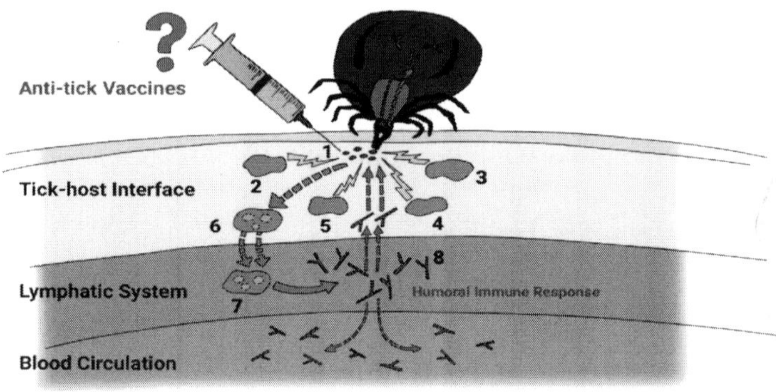

Figure 1: Tick-induced immune response: During feeding, ticks inoculate salivary proteins (1) which trigger recruitment of (2) antigen presenting cells (APC) which include macrophages and dendritic cells, (3) granulocytes which include neutrophils, basophils and eosinophils, (4) mast cells, (5) cytokines [16,17,18,19,20]. Salivary proteins are taken up by APCs (6), carried via the lymphatic system and presented to the lymph nodes to (7) induce a humoral immune response: antibody production (8). The antibodies move through the blood circulation and to the tick feeding site.

Additionally, although some concealed and non-concealed anti-tick vaccines have shown potential to interfere or block tick–pathogen transmission, researchers have explored the possibility of developing more specific vaccines [15]. The more specific tick–pathogen transmission-blocking vaccines are developed based on the pathogen genes expressed at the pathogen–vertebrate and/or the tick–pathogen interface. What this means is that unlike anti-tick vaccine-induced antibodies which bind to the tick proteins and/or cross-bind to the pathogen, transmission-blocking vaccines induce antibodies specifically that bind to the pathogen protein(s) to interfere with its development in ticks and/or in the vertebrate host. Beyond that, it is thinkable that the tick–pathogen transmission-blocking specific vaccines may cross-bind to the structurally related tick proteins. Whether that can impact on the tick biological parameters, remains an open question.

Through the described mechanism (Figure 1), tick salivary proteins can promote acquired tick animal resistance [21]. Note that the vertebrate host resistance is influenced by other factors such as the animal species, animal breed and tick species [22,23]. Evidently, building on the host acquired resistance phenomenon, Trager [24] initiated the concept of anti-tick vaccines in 1939. However, although salivary proteins are regularly inoculated during tick feeding, the proteins are insufficient to instantly induce a strong immune response [25]. Therefore, the inoculation of an anti-tick vaccine aims to instantly induce a strong humoral immune response. However, since the inception of the concept of anti-tick vaccines [24], only two vaccine derivatives of the Bm86 antigen (Gavac™ in Cuba and TickGARD^{PLUSTM} in Australia) have been commercialized, of which TickGARD^{PLUSTM} is no longer available on the market. Even though Gavac™ has shown success in controlling *R. microplus* in Cuba [26], the vaccine is shown not to equally induce protection against the *R. microplus* strains of South America [27]. This is attributed to the genetic diversity of *R. microplus* [28]. Nonetheless, the vaccine was shown to induce cross-protection against other one-host ticks, *R. annulatus* [29] and *R. decoloratus*, but not against three-host ticks, particularly, *R. appendiculatus* and *A. variegatum* [30]. It is unclear whether the low cross-protection was attributed to the difference in vaccine pre-exposure time against the one-host ticks compared to the pre-exposure time against three-host ticks. Unfortunately, while examining the effect against three-host ticks, the vaccine was assessed against only one developmental stage of the tick life cycle (adults). Although progress has been made in identifying tick antigens, the question remains: why is it that the new anti-tick vaccines have not reached the market? [31].

The key observations when assessing new antigens are that the factors which influence vaccine efficacy are less scrutinized and that the vaccines fall short of the goal, that is, to reduce the overall tick population. Therefore, the premise of this article is two-fold: (1) to highlight the bottlenecks to formulating efficacious anti-tick vaccines; and (2) to propose a model for conducting anti-tick vaccine effectiveness studies and a pipeline for assessing the vaccines from the bench to the field.

THE BASIC IMMUNOLOGICAL PRINCIPLE OF ANTI-TICK VACCINATION

World-wide vaccines are classified into two groups: T-cell-independent and T-cell-dependent vaccines [32,33]. Specifically, in contrast to T-cell-dependent vaccines, the T-cell-independent antigens do not require helper T cells to induce adaptive immune responses. Consequently, T-cell-independent vaccines induce a shorter lasting adaptive immune memory compared to the T-cell-dependent vaccines [34]. The mechanism of immune memory induction is illustrated clearly with human vaccines [35,36], but not with anti-tick vaccines. Nonetheless, it is likely that the mechanism is similar with anti-tick vaccines. For a deeper insight on the basics of immune responses, refer to Abbas et al. [37] In regard to memory induction, highly immunogenic tick antigens such as Bm86 [38] and GSTs [39] are reported to have T- and B-cell epitopes. What this suggests is that after obtaining omics data, researchers should examine the immunogenic potential of the target protein(s) based on the presence of B- and T-cell epitopes [40]. Fortunately, this can easily be achieved using immunoinformatics tools. For instance, based on tick GST protein coding sequences, Ndawula et al. [39] proposed a pipeline for predicting B- and T-cell epitopes.

Hypothetical Mechanism of Anti-Tick Immune Memory Induction

In brief, after inoculating the initial vaccine dose (priming dose), the antigen presenting cells (APCs), such as dendritic and macrophages [41], absorb, process, and present the antigen in the lymph node via the major histocompatibility complex (MHC) [42]. The activated dendritic cells interact with naive T cells which differentiate into effector T-helper cells. These cells migrate to the lymphatic germinal center where they interact with naive B cells which leads to the formation plasma and memory B cells. The plasma cells migrate to the bone marrow [43] leading to the production of antigen-specific antibodies, whereas the memory B cells sequester in the spleen and lymph nodes [44,45] whenceforth they await subsequent cognate antigen inoculation. The probable mechanism is illustrated in Figure 2. Refer to Abbas et al. [37] for further understanding of the fundamentals of basic immune responses.

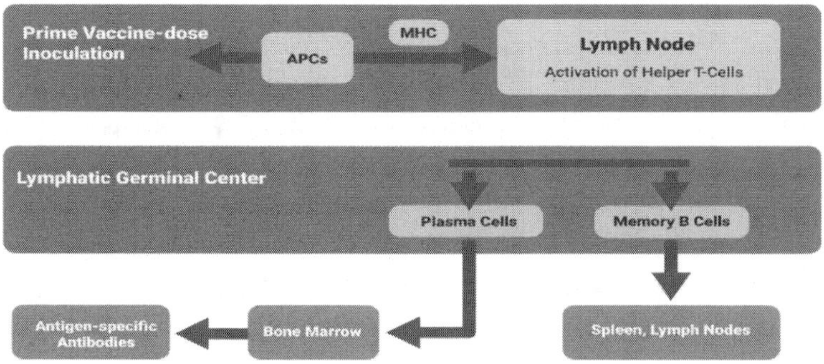

Figure 2: A schematic illustration of hypothetical mechanism of anti-tick immune-memory induction. Abbreviations: MHC (major histocompatibility complex). APCs (antigen presenting cells).

DETERMINATION OF ANTI-TICK VACCINE EFFICACY

Note that although the term anti-tick 'vaccine efficacy' is used interchangeably with 'anti-tick vaccine effectiveness′ [46], the two terms are distinct [47] (vaccine effectiveness is discussed in Section 6). Vaccine efficacy is the measure of the vaccine performance under ideal controlled experimental house/laboratory conditions. In line with conventional vaccines [48], anti-tick vaccine overall efficacy is the proportional measure of reduction in the reproduction success of ticks fed on vaccinated experimental vertebrate hosts compared to those fed on unvaccinated hosts under ideal controlled conditions.

Mathematically, anti-tick vaccine overall efficacy (E) is the relative effect on ticks fed on vaccinated and to ticks fed on unvaccinated animals. Note that in principle, tick reproduction success incorporates the male tick fertilization and the entire tick life cycle. However, often the vaccine-induced overall effect is assessed based on the reduction in blood acquisition, weight and viability of eggs, and the number of female engorged ticks only [49].

$$E\% = 100 \times \left[\frac{Tcontrol - Tvaccinated}{Tcontrol}\right] = 100 \times \left[1 - \frac{Tvaccinated}{Tcontrol}\right]$$

where $T_{Vaccinated}$ is the effect on ticks fed on vaccinated animals. $T_{Control}$ is the effect on ticks fed on control animals.

Currently, researchers use different methods to evaluate the efficacy of anti-tick vaccines [50,51,52]. For example, the formulae that are often used to assess the vaccine-induced effect on one-host and three-host ticks [50,51,52] only take into account the female tick engorgement, oviposition, and egg hatchability [49]. Although not often used, another formula has been proposed to assess the vaccine effect against three-host ticks [51]. In contrast, this method examines the effect induced on the weight and number of the ticks while at different stages of the life cycle [53]. What these differences indicate is that researchers are yet to assent on a uniform method of calculating anti-tick vaccine efficacy [54].

BOTTLENECKS TO DETERMINING ANTI-TICK VACCINE EFFICACY

Although the goal of the researchers is to attain anti-tick vaccines with a high vaccine protection efficacy, note that efficacy mainly depends on the induced humoral immune responses or antibodies. Therefore, hereafter we examine (1) the factors that could influence the antibody production, and (2) the limitations to the methods currently used to assess vaccine efficacy.

Determinants of Anti-Tick Vaccine Immune Response Induction

Adjuvant Anti-Tick Vaccine Formulation

Currently, anti-tick vaccines are most often expressed in *Escherichia coli* than in *Pichia pastoris* [55]. The drawback of *E. coli* expression systems is that the resulting proteins are not likely to undergo proper folding and post-translational modification [56]; hence, affecting their immunogenicity and bioactivity. However, this does not suggest that all proteins expressed in *E. coli* systems are of improper structure, that they lack bioactivity, or that they are not immunogenic. Rather, this suggests that *E. coli* expression systems are not always ideal for the expression of all tick proteins. Nonetheless, it is vital that adjuvants are added to all anti-tick vaccine formulations to augment their immunogenicity. Adjuvants can enhance immunogenicity mainly through two mechanisms: (A) antigen depot formation to ensure slow release and induction of cell-mediated immune responses which are a key precursor in stimulating a strong humoral immune response [57], and (B) inert delivery of antigen to the target immune cells [57]. Some adjuvants could confer a dual function [58]. Further, adjuvants may improve stability and prevent degradation of the antigen [59]. Although advances have been made in adjuvant development [60], a few adjuvants, namely Complete Freund's adjuvant (CFA), Incomplete Freund's adjuvant (FIA), and montanides have been exploited in the formulation of anti-tick vaccines [55]. Note that all of these are emulsion-based adjuvants which act through depot formation at the injection site and slow antigen release to stimulate the immune system [61]. Yet, other forms of adjuvants, such as mineral compounds (e.g., alum, calcium salts), liposomes, bacterial products (e.g., bacterial derived toxins and non-toxins), ISCOMs, tensoactive compounds (e.g., Quil-A), and nanoparticles, have been reported [62]. Nevertheless, like with other vaccines, the selection of adjuvants for the formulation of anti-tick vaccines should be based on the adjuvant potency and the minimal induction of side effects [63]. However, reports on comparison studies regarding the effect of adjuvants on the efficacy of anti-tick vaccines are scarce [63]. Therefore, this could be a key factor, but certainly not the only reason for the paucity of commercial anti-tick vaccines.

Vaccine Dose Administration

(A). Prime vaccine dose

There is a consensus that the antigen dose impacts proportionally on humoral immune responses and the vaccine protection outcome [64,65]. However, a higher antigen dose may not necessarily be better [65,66,67]. For example, unlike with a low antigen dose, when primed with a high antigen dose, the antibody response is higher and quicker, but the antibody titer levels off earlier [68]. Additionally, in contrast to low-dose-induced antibodies, the antibodies due to a high antigen dose are of higher affinity/avidity, yet such antibodies impair the antibody formation mechanism after immunization [68,69,70,71]. In other words, low-affinity antibodies enhance the initial immune response and B-cell activation. By definition, affinity is the binding strength between the antibody epitope and the antigen binding sites (parotype), whereas avidity is the measure of total antibody complexes. Ultimately, inoculating a high antigen concentration also induces vaccine immunotolerance, which in return reduces vaccine efficacy [69,72,73]. The optimum inoculation dose could vary depending on (A) the type of antigen and its ability to trigger different components of immune system [74], (B) the adjuvant used in the vaccine formulation, whereby each adjuvant has unique characteristics and action mechanisms [75], (C) route of administration [60,76], and (D) host related factors. Therefore, with regard to anti-tick vaccine administration, the question is: does the vaccine dose concentration increase with body size? In principle, an equal vaccine dose should be administered irrespective of the size, body weight or breed [77,78,79]. So, how is the vaccine dose determined? Contrary to the principle of drug dose administration [80] (which depends on the circulating drug concentration in blood or tissues), vaccine dose determination depends on the number of circulating lymphoid molecules. Inoculating a vaccine dose based on animal body weight could lead to undesirable immunological effects, and hence a reduction in vaccine efficacy. Therefore, to accurately determine the optimum anti-tick vaccine dose it is important to exploit the humoral immune response kinetics curve (Figure 3) that is based on data derived from pilot-dose response immunological studies performed with a wide range of antigen concentrations. Alternatively, the prime vaccine dose may be determined based on existing information about similar antigens. For instance, data generated from previous studies on protective vaccines, such as for Bm86 [81], could be of use in determining dose concentration of membrane antigen-based anti-tick vaccines. In addition to assessing the antibody titer after inoculating the prime vaccine dose, researchers ought to investigate the antibody avidity [82], given that it depicts antibody affinity.

(B). Booster vaccine dose

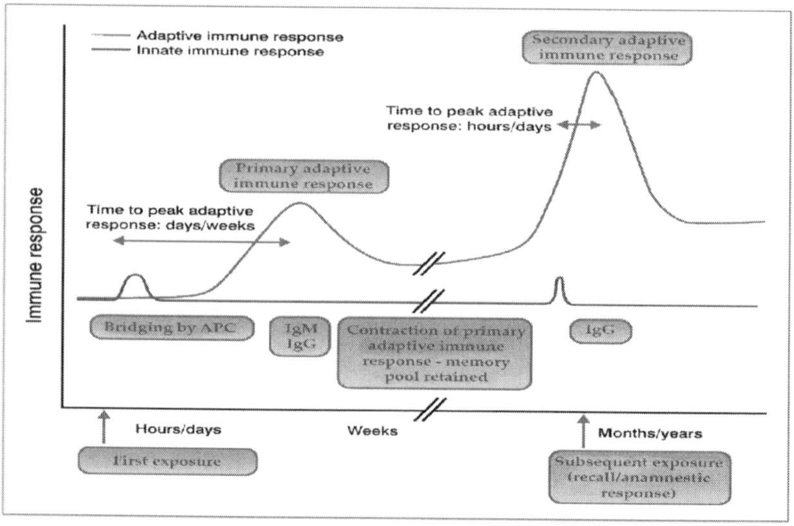

Figure 3: Illustration of the humoral immune response kinetics curve adapted from Leo et al. [104].

The aim of boost vaccination is to elicit a specific, robust, and long-lasting immune protection. Note that although booster vaccination also follows the earlier-described action mechanism (Figure 2), the response is faster [83]. This is because the booster vaccine exploits the prime vaccine-induced memory B cells which change into plasma B cells to produce antigen-specific antibodies. However, evidence shows that the secondary immune response is inversely proportional to primary immune response [84,85]. For instance, Hanna and Peters [85] illustrated that priming an animal with a low vaccine dose leads to a stronger secondary immune response than priming with a higher vaccine dose. However, until now the optimum anti-tick vaccine prime dose is not known. Rather, often tick researchers administer the same vaccine dose during prime-booster vaccination. Based on the discussion regarding the effect of the prime vaccine dose on avidity and immune tolerance, two questions arise: (1) could it be that the administration of a high booster vaccine dose affects the avidity of the secondary antibodies, hence the anti-tick vaccine efficacy?; and (2) does administering half the prime vaccine dose as a booster vaccine dose confer a stronger humoral response? In other words, it is important that the above factors are taken into account when formulating efficacious anti-tick vaccines.

Prime Boost Vaccination Strategy

The concept of prime boost dose anti-tick vaccination was conceived based on the observation that successive tick–animal feeding induces acquired immunity [24], hindering the subsequent tick batch from feeding to engorgement. What this observation suggests is that tick–animal infestation triggers immune memory (Figure 1). After inoculating hosts with tick larvae crude protein extracts, Trager [86] further observed antibody production. However, considering that subunit vaccines, (anti-tick vaccines inclusive) are less immunogenic [87,88], prime boost vaccination must be applied to ensure induction of a sufficient immune memory. There are two methods of prime boost vaccination:

(A). Homologous prime boost dose vaccination

This involves repeated host vaccination (priming and boosting) with the same type of vaccine or antigen [89]. The strategy is widely used for investigating vaccine efficacy. However, with regard to anti-tick vaccines, the strategy has shown mixed success. It is probable that the homologous prime boost vaccination strategy leads to a weak cellular immune response [90]. For instance, homologous prime boost vaccination tends to diminish release of antigen presentation cells and associated signals [89,90]. Therefore, it is likely that this strategy influences the efficacy of new anti-tick vaccines.

(B). Heterologous prime boost dose vaccination

This involves host repeated vaccination (priming and boosting) with different types of vaccines or antigens. For instance, DNA prime/protein booster vaccination [91,92]. This strategy has been adopted to circumvent the limitation of the homologous prime booster vaccination strategy. The benefit of heterologous prime boost vaccination is that it enhances cell-mediated responses, leading to a high humoral immune response [89,91,92,93,94,95,96] which influences vaccine efficacy [96]. However, with regard to anti-tick vaccination, the heterologous prime boost strategy is underexploited [97]. Yet, this vaccination strategy could enhance the efficacy of anti-tick vaccines.

Vaccination Interval and Frequency

The underlying tenet of anti-tick vaccines is to elicit a long-lasting, robust adaptive immune response to limit the animal–tick burden [98]. Immune protection depends on the ability of the memory B cells to generate effective antibodies after inoculation with the cognate antigen [99]. In humans, when there is continuous immune system exposure to the cognate antigen that was used for priming, the protection can last for decades [96]. It is possible that a similar phenomenon could occur with anti-tick vaccines, particularly with non-concealed antigens [12] or concealed/non-concealed (those expressed in saliva and other tissues) [99]. However, the response could be slower considering that a minute quantity of salivary proteins are inoculated during tick feeding [18]. Furthermore, the serum immunoglobulins have a short half-life of 1–3 weeks [100,101]. Therefore, to sustain the immune response, it is necessary to inoculate booster vaccine dose(s) by taking into account two fundamental considerations:

(A). Vaccination booster frequency

Evidence shows that booster vaccination at a short interval after the priming does not lead to induction of a stronger secondary immune response, rather it interferes with the primary response, which leads to immune tolerance [102]. For that reason, while determining the interval for administering the booster dose, researchers should consider the humoral immune-response kinetics curve (Figure 3). For instance, in humans, the humoral immune-response self-termination feedback mechanism occurs at least 2–3 months after the primary vaccination, which is optimal to induce a robust secondary adaptive immune response [103]. However, delays in booster vaccination beyond 3 months does not guarantee improved secondary adaptive immune response [103]. The benefit of allowing the primary immune response to wane is that it allows the completion of affinity maturation and the development of highly specific memory B cells which would subsequently lead to a stronger immune response. For a deeper insight on affinity maturation, refer to Abbas et al. [37]. By contrast, often the booster anti-tick vaccine doses are administered after two weeks [55]. This suggests that most of the anti-tick vaccines are administered before the prime immune response has waned which could affect the immune memory induction, and hence the vaccine protection efficacy.

(B). Number of administered booster vaccine doses

With regard to booster vaccination, it is common practice for tick researchers to administer at least two doses [55]. It is presumed that the more anti-tick vaccination booster doses, the stronger the immune protection. However, practically, this does not occur. Therefore, it is likely that the commonly used booster vaccination approaches [55] lead to interference between the booster-vaccine responses, and hence reduced anti-tick vaccine efficacy. Beyond that, administering a high number of booster doses has vaccine cost implications. Unlike with the anti-tick vaccines, the booster dose of some human vaccines is administered after a longer interval. For instance, the tetanus and diphtheria vaccine booster doses are administered after ten years [105], whereas the interval for hepatitis B is at least 20 years [106]. However, with regard to anti-tick vaccines, reports about booster vaccination doses are scarce. For instance, based on Bm86 (Gavac^{plus}) vaccination against *R. microplus*, Vagas et al. [107] demonstrated that one booster dose per year was sufficient to maintain humoral protection and that the booster dose should not be administered until the antibody titer drops below 1:640. Therefore, it is also possible that anti-tick vaccines, particularly those derived from proteins expressed in the tick saliva and other tissues, may induce a longer protection. The hypothesis is that after administering the prime dose of a concealed/non-concealed vaccine, regularly during blood ingestion, ticks will inoculate the cognate salivary proteins (antigens) in minute quantities, which will continue to sensitize the immune system, hence maintaining humoral immune responses.

Finally, it is prudent that while determining the number of booster doses, the researchers review the humoral immune response kinetics (Figure 3). Note that the booster dose could also be influenced by the booster vaccination strategy, route of administration, animal age, and the booster dose.

Route of Vaccination

Worldwide, vaccines are administered through two routes: mucosally and parenterally. The mucosal route entails administering the vaccine via the mucosal membrane (e.g., nasal, rectal, ocular, oral, vaginal), whereas the parenteral route entails injecting the vaccine through the skin. The skin layers are highly populated with immune cells, such as the granulocytes, antigen-presenting cells, and cytokines, which are key components for the induction of antigen-induced innate responses; a sine qua non for induction of adaptive immune responses [108,109,110,111]. Therefore, to access the skin's layers, parenteral vaccination can be executed via three routes: subcutaneous (SC), intradermal (ID), and intramuscular (IM) [112]. Of the three routes, the parenteral SC route is commonly used to administer anti-tick vaccines.

By contrast, other vaccines (for example, human vaccines) are commonly administered via the parenteral IM route. Generally, vaccination via the IM route induces higher and quicker adaptive immune responses than vaccination via the SC or ID route [113,114,115]. This phenomenon can be explained as follows: unlike via the ID or SC route, when vaccines are inoculated via the IM route, they drain directly into the lymphatic system [114]. In addition, IM vaccine inoculation induces fewer adverse effects than the SC route. For example, in comparison to the IM inoculation, often reactogenicity occurs at the site of injection after ID and SC inoculation [116,117]. Moreover, vaccination via the ID or SC route requires a higher dose compared to vaccination via the IM route [117,118,119]. To enhance the anti-tick vaccine immune protection, researchers may consider using the IM route for prime dose inoculation and other routes for booster dose inoculation. For instance, McCluskie et al. [120] illustrate that parenteral prime mucosal boost vaccination elicits a diverse immune response.

In addition to the parenteral route, recently, Contreras et al. [121] demonstrated the possibility of administering anti-tick vaccines via the mucosal route. Notably, rather than combining the subolesin with the commonly used emulsion adjuvants [55], the vaccine was combined with an immunostimulant: heat-inactivated *Mycobacterium bovis*. It was demonstrated that when anti-tick vaccines are administered via the oral route, it is possible to still attain a high protection efficacy while yet minimizing the stress inflicted to cattle during vaccination [121]. Simply, this is because via the mucosal route, vaccines are administered non-invasively, which does not require the use a of needle. The rationale is that, unlike the skin, which is stratified and keratinized, the mucosal membrane is comprised of a thin single-layered epithelia which permits rapid permeability of the vaccine to the underlying immune cells [122]. Although the mucosal route is attractive, administering anti-tick vaccines (immunogenic tick proteins) orally poses a risk of antigen/protein degradation due to digestive enzymes and bacterial proteases [123].

Nevertheless, to enhance the anti-tick vaccine animal immune protection, researchers may consider using the IM route for prime dose inoculation and other routes for booster dose inoculation. For instance, McCluskie et al. [120] demonstrates that parenteral prime and mucosal boost vaccination elicits a diverse immune response. Finally, it is worthwhile to further exploit the benefits conferred by adjuvants when combined with vaccines [118], and to remember that the route could be influenced by adjuvants [124,125].

Vaccination Animal Model

Until now, naive small animal models (e.g., rabbits, guinea pigs, and mice) or large (sheep, and cattle) are used to assess the efficacy of anti-tick vaccines [55,91]. It should be noted that animal models exhibit varying immunocompetence, which significantly influences the vaccine-induced humoral response. The difference in immunocompetence is higher in mice [126] than in rabbits [127]. Similarly, different cattle breeds, exhibit variable immunocompetence and resistance against diseases or vectors [128,129,130]. The difference in resistance is attributed to the variation in T-cell responses among cattle breeds [130,131]. On the contrary, evidence shows that after anti-tick vaccination, no significant variation in humoral immune response was exhibited among African and European cross-bred cattle [53,132]. However, these studies did not investigate T-cell responses.

Aside from the animal breed, age influences the vaccine immune response. For example, in humans [133], the immune system wanes with age. By contrast, reports on whether the same phenomenon occurs in animals are scarce. Nonetheless, from the field point of view, the older the animal, the more it is exposed to ticks and tick-borne diseases, hence acquiring partial immunity. However, this does not imply that younger animals should be administered with higher vaccine doses than the older animals. The implications are elucidated above. Therefore, in an effort to ensure uniform protection among all animals, researchers ought to consider the following:

(A) The immunocompetence background of the animal model; (B) prime booster vaccine dose administration intervals—in comparison to younger animals, older animals may require a longer interval after the priming dose and between the booster doses; (C) route of administration; and (D) number of booster vaccine doses. Unlike older animals, younger animals may require regular booster vaccine doses. In summary, irrespective of animal breed or age, anti-tick vaccines should confer insignificant variations in the induced immune protection efficacy.

APPROACH OF ASSESSING VACCINE EFFICACY

In general, the current approach of assessing the efficacy of tick vaccines is based on one-host tick species. This entails determining the relative weight reduction of engorged female ticks, eggs, and viable larvae from vaccinated and non-vaccinated hosts [50,51]. However, using this approach, it is not possible to deduce whether the vaccine could reduce the tick population. For example, in the current protocols [50,51], after feeding on experimental animals, the weight of engorged female ticks is recorded to assess the vaccine effect on tick feeding success. In principle, the weight of the tick is directly proportional to the number of eggs [134,135]. Therefore, it is anticipated that ticks fed on vaccinated animals will ingest less blood compared to those fed on unvaccinated animals. It is expected that the ticks which fed on vaccinated animals will lay fewer eggs compared to the eggs from ticks which fed on the unvaccinated animals. Further it is anticipated that the eggs from the ticks of the vaccinated group may fail to hatch, or that the larvae hatched from the eggs of the ticks which fed on vaccinated animals will weigh less compared to the larvae hatched from eggs of the ticks which fed on unvaccinated animals. However, to establish the weight, the larvae are sterilized by freezing. The drawback to this approach is that the viability of the larvae hatched from the eggs from ticks which fed on vaccinated or unvaccinated animals cannot be examined (i.e., nymph development). With this approach, the following questions arise: will the larvae from eggs of lower weight imbibe less blood? If the answer is no, does that imply that larvae cannot molt into nymphs? Additionally, if the larvae do molt, what is the fate of the nymphs? In other words, using the current approach [50,51], it is not possible to extensively assess the anti-tick vaccine efficacy and effectiveness. Therefore, to develop effective universal anti-tick vaccines, there is need for a model for assessing population reduction and mortality of single- and multiple-host ticks. Herein, an alternative approach to determining anti-tick vaccine efficacy while considering the above questions is proposed.

Assessing the Anti-Tick Vaccine Efficacy against Three-Host Ticks

This model is based on *R. appendiculatus* which feeds on three different hosts throughout its life cycle. Note that although rabbits are mentioned in subsequent discussions, they are not the best model for vaccine efficacy studies for one-host ticks [136].

- Step 1: Assess the humoral immune response kinetics

Collect serum after prime dose vaccine inoculation, before and throughout the tick infestation experiments. Preserve the serum at −20 °C. It will later be used to assess the vaccine induced humoral response kinetics. The assumption is that earlier independent vaccine immunogenicity studies were performed in rabbits or mice and that the different vaccine parameters were established. These are the vaccine dose, vaccination strategy, booster vaccination interval, and route of vaccine administration.

- Step 2: Assess adult tick engorgement

Divide naive rabbits into two groups (one unvaccinated control and one vaccine treatment) each made up of at least three rabbits. Infest each rabbit with 60 adult ticks (30 males and 30 females) obtained from the same tick colony batch. Monitor treatment groups daily for 5–7 days [137]. Determine the number of ticks successfully attached after 24 h. Collect male and female ticks as they detach from the host after engorgement. Note the term 'engorgement′ refers to when a tick has fed to repletion and detached from the host. Ticks can feed to partial or full engorgement. Determine the number and weight of female ticks that successfully fed to engorgement. Calculate the difference in weight between the female engorged ticks collected from the control and vaccine-treated rabbits. However, engorgement does not apply to male ticks. There are two reasons: (1) the male ticks ingest less blood than the female ticks. Rather, they feed to ensure copulation and promote female tick feeding [138]; and (2) The male ticks detach and reattach probably to copulate with different female ticks. It is possible, based on these data, to deduce whether the vaccine affects tick feeding, although this does not imply that the harvested ticks cannot lay eggs or that the eggs cannot hatch and develop into nymphs and adults.

- Step 3: Vaccine effect on adult tick egg laying (oviposition)

Incubate the female ticks independently under the appropriate conditions [137]. Collect data on the number ticks laying eggs, and number/weight of eggs per tick.

- Determine the number of eggs by weight

Weigh a portion of the eggs of at least three ticks of equal weight that were fed on the same rabbit. Count the number of eggs in each portion. The eggs can be counted under a magnifying glass. Return the counted eggs to the egg cluster and weigh the entire eggs per tick. Based on the proportionality between the total egg weight and the number of eggs, determine the total number of eggs in the entire egg cluster per tick. Calculate the average number of eggs from the three ticks. Weigh separately the eggs from the remaining ticks and, based on the tick weight and average egg count, calculate the number of eggs from each of the remaining ticks. Finally, calculate the average and standard deviation of the number of eggs per group. Take care not to damage the eggs while handling and counting.

Mathematical example:

Presume that the engorged tick 1, 2, and 3 each weigh 0.5 g.

A portion of eggs from tick 1, 2, and 3 weighs 0.1 g, 0.14 g, and 0.15 g and the corresponding egg number is 450, 480, and 500, respectively. The total egg weight cluster per tick is 0.31 g, 0.33 g, and 0.35 g.

Therefore, the total number of eggs per tick can be calculated as below:

If 0.1 g = 450 eggs
0.31 g = X eggs
X = (0.31 × 450)/0.1
X= 1395 eggs

If 0.14 g = 480 eggs
0.33 g = X eggs
X = (0.33 × 480)/0.14
X= 1131.4 eggs

If 0.15 g = 500 eggs
0.35 g = X eggs
X = (0.35 × 500)/0.15
X= 1166.6 eggs

Batch average tick number: (1395 + 1131.4 + 1166.6)/3 = 1231 eggs.

Batch average egg weight: (0.31 + 0.33 + 0.35)/3 = 0.33 g.

- Step 4: Vaccine effect on egg hatchability

(A)Independently pool and mix (e.g., using a paper) the eggs of ticks from each rabbit. Then, weigh a portion of eggs equivalent to the average cluster egg weight per tick (calculated in Step 2) and dispense the egg clusters into separate microtubes. Place the microtubes into separate bottles and incubate while slanted under appropriate conditions [137].

(B)After the eggs have hatched, expose the molting larvae to white light [139] to ensure larvae migration from the egg shells to the sides of the bottle.

(C)Randomly select one bottle of larvae from the corresponding rabbit within treatment groups to be used in subsequent infestation. Place the other bottles of larvae at −20 °C to freeze sterilize the larvae for at least 3 h. Remove the microtube with egg shells, pour the larvae on a Petri dish, weigh and count the larvae. Make a larvae count for at least three egg batches and calculate the average number of larvae. Then, weigh the other larvae. Based on the proportionality between the average larvae count and the weight of larvae from the remaining eggs, calculate the larvae count per egg batch. Calculate the standard deviation and the average larvae count per group. Given that an equal weight of eggs were incubated to obtain larvae, the average larvae count is representative of all egg clusters. Therefore, it is possible to extrapolate the number of larvae which is used for subsequent infestation.

- Step 5: Vaccine effect on larval feeding

Infest the earlier-vaccinated cognate rabbits with larvae hatching from at least one egg batch and determine the engorged larvae recovery. Observe the time taken for the larvae to feed to engorgement. Incubate the recovered engorged larvae under appropriate conditions [137].

- Step 6: Vaccine effect on larval molting

After larval molting, count the number of nymphs that were recovered. It is possible to also take note of the time taken for the larvae to molt into nymphs. Incubate the nymphs under the appropriate conditions.

- Step 7: Vaccine effect on nymph feeding

Feed the nymphs on the earlier-vaccinated cognate rabbits. Assess the nymph feeding success based on the nymph recovery. Incubate the recovered nymph under appropriate conditions. Afterwards, assess whether the vaccine affects nymph molting.

- Step 8: Vaccine effect on adult tick engorgement and egg laying

Repeat Step 1 and Step 2, but with slight modifications. In brief, infest the vaccinated cognate rabbits with 60 ticks. Monitor and assess the following parameters: (1) the attachment success after 24 h and feeding time interval to engorgement; (2) the number of female ticks which feed to engorgement; and (3) the egg laying capacity of the ticks.

- Step 9: Determine vaccine efficacy

Calculate the vaccine overall efficacy based on the vaccine-induced effect on different tick biological parameters (e.g., tick feeding, egg laying, egg hatching, larvae and nymph molting).

Although other data may be collected, the key parameters below should be used for assessing anti-tick vaccine efficacy. The rationale is that these are the visible indicators/markers of tick population growth under field conditions.

(A)Assess the vaccine effect on the feeding of female ticks

Determine the standard deviation and mean of female ticks harvested in Step 2 and 8. Using the mean value, calculate the percentage effect on tick feeding.

Therefore, % effect on the feeding of female ticks

$$E_{TF} = 100 \times \left[\frac{TFcontrol - TFvaccinated}{TFcontrol}\right] = 100 \times \left[1 - \frac{TFvaccinated}{TFcontrol}\right]$$

where $TF_{Control}$ = mean of ticks from the control group, $TF_{Vaccinated}$ = mean ticks from the vaccinated group.

(B)Assess the vaccine effect on egg laying

Determine the standard deviation and mean of eggs from the ticks harvested in Step 3 and Step 8. Using the mean value, calculate percentage effect on egg laying.

Therefore, % effect on egg laying

$$E_{TE} = 100 \times \left[\frac{TEcontrol - TEvaccinated}{TEcontrol}\right] = 100 \times \left[1 - \frac{TEvaccinated}{TEcontrol}\right]$$

where $TE_{Control}$ = mean of eggs from tick of the control group, $TE_{Vaccinated}$ = mean of eggs from ticks of the vaccinated group.

(C)Assess the vaccine effect on egg hatchability

Determine the standard deviation and mean of larvae hatching from eggs harvested in Step 4. Using the mean value, calculate percentage effect on egg hatching.

Therefore, % effect on egg hatchability

$$E_{TH} = 100 \times \left[\frac{THcontrol - THvaccinated}{THcontrol}\right] = 100 \times \left[1 - \frac{THvaccinated}{THcontrol}\right]$$

where $TH_{Control}$ = mean of larvae hatched from the control group, $TH_{Vaccinated}$ = mean of larvae hatched from the vaccinated group.

(D)Assess vaccine effect on larvae feeding

Determine the standard deviation and mean number of engorged larvae harvested in Step 5. Using the mean value, calculate percentage effect on larvae feeding.

Therefore, % effect on larvae feeding

$$E_{TL} = 100 \times [\frac{TLcontrol - TLvaccinated}{TLcontrol}] = 100 \times [1 - \frac{TLvaccinated}{TLcontrol}]$$

where $TL_{Control}$ = mean of engorged larvae from the control group, $TL_{Vaccinated}$ = mean of engorged larvae from the vaccinated group.

(E)Assess vaccine effect on nymph feeding

Determine the standard deviation and mean of engorged nymph harvested in Step 7. Using the mean value, calculate percentage effect on nymph feeding.

Therefore, % efficacy on nymph feeding

$$E_{TN} = 100 * [\frac{TNcontrol - TNvaccinated}{THcontrol}] = 100 * [1 - \frac{TNvaccinated}{TNcontrol}]$$

where $TN_{Control}$ = mean of engorged nymph from the control group, $TN_{Vaccinated}$ = mean of engorged nymph from the vaccinated group.

(F)Calculate vaccine overall efficacy

The vaccine overall efficacy is a product of the vaccine relative effect on different key biological parameters among ticks from the vaccinated (Vac) and control (Cont) group. The parameters are: TF (Tick feeding), TE (egg laying), TH (egg hatchability), TL (larvae feeding), and TN (nymph feeding).

Therefore, percentage overall efficacy (E%):

$$E\% = 100 \times [1 - [[\frac{TFvac}{TFcont}] \times [\frac{TEvac}{TEcont}] \times [\frac{THvac}{THcont}] \times [\frac{TLvac}{TLcont}] \times [\frac{TNvac}{TNcont}]]]$$

$$= 100 * [1 - [ETF * ETE * ETH * ETL * ETN]]$$

Can the Approach Be Used to Assess Vaccine Efficacy against the One-Host Ticks?

This approach could be also used to assess the efficacy of anti-tick vaccines against one-host ticks. However, the difference between the feeding behaviors of the one-host and three-host ticks influences how the method is applied. Unlike three-host ticks, one-host ticks (e.g., *R. microplus*, *R. decoloratus*, *R. annulatus*) undergo their entire lifecycle on the same host [140]. Note that, unlike with the three-host ticks, one-host ticks do not successfully feed on rabbits [135] or even on other small animal models such as guinea pigs, mice, and rats. Rather, one-host ticks thrive best on cattle. Considering that cattle are more costly, researchers may consider using sheep or goats given that they are closely related to cattle [141]. Contrary to three-host ticks, assessing vaccine efficacy against one-host ticks should include the number of adults obtained after the initial infestation, the number of eggs laid per tick and the number of hatched larvae per egg batch.

- Step 1: Assess vaccine immunogenicity and prepare study ticks

Administer animals with the prime booster vaccine dose. Use ELISA tests to determine vaccine immunogenicity before and during the tick infestation experiments. Select at least 10 engorged colony ticks (of equal weight and from the same batch) and incubate them independently under appropriate conditions to allow egg laying. Select the number of ticks based on the number of animals to be used in the experiment. Determine the average number of eggs laid per tick. For example, using a paper, pool and carefully mix the eggs from all the ticks. Dispense one average weight per tick into independent microtubes and place the microtube in another bottle and incubate under appropriate conditions. After hatching, expose the emerged larvae to white light to enhance and hasten larvae migration [139] from the microtube to the bottle. Determine the number of larvae per average tick egg weight as follows:

Randomly select at least three bottles and keep the bottles at −20 °C for at least 12 h. The aim is to sterilize the larvae. Thereafter, remove the microtube which contains the egg shells, pour out the larvae, weigh and count. Considering that egg batches were of equal count and number, the established average larvae count applies to all the obtained larvae.

- Step 2: Assess the vaccine effect on adult tick engorgement

After prime-boost vaccination, infest experimental animals with a predetermined number of larvae, collect and count the self-detached (full or partially engorged) adult female ticks, and weigh each. Note that although the resulting ticks may be of lower weight, this does not imply that such ticks cannot lay or that the eggs will not hatch.

- Step 3: Assess vaccine effect on adult tick egg laying (oviposition).

Randomly select 30 engorged female ticks per animal and incubate each tick independently under appropriate conditions. Calculate the average tick number per animal as earlier illustrated in Step 3. Pool and mix the eggs as described in the tick preparation stage (Step 1).

- Step 4: Assess vaccine effect on egg hatchability.

After egg hatching, process the emerged larvae per egg batch and count as described in Step 4 (c).

- Step 5: Assess vaccine effect on larvae to adult development.

Infest the vaccinated animals with a predetermined number of the hatched larvae. After, assess the tick engorgement, oviposition, and egg hatchability as in Steps 2, 3, and 4.

- Step 6: Vaccine efficacy assessment

Calculate the mean of ticks dropping off the host, laid eggs, hatched larvae, and the percentage effect (as described in Step 9, A, B, and C, respectively) on tick feeding (TF), egg laying (TL), and egg hatching (TH). Therefore, calculate the percentage overall efficacy (E%) as follows:

$$E\% = 100 \times [1-[[\frac{TFvac}{TFcont}] \times [\frac{TEvac}{TEcont}] \times [\frac{THvac}{THcont}]]]$$

$$= 100 \times [1-[ETF \times ETE \times ETH]]$$

In comparison to the previous methods [50], the proposed method reduces the number of tasks such as sorting larvae from eggs shells, over infesting an animal, and the subsequent challenge of tick incubation during oviposition. Note, however, that the findings only depict the vaccine efficacy, not effectiveness (the reasons are discussed subsequently)

Foreseen Limitations and Opportunities of the Vaccine Efficacy Assessment Approach

In summary, the above-proposed method examines the potential of anti-tick vaccines to reduce tick populations. However, two questions may arise: (1) during the experiments, will the experimental animals, e.g., rabbits or sheep, not acquire humoral resistance against ticks?, and (2) Will that not affect the vaccine efficacy assessment?

In response, first with reference to vaccine efficacy against three-host ticks, it should be remembered that the life cycle entails ticks feeding at three times and on different vertebrate hosts. This, however, does not indicate that three-host ticks cannot feed on the same host at all stages of the tick life cycle. In other words, the ticks do not have a choice; rather, what matters is acquiring a blood meal for continuity of the life cycle. For instance, under the controlled farm system, there is a high probability that at different stages of the life cycle, the ticks may feed on the same animal. This described scenario is similar to what transpires with one-host ticks that feed on one animal throughout its life cycle.

Against that background, the possibility that the experimental animals will acquire tick resistance cannot be ruled out completely. However, evidence indicates that for an animal to acquire tick resistance, multiple tick infestations are required [142] and that the degree of resistance is directly proportional to the number of infesting ticks. Particularly for the three-host ticks, the degree of resistance also varies depending on the preceding tick stage of feeding [24] and the interval between infestations [142]. Specifically, although the animals may acquire some resistance, this is not likely to significantly affect the vaccine efficacy assessment. The reasons are that in the proposed protocol: (1) for the three-host vaccine efficacy studies, rabbits are infested once with the nymph, larval, and adult stages; (2) for vaccine efficacy studies against one- or three-host ticks, a few ticks are used during feeding; and (3) there is an interval between each stage of feeding as the ticks molt to the next state. Note that to improve of the quality of vaccine efficacy data, prior to the experiment, the animals must be naive to ticks.

Another question may arise, particularly with vaccine efficacy studies against one-host ticks: Will the vaccine efficacy assessment against tick feeding not be skewed given that the number of male and female ticks on the animal is not known? The rationale would be that at the start of the experiment, the researcher cannot determine how many females will molt from the nymphs. Nor can he determine the number of females while they are on the animal. Unfortunately, the information regarding the ratio of female and male tick molting from nymphs from a batch of eggs of one-host ticks is still scarce. For that reason, unless there is information on the male/female tick ratio, the researcher may consider focusing on assessing the vaccine effect on egg laying (TE) and egg hatching (TH). In this case, the protocol can be modified accordingly. By contrast, with the three-host ticks, the researcher determines the exact number of ticks to attach and it is possible to keep track of all the subsequent stages especially if engorged ticks are collected from the animals. In other words, the modification proposed in the vaccine efficacy assessments against the one-host ticks does not apply to assessments against three-host ticks. Fortunately, under lab conditions, it is practically possible to conduct independent investigations to establish the male/female tick ratio per egg batch of one-host or three-host ticks.

Nonetheless, through this approach it is possible to (1) determine the degree of resistance based both on the humoral response of the control rabbits and the effect on ticks, (2) establish the span of the humoral immune response or antibody production following the humoral immune response kinetics curve (Figure 3)—this helps the researcher to further investigate the question on the number of vaccine booster doses required, (3) establish whether the vaccine can induce protection against the different stages of tick development, and (4) establish whether natural immunity augments anti-tick vaccine efficacy.

APPROACH TO ASSESSING VACCINE EFFECTIVENESS

Often, vaccine 'efficacy′ is mistaken for vaccine 'effectiveness′ and vice versa [46], yet the terms are distinct [47]. Vaccine effectiveness is defined as the measure of the vaccine performance under uncontrolled field conditions. An alternative defination is that vaccine effectiveness is the measure of how a vaccine that was earlier tested under ideal experimental house/laboratory condition performs under natural conditions. What this means is, unlike the vaccine efficacy data, the vaccine effectiveness assessment data could influence decision making or policy before the vaccine is launched for public livestock health.

Although steady progress has been made in anti-tick vaccine research [142,143,144], only Bm86 is reported to be effective under field conditions [145,146,147]. Specifically, the reports on the effectiveness of Bm86 anti-tick vaccines were based on basic assessments of tick burden and the prevalence of tick-borne disease (anaplasmosis and babesiosis) infection status among cattle that were consistently vaccinated for at least a decade. Note, however, that under field assessments, the effects of tick predictors such as the Oxpeckers, cattle egrets (*Bubulcus ibis*, *Ardeola ibis*) [2], acaricides [148], or pasture conditions [149] come into play. Such confounding factors are likely to significantly influence the interpretation of data collected for vaccine effectiveness under field clinical trial conditions. Therefore, while assessing the effectiveness of anti-tick vaccines, it is relevant to limit the confounding factors. For instance, a semi-field system has been designed

[150] and used for investigating novel malaria mosquito control methods [151]. By contrast, a similar SFS has not been developed for anti-tick vaccine effectiveness assessment clinical trials.

Semi Field System Model for Conducting Anti-Tick Vaccine Effectiveness Studies

As illustrated in Figure 4, the system comprises two cattle shelter houses which can independently accommodate at least 12 cattle. The houses are each surrounded by a trench with water in oil to prevent tick movement from one cattle population to another [52] or from the environment. The water level in the trench is controlled by a safety float value. Adjacent to each house is a paddock with forage that does not inflict anti-tick properties [152,153]. The pasture serves as a breeding ground for ticks, and less so for feeding. Each paddock could be approximately 0.5 hectares. The aim is to limit the cost of fencing for the area (around and above) and increases tick–host encounters. After each tick feeding cycle, the animals should be moved to the houses to allow the grass to grow and ticks to hatch or molt. Experimental animals are returned to the paddocks after sufficient time has passed (depending on the tick life cycle) to allow feeding of the molted ticks.

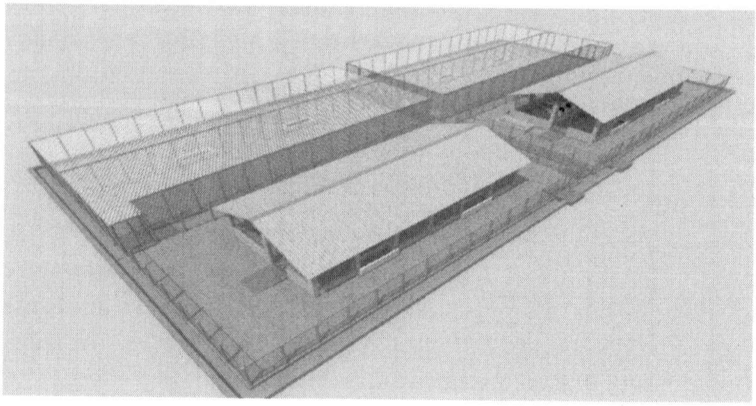

Figure 4: Illustration of the SFS model for conducting anti-tick vaccine effectiveness experiments [154].

Assessment of Vaccine Effectiveness

Following the protocol that was used for determining vaccine efficacy above, but with slight modifications below:

(A) Conduct experiments against all ixodid ticks (one-, two-, or three-host tick) using cattle.

(B) In comparison to efficacy studies, conduct vaccine effectiveness studies with a greater number of ticks.

(C) During each stage of tick infestation, collect and handle samples accordingly. This should also aid in monitoring the development of detached field ticks.

(D) After each infestation stage, let the ticks to drop into the pasture and return the cattle to the shelter house.

(E) Assess the vaccine effectiveness using two cattle populations. Figure 5 illustrates the vaccine effectiveness experimental design.

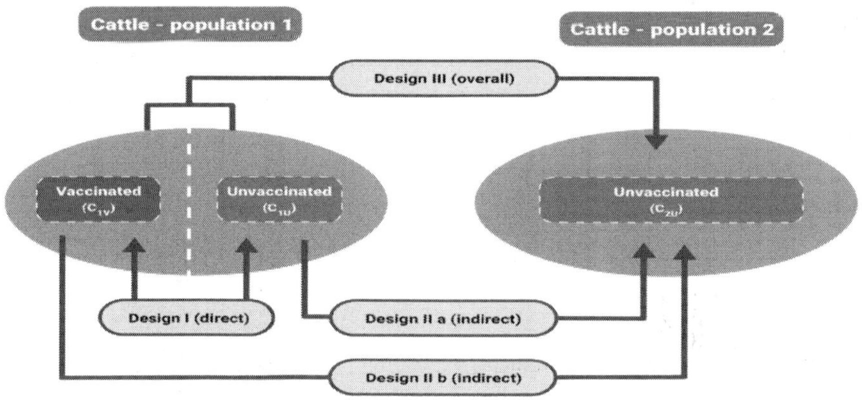

Figure 5: Illustration of the experimental design for assessing anti-tick vaccine effectiveness (adopted from [155]). The assessment is performed using two cattle populations. In Population 1, both the vaccinated (C_{1V}) and unvaccinated (C_{1U}) cattle are kept together, while in Population 2, only unvaccinated (C_{2U}) cattle are present. The two populations are separated from each other. The vaccine-induced effect on ticks from the cattle populations is determined as: Design I: a comparison between effects on ticks from vaccinated and unvaccinated cattle in Population 1. Design II a: a comparison between effects on ticks from unvaccinated cattle in Population 1 and unvaccinated cattle in Population 2. Design II b: a comparison between effects on ticks from vaccinated cattle in Population 1 and unvaccinated cattle in Population 2. Design III: a comparison between effects on ticks from both vaccinated and unvaccinated cattle in Population 1 and unvaccinated cattle in Population 2.

Determine Vaccine Effectiveness

Both anti-tick vaccine efficacy and effectiveness are calculated using the aforementioned formulae [48]. Vaccine effectiveness against three-host ticks assesses the effect on egg laying (TE), egg hatchability (TH), larvae to nymph molting (TL), and nymph to adult molting (TN). By contrast, the assessment of vaccine effectiveness against one-host ticks is based on two parameters: egg laying (TE) and egg hatchability (TH). This is attributed to the differences in tick life cycles. Contrary to the proposed protocol for assessing vaccine efficacy, it is not possible to assess the vaccine effectiveness against tick feeding (TH). The reason is that unlike in the vaccine efficacy assessment, under the vaccine effectiveness assessment, ticks are allowed to drop to the ground into the grass where they will continue their life cycle. In other words, it is not possible to count how many ticks have engorged and dropped onto the grass. Nonetheless, the parameters selected (TE, TH, TL, and TN) are clear indicators of tick burden under field conditions. The assessment is based on the representative tick samples collected from specific cattle in a particular population. After assessing the effect on each parameter, the ticks may be returned to the respective pasture, although not returning the collected samples will not significantly affect the outcome.

Finally, to determine vaccine effectiveness, examine the vaccine-induced effects for at least one year. The objective of conducting vaccine effectiveness studies for at least a year is to incorporate the tick infestation cycles during rainy and sunny season.

The calculations are based on the effect induced on ticks collected from cattle denoted as follows.

- C_{1U}: Ticks from unvaccinated cattle in population 1.
- C_{2U}: Ticks from unvaccinated cattle in population 2.
- C_{1V}: Ticks from vaccinated cattle in population 1.
- C_{OVR}: Overall tick from vaccinated and unvaccinated cattle population in 1.

Determining Vaccine Effectiveness against Three-Host Ticks

- What is the vaccine effectiveness on egg laying?

Collect a predetermined number of engorged adult female ticks from each cattle in a specific population. Based on the mean egg count, determine whether the vaccine affects tick egg-laying ability (for guidance on how to count tick eggs, see the aforementioned mathematical illustrations 5.1: Step

3). The assumption is that after imbibing the vaccine-induced antibodies, ticks are not able to feed to full engorgement, and hence will lay less eggs than ticks which fed on unvaccinated cattle.

Therefore, % effectiveness on egg laying (TE)

Design I (direct)

$$100 \times [\frac{TEC1U - TEC1V}{TEC1U}] = 100 \times [1 - \frac{TEC1V}{TEC1U}]$$

Design II a (indirect)

$$100 \times [\frac{TEC2U - TEC1U}{TEC2U}] = 100 \times [1 - \frac{TEC1U}{TEC2U}]$$

Design II b (indirect)

$$100 \times [\frac{TEC2U - TEC1V}{TEC2U}] = 100 \times [1 - \frac{TEC1V}{TEC2U}]$$

Design III (Overall)

$$100 \times [\frac{TEC2U - TECOVR}{TEC2U}] = 100 \times [1 - \frac{TECOVR}{TEC2U}]$$

- What is the vaccine effectiveness on egg hatching?

The aim is to assess the whether the vaccine affects egg larvae hatching. The larvae used herein are obtained after incubation of the eggs predetermined above. The assumption is that not all eggs laid by the ticks fed on vaccinated animals will hatch to larvae.

Therefore, % effectiveness on egg hatching (TH)=

Design I (direct)

$$100 \times [\frac{THC1U - THC1V}{THC1U}] = 100 \times [1 - \frac{THC1V}{THC1U}]$$

Design II a (indirect)

$$100 \times [\frac{THC2U - THC1U}{THC2U}] = 100 \times [1 - \frac{THC1U}{THC2U}]$$

Design II b (indirect)

$$100 \times [\frac{THC2U - THC1V}{THC2U}] = 100 \times [1 - \frac{THC1V}{THC2U}]$$

Design III (overall)

$$100 \times [\frac{THC2U - THCOVR}{THC2U}] = 100 \times [1 - \frac{THCOVR}{THC2U}]$$

- What is the vaccine effectiveness on larvae feeding?

The aim is to assess whether, under field conditions, the vaccine affects larval development. To attain this goal, collect predetermined larvae from the cattle populations, incubate and compare the larval molting success. The assumption is that if the vaccine-induced antibodies affecting larval blood ingestion (e.g., by impeding larval full engorgement), this could reciprocate to reduced success in larval molting.

Therefore, % effectiveness on larval feeding (TL)

Design I (direct)

$$100 \times [\frac{TLC1U - TLC1V}{TLC1U}] = 100 \times [1 - \frac{TLC1V}{TLC1U}]$$

Design II a (indirect)

$$100 \times [\frac{TLC2U - TLC1U}{TLC2U}] = 100 \times [1 - \frac{TLC1U}{TLC2U}]$$

Design II b (indirect)

$$100 \times [\frac{TLC2U - TLC1V}{TLC2U}] = 100 \times [1 - \frac{TLC1V}{TLC2U}]$$

Design III (overall)

$$100 \times [\frac{TLC2U - TLCOVR}{TLC2U}] = 100 \times [1 - \frac{TLCOVR}{TLC2U}]$$

- What is the vaccine effectiveness on nymph feeding?

The aim is to assess whether, under field conditions, the vaccine affects nymph development. Similarly, collect predetermined nymph from the cattle populations, incubate and compare nymph molting success. The assumption is that if the vaccine-induced antibodies affect nymph blood ingestion (e.g., by impeding larval full engorgement), this could lead to reduced nymph molting success.

Therefore, % effectiveness on nymph feeding (TN)

Design I (direct)

$$100 \times [\frac{TNC1U - TNC1V}{TNC1U}] = 100 \times [1 - \frac{TNC1V}{TNC1U}]$$

Design II a (indirect)

$$100 \times [\frac{TNC2U - TNC1U}{TNC2U}] = 100 \times [1 - \frac{TNC1U}{TNC2U}]$$

Design II b (indirect)

$$100 \times [\frac{TNC2U - TNC1V}{TNC2U}] = 100 \times [1 - \frac{TNC1V}{TNC2U}]$$

Design III (overall)

$$100 \times [\frac{TNC2U - TNCOVR}{TNC2U}]$$

Calculate the Vaccine Overall Effectiveness

This entails calculating the overall vaccine effectivity per design, based on which the vaccine effectiveness on the entire cattle population is determined.

Therefore, the vaccine effectiveness (E)%

Design I effectiveness (E_I)

$$E_I = 100 \times [1-[[\frac{TEC1V}{TEC1U}] \times [\frac{THC1V}{THC1U}] \times [\frac{TLC1V}{TLC1U}] \times [\frac{TNC1V}{TNC1U}]]]$$

Design II a effectiveness (E_{II-a})

$$100 \times [1-[[\frac{TEC1U}{TEC2U}] \times [\frac{THC1U}{THC2U}] \times [\frac{TLC1U}{TLC2U}] \times [\frac{TNC1U}{TNC2U}]]]$$

Design II b effectiveness (E_{II-b})

$$100 \times [1-[[\frac{TEC1V}{TEC2U}] \times [\frac{THC1V}{THC2U}] \times [\frac{TLC1V}{TLC2U}] \times [\frac{TNC1V}{TNC2U}]]]$$

Design III effectiveness (E_{III})

$$100 \times [1-[[\frac{TECOVR}{TEC2U}] \times [\frac{THCOVR}{THC2U}] \times [\frac{TLCOVR}{TLC2U}] \times [\frac{TNCOVR}{TNC2U}]]]$$

Therefore, vaccine overall effectiveness in the semi field system

$$E_{SFS} = 100 \times [1 - [EI \times EII - a \times EII - b \times EIII]]$$

- Determining vaccine effectiveness against one-host ticks

The approach described above could also be used for the one-host ticks, but with a slight modification. In this case, calculation is based on egg laying and egg hatching. The rationale is that unlike three-host ticks, during feeding, the nymph and larvae do not drop off the cattle.

Therefore, the vaccine overall effectiveness in the semi field system ($E_{SFS\%}$) is derived from the following vaccine effectiveness (E) calculations:

Design I effectiveness (E_I)

$$100 \times [1-[[\frac{TEC1V}{TEC1U}] \times [\frac{THC1V}{THC1U}]]]$$

Design II a effectiveness ($E_{II\text{-}a}$)

$$100 \times [1-[[\frac{TEC1U}{TEC2U}] \times [\frac{THC1U}{THC2U}]]]$$

Design II b effectiveness ($E_{II\text{-}b}$)

$$100 \times [1-[[\frac{TEC1V}{TEC2U}] \times [\frac{THC1V}{THC2U}]]]$$

Design III effectiveness (E_{III})

$$100 \times [1-[[\frac{TECOVR}{TEC2U}] \times [\frac{THCOVR}{THC2U}]]]$$

The vaccine overall effectiveness in the semi-field system

$$(E_{SFS}) = 100 \times [1 - [EI \times EII - a \times EII - b \times EIII]] \quad (1)$$

Foreseen Limitations and Opportunities of the Vaccine Effectiveness Assessment Approach

The limitation to the proposed model is that the experiment is costly in terms acquiring experimental cattle. Yet, in temperate areas, for instance in Africa, it may be difficult to find cattle with no or minimal tick exposure. The merit of conducting vaccine-effectiveness studies under the proposed SFS model is that the researcher is able to examine the potential of the vaccine to confer "herd immunity". For a deeper insight on "herd immunity", see the report by John and Samuel [156]. Specifically, the whole principle of anti-tick vaccine-induced herd immunity is that if you inoculate a high percentage of cattle with an anti-tick vaccine, you will reduce the subsequent tick infestation burden, but also prevent ticks from infesting unvaccinated cattle. Ultimately, a reduction in infestation tick burden reciprocates into low transmission of tick-borne pathogens, and hence a reduction in tick-borne diseases. Furthermore, using the SFS model, it is possible to concurrently assess the vaccine effectiveness against multiple tick species which is a common phenomenon in Africa, e.g., in Uganda.

A PIPELINE/MAP FOR DEVELOPMENT OF ANTI-TICK VACCINES

Even though a map/pipeline to guide in the development of general veterinary vaccines is established [157], until now, there has been no defined pipeline specific for the development of anti-tick vaccines globally [54]. So far, the available pipeline was based on steps for anti-tick vaccine development in Thailand [46]. By contrast, however, the pipeline described by Jittapalapong et al. [46] is far less elaborate than that presented by Francis [157], and it specifically guides development of a vaccine against *R. microplus*. Therefore, building on the existing pipelines [46,157], and taking into account the bottlenecks in assessing vaccine efficacy and effectiveness, a pipeline (Figure 6) is proposed to guide the development of anti-tick vaccines against one host and multiple-host ticks.

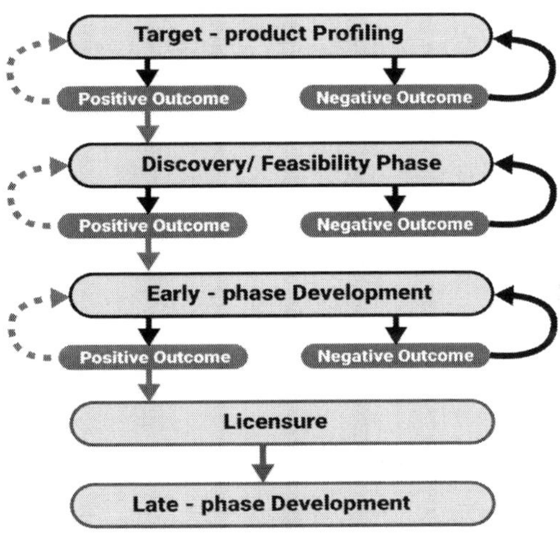

Figure 6: A flow chat for anti-tick vaccine development. Depending on whether the outcome is positive or negative, the phase studies may be repeated as shown in the curved arrow.

Phase I: Target Product Profiling

Similarly, this is the first phase for the development of veterinary vaccines. In this case, though, the phase mainly entails: (1) development of the desired product following the basic steps: gene identification/antigen discovery [55]—the steps include: cloning, expression, and protein purification; (2) assess the biochemical functionality of the protein to ensure proper structural conformation which can influence immunogenicity [88]; and (3) assess the immunogenicity of the candidate vaccine. The main aspects of concern include the adjuvant vaccine formulation, dose concentration, vaccination interval, route of administration, optimum or minimum target, and the number of booster doses. Immunogenicity studies can be conducted using small animal models such as rabbits, guinea pigs, and mice. The generated data will guide the subsequent phase studies and it is crucial for product registration.

Phase II: Discovery/Feasibility Studies

This precedes establishment of the vaccine formulation. Specifically, the aim is to show (1) proof of concept of the antigen immunogenicity, and (2) anti-tick vaccine efficacy. Although rabbits are often used due to their immunocompetence [127], they are not compatible hosts for all tick species. For example, the recovery of *R. microplus* ticks fed on rabbits is low [136]. Therefore, to ensure consistency, phase II studies against either one or three-host ticks should be conducted using a universal model, preferably a large animal model. In principle, cattle are the best model; however, they are costly; hence, sheep or goats could be adopted as an alternative. The reason is that in comparison to cattle, sheep and goats are less expensive, yet they are closely related to cattle [81,141]. For example, Jackson and Opedeeck [81] demonstrated the option of using sheep as an alternative model for examining vaccine efficacy. It is also important that the efficacy correlates with the induced humoral response, and in this way an optimum vaccination strategy can be developed. Although the immune response in sheep or goats could vary slightly from that of cattle, the data can demonstrate that the vaccine is suited for vaccine effectiveness assessment. Additionally, the data can suggest that the vaccine could confer protection to sheep or goats especially as both species are hosts under field conditions for some stages of the tick life cycle. Again, note that the vaccine efficacy studies do not substitute vaccine effectiveness studies. Therefore, high vaccine efficacy does not warrant field success.

Phase III: Early Phase Development

This is conducted to establish whether the vaccine can effectively induce protection in cattle. Firstly, following good laboratory practice (GLP), the pre-master seed (vaccine trial batch) is prepared and safely stored. It is important that the source material required for preparation of the antigen is traceable.

Secondly, unlike other vaccines (where the vaccine effectiveness studies are conducted directly in the field), the effectiveness of anti-tick vaccines should be investigated under a semi field system (SFS) (Figure 5). The merits of using this system have been discussed. The other aspects examined under the vaccine effectiveness studies are vaccine biosafety (e.g., antibody traces in milk), variation of immune response (according to animal age, breed, and in pregnant cattle), vaccine adverse effects (e.g., autoimmune response), and immune response duration.

A question may arise whether it is necessary to conduct vaccine effectiveness studies in different geographic areas which would be ideal. However, it would be costly to establish SFS in different geographic regions. Rather, the suggestion is to collect and assess ticks from different regions on one trial site. However, this may also prove costly in terms of the number of cattle required per experiment. Nonetheless, at least one phase-III trial should reflect the vaccine protective effect against ticks from different regions. Before vaccination, it is important to ensure that the cattle are naive to ticks and free from tick-borne diseases and other diseases, as ticks are capable of harboring a plethora of pathogens including bacteria, viruses, fungi, and protozoa [1,2]. If present, the pathogens could be transmitted between cattle during the study. On the other hand, the ticks used in the phase-III studies should be pathogen free. In the event that diseases manifest in experimental cattle, for ethical reasons, these should be immediately treated; however, no acaricides should be applied in the study.

Licensure

On the basis of data generated under Phase I, II and III, an application for vaccine licensure can be compiled and submitted. Furthermore, this data may be sufficient to address the key elements of the regulatory dossier. Upon licensure, upscale vaccine production and commercialization may ensue.

Phase IV: Post Licensure/Late Phase Development

Although these studies precede licensure, they are a continuation of phase III studies. On the contrary, the phase IV studies should be conducted directly under field conditions, such as on farm vaccine trials. Again, the aim is to ascertain the effectiveness of the vaccine in question. In this case, an application to conduct field trials should be submitted to the national drug regulatory body. The application is supported by phase III data. While conducting phase IV studies, it important to further examine aspects such as vaccine biosafety, duration of the humoral immune response, and vaccine safety. For ethical reasons, one may consider conducting the study against one population which consists of vaccinated and unvaccinated cattle. Under these studies, it also important to assess tick-borne disease burden among the study cattle population. However, in the event that tick-borne diseases manifest among experimental cattle, for ethical reasons, these should be immediately treated. However, no acaricides should be applied in the study. An alternative is to use recombinant antigens to not-live pathogens to vaccinate cattle against relevant tick-borne diseases prior to commencement of the trial.

CONCLUDING REMARKS

Currently, although advances has been made in identifying tick antigens, far less effort has been devoted to exploring the immunological related factors that influence vaccine efficacy. Furthermore, the methods for determining vaccine efficacy are still questionable, and there is not a defined protocol for assessing anti-tick vaccine effectiveness.

The approaches proposed herein for assessing vaccine efficacy or effectiveness aims to complement earlier methods [50,51], and to arouse the curiosity of tick researchers concerning how efficacy and effectiveness evaluation studies could be executed while taking into account the ultimate question: can the anti-tick vaccines reduce the tick population? Additionally, given the differences in tick life cycles, it could be erroneous to directly apply the formula used to determine vaccine efficacy or effectiveness against one-host ticks for vaccine assessment against three-host ticks.

The bottlenecks to efficacious anti-tick vaccines discussed herein and the proposed pipeline do not only apply to ixodid anti-tick vaccines, but also in determining argasid anti-tick vaccine efficacies. However, on the contrary, the model proposed for determining anti-tick vaccine efficacy and effectiveness suits mainly ixodid ticks, but not the argasid ticks. This is attributed to the difference between the ixodid and argasid tick life cycles [4].

Similar to development of other veterinary vaccines [157], anti-tick vaccines could take up to 5 to 6 years, yet by contrast, human vaccines require up to 10–20 years [157], except for COVID-19 vaccines. In other words, given the public importance of livestock, tick researchers ought to rigorously scrutinize the candidate anti-tick vaccines before field-application use of anti-tick vaccines. For this reason, a uniform pipeline/

map to assist in assessing the efficacy and effectiveness of anti-tick vaccines globally is urgently needed as earlier highlighted at the CATVAC meeting [54]. Fortunately, for instance, the semi-field system model for vaccine effectiveness studies (Figure 5) proposed herein is simple, practical, and minimizes the number of animals. Yet, it helps the researcher to fast-track the potency of the vaccine compared to when the studies are directly conducted in the field.

With slight modifications, the pipeline could also be followed, in assessing the efficacy and effectiveness of cocktail anti-tick vaccines [55] and tick pathogen transmission-blocking vaccines [15]. Furthermore, the factors that influence vaccine humoral immune response induction may also apply towards enhancing the efficacy and effectiveness of cocktail vaccines [55]. The SFS model could also be explored while assessing other methods of tick control, for example, biological and chemical acaricides.

ACKNOWLEDGMENTS

Special thanks to Stephen Wikel, Wes Watson and Sayan Dasgupta for perusing the draft and for providing criticism. Thanks to Robert Seremba and Leonard Semwogerere Katetemera for the illustrations. Finally, thanks to the administration at National Agricultural Research Organization secretariate and National Livestock Resources Research Institute for providing financial support toward APC.

REFERENCES

1. Brites-Neto J., Duarte K.M., Martins T.F. Tick-borne infections in human and animal population worldwide. *Vet. World.* 2015;8:301–315. doi: 10.14202/vetworld.2015.301-315.
2. Samish M., Rehacek J. Pathogens and predators of ticks and their potential in biological control. *Annu. Rev. Entomol.* 1999;44:159–182. doi: 10.1146/annurev.ento.44.1.159.
3. Horak I.G., Camicas J.L., Keirans J.E. The Argasidae, Ixodidae and Nuttalliellidae (Acari: Ixodida): A world list of valid tick names. *Exp. Appl. Acarol.* 2002;28:27–54. doi: 10.1023/A:1025381712339.
4. Walker A.R., Bouattour A., Camicas J.L., Estrada-Pena A., Horak I.G., Latif A.A., Pegram R.G., Preston P.M. *Ticks of Domestic Animals in Africa: A Guide to Identification of Species.* Bio-Science Reports; Edinburgh, UK: 2003.
5. Guglielmone A.A., Robbins R.G., Apanaskevich D.A., Petney T.N., Estrada-Peña A., Horak I.G., Shao R., Barker S.C. The Argasidae, Ixodidae and Nuttalliellidae (Acari: Ixodida) of the world: A list of valid species names. *Zootaxa.* 2010;25:1–28. doi: 10.11646/zootaxa.2528.1.1.
6. McCoy B.N., Raffel S.J., Lopez J.E., Schwan T.G. Bloodmeal size and spirochete acquisition of *Ornithodoros hermsi* (Acari: Argasidae) during feeding. *J. Med. Entomol.* 2010;47:1164–1172. doi: 10.1603/ME10175.
7. Walker A.R. Ticks and associated diseases: A retrospective review. *Med. Vet. Entomol.* 2014;28:1–5. doi: 10.1111/mve.12031.
8. Abbas R.Z., Zaman M.A., Colwell D.D., Gilleard J., Iqbal Z. Acaricide resistance in cattle ticks and approaches to its management: The state of play. *Vet. Parasitol.* 2014;203:6–20. doi: 10.1016/j.vetpar.2014.03.006.
9. Graf J.F., Gogolewski R., Leach-Bing N., Sabatini G.A., Molento M.B., Bordin E.L., Arantes G.J. Tick control: An industry point of view. *Parasitology.* 2004;129:S427–S442. doi: 10.1017/S0031182004006079.
10. Manjunathachar H.V., Saravanan B.C., Kesavan M., Karthik K., Rathod P., Gopi M., Tamilmahanf P., Balarajud B.L. Economic importance of ticks and their effective control strategies. *Asian Pac. J. Trop. Dis.* 2014;4:770–779. doi: 10.1016/S2222-1808(14)60725-8.

11. De la Fuente J., Kocan K.M. Strategies for development of vaccines for control of Ixodid tick species. *Parasite Immunol.* 2006;28:275–283. doi: 10.1111/j.1365-3024.2006.00828.x.
12. Nuttall P.A., Trimnell A.R., Kazimirova M., Labuda M. Exposed and concealed antigens as vaccine targets for controlling ticks and tick-borne diseases. *Parasite Immunol.* 2006;28:155–163. doi: 10.1111/j.1365-3024.2006.00806.x.
13. Bensaci M., Bhattacharya D., Clark R., Hu L.T. Oral vaccination with vaccinia virus expressing the tick antigen subolesin inhibits tick feeding and transmission of *Borrelia burgdorferi. Vaccine.* 2012;30:6040–6046. doi: 10.1016/j.vaccine.2012.07.053.
14. Labuda M., Trimnell A.R., Licková M., Kazimírová M., Davies G.M., Lissina O., Hails R.S., Nuttall P.A. An antivector vaccine protects against a lethal vector-borne pathogen. *PLoS Pathog.* 2006;2:e27. doi: 10.1371/journal.ppat.0020027.
15. Neelakanta G., Sultana H. Transmission-blocking vaccines: Focus on anti-vector vaccines against tick-borne iseases. *Arch. Immunol. Ther. Exp.* 2015;63:169–179. doi: 10.1007/s00005-014-0324-8.
16. Wikel S.K., Ramachandra R.N., Bergman D.K. Tick-induced modulation of the host immune response. *Int. J. Parasitol.* 1994;24:59–66. doi: 10.1016/0020-7519(94)90059-0.
17. Wikel S.K. Host immunity to ticks. *Annu. Rev. Entomol.* 1996;41:1–22. doi: 10.1146/annurev.en.41.010196.000245.
18. Brossard M., Wikel S.K. Immunology of interactions between ticks and hosts. *Med. Vet. Entomol.* 1997;11:270–276. doi: 10.1111/j.1365-2915.1997.tb00406.x.
19. Brossard M., Wikel S.K. Tick immunobiology. *Parasitology.* 2004;129:S161–S176. doi: 10.1017/S0031182004004834.
20. Robbertse L., Richards S.A., Maritz-Olivier C. Bovine immune factors underlying tick resistance: Integration and future directions. *Front. Cell. Infect. Microbiol.* 2017;7:522. doi: 10.3389/fcimb.2017.00522.
21. Kurokawa C., Narasimhan S., Vidyarthi A., Booth C.J., Mehta S., Meister L., Diktas H., Strank N., Lynn G.E., DePonte K., et al. Repeat tick exposure elicits distinct immune responses in guinea pigs and mice. *Ticks Tick-Born Dis.* 2020;11:101529. doi: 10.1016/j.ttbdis.2020.101529.

22. Wambura P.N., Gwakisa P.S., Silayo R.S., Rugaimukamu E.A. Breed-associated resistance to tick infestation in *Bos indicus* and their crosses with *Bos taurus*. *Vet. Parasitol.* 1998;77:63–70. doi: 10.1016/S0304-4017(97)00229-X.
23. Rechav Y. Naturally acquired resistance to ticks- A global view. *Int. J. Trop. Insect. Sci.* 1992;13:495–504. doi: 10.1017/S1742758400016076.
24. Trager W. Acquired immunity to ticks. *J. Parasitol.* 1939;25:57–81. doi: 10.2307/3272160.
25. Anderson J.M., Moore I.N., Nagata B.M., Ribeiro J., Valenzuela J.G., Sonenshine D.E. Ticks, *Ixodes scapularis*, feed repeatedly on white-footed mice despite strong inflammatory response: An expanding paradigm for understanding tick-host interactions. *Front Immunol.* 2017;8:1784. doi: 10.3389/fimmu.2017.01784.
26. Valle M.R., Mèndez L., Valdez M., Redondo M., Espinosa C.M., Vargas M., Cruz R.L., Barrios H.P., Seoane G., Ramirez E.S., et al. Integrated control of *Boophilus microplus* ticks in Cuba based on vaccination with the anti-tick vaccine GavacTM. *Exp. Appl. Acarol.* 2004;34:375–382. doi: 10.1007/s10493-004-1389-6.
27. Andreotti R. Performance of two Bm86 antigen vaccine formulation against tick using crossbreed bovines in stall test. *Rev. Bras. Parasitol. Vet.* 2006;15:97–100.
28. García-García J.C., Gonzalez I.L., González D.M., Valdés M., Méndez L., Lamberti J., D'Agostino B., Citroni D., Fragoso H., Ortiz M., et al. Sequence variations in the *Boophilus microplus* Bm86 locus and implications for immunoprotection in cattle vaccinated with this antigen. *Exp. Appl. Acarol.* 1999;23:883–895. doi: 10.1023/A:1006270615158.
29. Fragoso H., Rad P.H., Ortiz M., Rodríguez M., Redondo M., Herrera L., de la Fuente J. Protection against *Boophilus annulatus* infestations in cattle vaccinated with the *B. microplus* Bm86-containing vaccine Gavac. *Vaccine.* 1998;16:1990–1992. doi: 10.1016/S0264-410X(98)00116-9.
30. De Vos S., Zeinstra L., Taoufik A., Willadsen P., Jongejan F. Evidence for the utility of the Bm86 antigen from *Boophilus microplus* in vaccination against other tick species. *Exp. Appl. Acarol.* 2001;25:245–261. doi: 10.1023/A:1010609007009.
31. De la Fuente J., Estrada-Peña A. Why new vaccines for the control of ectoparasite vectors have not been registered and commercialized? *Vaccines.* 2019;7:75. doi: 10.3390/vaccines7030075.

32. Waldmann H., Munro A.T. Cell-dependent mediator in the immune response. *Nature.* 1973;243:356–357. doi: 10.1038/243356a0.

33. Lesinski G.B., Westerink M.A. Novel vaccine strategies to T-independent antigens. *J. Microbiol. Methods.* 2001;47:135–149. doi: 10.1016/S0167-7012(01)00290-1.

34. Sallusto F., Lanzavecchia A., Araki K., Ahmed R. From vaccines to memory and back. *Immunity.* 2010;33:451–463. doi: 10.1016/j.immuni.2010.10.008.

35. Shetty V.U., Chaudhuri P., Sabella C. Rationale for the immunization schedule: Why is it the way it is? *Pediat. Rev.* 2019;40:26–36. doi: 10.1542/pir.2018-0033.

36. Akkaya M., Kwak K., Pierce S.K. B cell memory: Building two walls of protection against pathogens. *Nat. Rev. Immunol.* 2020;20:229–238. doi: 10.1038/s41577-019-0244-2.

37. Abbas A.K., Lichtman A.H., Pilli S. *Cellular and Molecular Immunology.* 9th ed. Elsevier; Philadelphia, PA, USA: 2018. p. 565.

38. Patarroyo S.J.H., de Sousa Neves E., Fidelis C.F., Tafur-Gómez G.A., de Araujo L., Vargas M.I., Sossai S., Prates-Patarroyo P.A. Bovine immunisation with a recombinant peptide derived from synthetic SBm7462® (Bm86 epitope construct) immunogen for *Rhipicephalus microplus* control. *Ticks Tick-Borne Dis.* 2020;11:101461. doi: 10.1016/j.ttbdis.2020.101461.

39. Ndawula C., Jr., Amaral Xavier M., Villavicencio B., Cortez Lopes F., Juliano M.A., Parizi L.F., Verli H., da Silva Vaz I., Jr., Ligabue-Braun R. Prediction, mapping and validation of tick glutathione S-transferase B-cell epitopes. *Ticks Tick-Borne Dis.* 2020;11:101445. doi: 10.1016/j.ttbdis.2020.101445.

40. Couto J., Seixas G., Stutzer C., Olivier N.A., Maritz-Olivier C., Antunes S., Domingos A. Probing the *Rhipicephalus bursa* sialomes in potential anti-tick vaccine candidates: A reverse vaccinology approach. *Biomedicines.* 2021;9:363. doi: 10.3390/biomedicines9040363.

41. Reichardt P., Dornbach B., Gunzer M. APC, T cells, and the immune synapse. *Curr. Top. Microbiol. Immunol.* 2010;340:229–249. doi: 10.1007/978-3-642-03858-7_12.

42. Reeves E., James E. Antigen processing and immune regulation in the response to tumours. *Immunology.* 2017;150:16–24. doi: 10.1111/imm.12675.

43. Manz R.A., Thiel A., Radbruch A. Lifetime of plasma cells in the bone marrow. *Nature.* 1997;388:133–134. doi: 10.1038/40540.
44. Phan T.G., Tangye S.G. Memory B cells: Total recall. *Curr. Opin. Immunol.* 2017;45:132–140. doi: 10.1016/j.coi.2017.03.005.
45. Moran I., Nguyen A., Khoo W.H., Butt D., Bourne K., Young C., Hermes J.R., Biro M., Gracie G., Ma C.S., et al. Memory B cells are reactivated in subcapsular proliferative foci of lymph nodes. *Nat. Commun.* 2018;9:3372. doi: 10.1038/s41467-018-05772-7.
46. Jittapalapong S., Thanasilp S., Kaewmongkol G., Sirinarukmitr T. The progress and process of the development of anti-tick vaccine against cattle ticks (*Boophilus microplus*) in Thailand. *Kasetsart J.* 2006;40:276–283.
47. Shim E., Galvani A.P. Distinguishing vaccine efficacy and effectiveness. *Vaccine.* 2012;30:6700–6705. doi: 10.1016/j.vaccine.2012.08.045.
48. Greenwood M., Yule G.U. The statistics of anti-typhoid and anti-cholera inoculations, and the interpretation of such statistics in general. *Proc. R Soc. Med.* 1915;8:113–194. doi: 10.1177/003591571500801433.
49. Wikel S.K. Immunomodulation of host responses to ectoparasite infestation-an overview. *Vet. Parasitol.* 1984;14:321–339. doi: 10.1016/0304-4017(84)90099-2.
50. Cunha R.C., Andreotti R., Garcia M.V., Aguirre A.A.R., Leitão A. Calculation of the efficacy of vaccines against tick infestations on cattle. *Rev. Bras. Parasitol. Vet.* 2013;22:571–578. doi: 10.1590/S1984-29612013000400019.
51. Aguirre A., Garcia M.V., Szabó M.P., Barros J.C., Andreotti R. Formula to evaluate efficacy of vaccines and systemic substances against three-host ticks. *Int. J. Parasitol.* 2015;45:357–359. doi: 10.1016/j.ijpara.2015.02.003.
52. Szabó M.P.J., Bechara G.H. Immunisation of dogs and guinea pigs against *Rhipicephalus sanguineus* ticks using gut extract. *Vet. Parasitol.* 1997;68:283–294. doi: 10.1016/S0304-4017(96)01079-5.
53. Kasaija P.D., Contreras M., Kabi F., Mugerwa S., de la Fuente J. Vaccination with recombinant Subolesin antigens provides cross-tick species protection in *Bos indicus* and crossbred cattle in Uganda. *Vaccines.* 2020;8:319. doi: 10.3390/vaccines8020319.
54. Schetters T., Bishop R., Crampton M., Kopáček P., Lew-Tabor A., Maritz-Olivier C., Miller R., Mosqueda J., Patarroyo J., Rodriguez-

Valle M., et al. Cattle tick vaccine researchers join forces in CATVAC. *Parasite Vectors.* 2016;9:105. doi: 10.1186/s13071-016-1386-8.

55. Ndawula C., Jr., Tabor A.E. Cocktail anti-tick vaccines: The unforeseen constraints and approaches toward enhanced efficacies. *Vaccines.* 2020;8:457. doi: 10.3390/vaccines8030457.
56. Simionatto S., Marchioro S.B., Galli V., Hartwig D.D., Carlessi R.M., Munari F.M., Laurino J.P., Conceição F.R., Dellagostin O.A. Cloning and purification of recombinant proteins of *Mycoplasma hyopneumoniae* expressed in *Escherichia coli. Protein Expr. Purif.* 2010;69:132–136. doi: 10.1016/j.pep.2009.09.001.
57. Christensen D. Vaccine adjuvants: Why and how. *Hum. Vaccin. Immunother.* 2016;12:2709–2711. doi: 10.1080/21645515.2016.1219003.
58. Gallovic M.D., Montjoy D.G., Collier M.A., Do C., Wyslouzil B.E., Bachelder E.M., Ainslie K.M. Chemically modified inulin microparticles serving dual function as a protein antigen delivery vehicle and immunostimulatory adjuvant. *Biomater. Sci.* 2016;4:483–493. doi: 10.1039/C5BM00451A.
59. Mohan T., Verma P., Rao D.N. Novel adjuvants and delivery vehicles for vaccines development: A road ahead. *Ind. J. Med. Res.* 2013;138:779–795.
60. Apostólico J., Lunardelli V.A., Coirada F.C., Boscardin S.B., Rosa D.S. Adjuvants: Classification, modus operandi, and licensing. *J. Immunol. Res.* 2016;2016:1459394. doi: 10.1155/2016/1459394.
61. Herbert W.J. The mode of action of mineral-oil emulsion adjuvants on antibody production in mice. *Immunology.* 1968;14:301–318.
62. Chauhan N., Tiwari S., Iype T., Jain U. An overview of adjuvants utilized in prophylactic vaccine formulation as immunomodulators. *Expert Rev. Vaccines.* 2017;16:491–502. doi: 10.1080/14760584.2017.1306440.
63. Brown S.J., Shapiro S.Z., Askenase P.W. Characterization of tick antigens inducing host immune resistance. I. Immunization of guinea pigs with *Amblyomma americanum*-derived salivary gland extracts and identification of an important salivary gland protein antigen with guinea pig anti-tick antibodies. *J. Immunol.* 1984;33:3319–3325.
64. Marois I., Cloutier A., Garneau É., Richter M.V. Initial infectious dose dictates the innate, adaptive, and memory responses to influenza in the respiratory tract. *J. Leukoc. Biol.* 2012;92:107–121. doi: 10.1189/jlb.1011490.

65. Handel A., Li Y., McKay B., Pawelek K.A., Zarnitsyna V., Antia R. Exploring the impact of inoculum dose on host immunity and morbidity to inform model-based vaccine design. *PLoS Comput. Biol.* 2018;14:e1006505. doi: 10.1371/journal.pcbi.1006505.

66. Taborda C.P., Rivera J., Zaragoza O., Casadevall A. More is not necessarily better: Prozone-like effects in passive immunization with IgG 1. *J. Immunol.* 2003;170:3621–3630. doi: 10.4049/jimmunol.170.7.3621.

67. Billeskov R., Lindenstrom T., Woodworth J., Vilaplana C., Cardona P.J., Cassidy J.P., Mortensen R., Agger E.M., Andersen P. High antigen dose is detrimental to post-exposure vaccine protection against tuberculosis. *Front. Immunol.* 2017;8:1973. doi: 10.3389/fimmu.2017.01973.

68. Andersson B. Studies on the regulation of avidity at the level of the single antibody-forming cell the effect of antigen dose and time after immunization. *J. Exp. Med.* 1970;132:77–88. doi: 10.1084/jem.132.1.77.

69. Siskind G.W., Dunn P., Walker J.G. Studies on the control of antibody synthesis. II. Effect of antigen dose and of suppression by passive antibody on the affinity of antibody synthesized. *J. Exp. Med.* 1968;127:55–66. doi: 10.1084/jem.127.1.55.

70. Goidl E.A., Paul W.E., Siskind G.W., Benacerraf B. The effect of antigen dose and time after immunization on the amount and affinity of anti-hapten antibody. *J. Immunol.* 1968;100:371–375.

71. Eisen H.N., Siskind G.W. Variations in affinities ofantibodies during the immune response. *Biochemistry.* 1964;3:996–1008. doi: 10.1021/bi00895a027.

72. Theis G.A., Siskind G.W. Selection of cell populations in induction of tolerance: Affinity of antibody formed in partially tolerant rabbits. *J. Immunol.* 1968;100:138–141.

73. Paul W.E., Siskind G.W., Benacerraf B. A study of the "termination" of tolerance to BSA with DNP-BSA in rabbits: Relative affinities of the antibodies for the immunizing and paralyzing antigens. *Immunology.* 1967;13:147.

74. Hu Z., Molloy M.J., Usherwood E.J. CD4(+) T-cell dependence of primary CD8(+) T-cell response against Vaccinia virus depends upon route of infection and viral dose. *Cell. Mol. Immunol.* 2016;13:82–93. doi: 10.1038/cmi.2014.128.

75. Awate S., Babiuk L.A., Mutwiri G. Mechanisms of action of adjuvants. *Front. Immunol.* 2013;4:114. doi: 10.3389/fimmu.2013.00114.
76. Singh M., O'Hagan D.T. Recent advances in vaccine adjuvants. *Pharm. Res.* 2002;19:715–728. doi: 10.1023/A:1016104910582.
77. Hum B.A.L., Chantler S.M. Production of reagent antibodies. *Methods Enzym.* 1980;70:104–142. doi: 10.1016/s0076-6879(80)70044-7.
78. Hanly W.C., Artwohl J.E., Bennett B.T. Review of polyclonal antibody production procedures in mammals and poultry. *ILAR J.* 1995;37:93–118. doi: 10.1093/ilar.37.3.93.
79. What's the Right "Dose" of a Vaccine for Small-Breed Dogs? [(accessed on 23 December 2020)]. Available online: https://skeptvet.com/Blog/2016/01/whats-the-right-dose-of-a-vaccine-for-small-breed-dogs/
80. Benet L.Z., Zia-Amirhosseini P. Basic principles of pharmacokinetics. *Toxicol. Pathol.* 1995;23:115–123. doi: 10.1177/019262339502300203.
81. Jackson L.A., Opdebeeck J.P. The effect of antigen concentration and vaccine regimen on the immunity induced by membrane antigens from the midgut of *Boophilus microplus*. *Immunology.* 1989;68:272–276.
82. Ndawula C.J., Sabadin G.A., Parizi L.F., da Silva Vaz I.J. Constituting a glutathione S-transferase-cocktail vaccine against tick infestation. *Vaccine.* 2019;37:1918–1927. doi: 10.1016/j.vaccine.2019.02.039.
83. Maruyama M., Lam K.P., Rajewsky K. Memory B-cell persistence is independent of persisting immunizing antigen. *Nature.* 2000;407:636–642. doi: 10.1038/35036600.
84. Svehag S.E., Mandel B. The formation and properties of poliovirus-neutralizing antibody. II. 19S and 7S antibody formation: Differences in antigen dose requirement for sustained synthesis, anamnesis, and sensitivity to x-irradiation. *J. Exp. Med.* 1964;119:21–39. doi: 10.1084/jem.119.1.21.
85. Hanna M.G., Jr., Peters. L.C. Requirement for continuous antigenic stimulation in the development and differentiation of antibody-forming cells: Effect of antigen dose. *Immunology.* 1971;20:707–718.
86. Trager W. Further observations on acquired immunity to the tick *Dermacentor variabilis* Say. *J. Parasitol.* 1939;25:137–139. doi: 10.2307/3272354.

87. Baxter D. Active and passive immunity, vaccine types, excipients and licensing. *Occup. Med.* 2007;57:552–556. doi: 10.1093/occmed/kqm110.
88. Clark T.G., Cassidy-Hanley D. Recombinant subunit vaccines: Potentials and constraints. *Dev. Biol.* 2005;121:153–163.
89. Woodland D.L. Jump-starting the immune system: Prime-boosting comes of age. *Trends Immunol.* 2004;25:98–104. doi: 10.1016/j.it.2003.11.009.
90. Nolz J.C., Harty J.T. Strategies and implications for prime-boost vaccination to generate memory CD8 T cells. *Adv. Exp. Med. Biol.* 2011;780:69–83. doi: 10.1007/978-1-4419-5632-3_7.
91. Ramshaw I.A., Ramsay A.J. The prime-boost strategy: Exciting prospects for improved vaccination. *Immunol. Today.* 2000;21:163–165. doi: 10.1016/S0167-5699(00)01612-1.
92. Lu S. Heterologous prime-boost vaccination. *Curr. Opin. Immunol.* 2009;21:346–351. doi: 10.1016/j.coi.2009.05.016.
93. Newman M.J. Heterologous prime-boost vaccination strategies for HIV-1: Augmenting cellular immune responses. *Curr. Opin. Investig. Drugs.* 2002;3:374–378.
94. Kardani K., Bolhassani A., Shahbazihttp S. Prime-boost vaccine strategy against viral infections: Mechanisms and benefits. *Vaccine.* 2016;34:413–423. doi: 10.1016/j.vaccine.2015.11.062.
95. Agrawal B. Heterologous Immunity: Role in natural and vaccine-induced resistance to Infections. *Front. Immunol.* 2019;10:2631. doi: 10.3389/fimmu.2019.02631.
96. Ahmed R., Gray D. Immunological memory and protective immunity: Understanding their relation. *Science.* 1996;272:54–60. doi: 10.1126/science.272.5258.54.
97. Hassan I.A., Wang Y., Zhou Y., Cao J., Zhang H., Zhou J. Cross protection induced by combined Subolesin-based DNA and protein immunizations against adult *Haemaphysalis longicornis. Vaccine.* 2020;38:907–915. doi: 10.1016/j.vaccine.2019.10.076.
98. Willadsen P., Riding G.A., McKenna R.V., Kemp D.H., Tellam R.L., Nielsen J.N., Lahnstein J., Cobon G.S., Gough J.M. Immunologic control of a parasitic arthropod. Identification of a protective antigen from *Boophilus microplus. J. Immunol.* 1989;143:1346–1351.

99. Hernandez E.P., Kusakisako K., Talactac M.R., Galay R.L., Hatta T., Matsuo T., Fujisaki K., Tsuji N., Tanaka T. Characterization and expression analysis of a newly identified glutathione S-transferase of the hard tick *Haemaphysalis longicornis* during blood-feeding. *Parasit. Vectors.* 2018;11:91. doi: 10.1186/s13071-018-2667-1.
100. Vieira P., Rajewsky K. The half-lives of serum immunoglobulins in adult mice. *Eur. J. Immunol.* 1988;18:313–316. doi: 10.1002/eji.1830180221.
101. Vieira P., Rajewsky K. The bulk of endogenously produced IgG2a is eliminated from the serum of adult C57BL/6 mice with a half-life of 6–8 days. *Eur. J. Immunol.* 1986;16:871–874. doi: 10.1002/eji.1830160727.
102. Albas A., Fontolan O.l., Pardo P.E., Bremer Neto H., Sartori A. Interval between first dose and booster affected antibody production in cattle vaccinated against rabies. *J. Venom. Anim. Toxins Incl. Trop. Dis.* 2006;12:476–486. doi: 10.1590/S1678-91992006000300010.
103. Castiglione F., Mantile F., De Berardinis P., Prisco A. How the interval between prime and boost injection affects the immune response in a computational model of the immune system. *Comput. Math. Methods Med.* 2012;2012:842329. doi: 10.1155/2012/842329.
104. Leo O., Cunningham A., Stern P.L. Vaccine immunology. In: Garcon N., Stern P., Cunningham T., Stanberry L., editors. *Understanding Modern Vaccines.* Volume 25. Elsevier; Philadelphia, PA, USA: 2011. p. 29.
105. Wang H.L. Revisiting the adult vaccination schedule for tetanus and diphtheria. *AJN Am. J. Nurs.* 2016;116:15. doi: 10.1097/01.NAJ.0000484919.12660.0e.
106. Ni Y.H., Huang L.M., Chang M.H., Yen C.J., Lu C.Y., You S.L., Kao J.H., Lin Y.C., Chen H.L., Hsu H.J., et al. Two decades of universal hepatitis B vaccination in Taiwan: Impact and implication for future strategies. *Gastroenterology.* 2007;132:1287–1293. doi: 10.1053/j.gastro.2007.02.055.
107. Vargas M., Montero C., Sánchez D., Pérez D., Valdés M., Alfonso A., Joglar M., Machado H., Rodríguez E., Méndez L., et al. Two initial vaccinations with the Bm86-based Gavac^{plus} vaccine against *Rhipicephalus* (*Boophilus*) *microplus* induce similar reproductive suppression to three initial vaccinations under production conditions. *BMC Vet. Res.* 2010;6:43. doi: 10.1186/1746-6148-6-43.

108. Mathers A.R., Larregina A.T. Professional antigen-presenting cells of the skin. *Immunol. Res.* 2006;36:127–136. doi: 10.1385/IR:36:1:127.

109. Pasparakis M., Haase I., Nestle F.O. Mechanisms regulating skin immunity and inflammation. *Nat. Rev. Immunol.* 2014;14:289–301. doi: 10.1038/nri3646.

110. Dubois B., Bridon J.M., Fayette J., Barthélémy C., Banchereau J., Caux C., Brière F. Dendritic cells directly modulate B cell growth and differentiation. *J. Leukoc. Biol.* 1999;66:224–230. doi: 10.1002/jlb.66.2.224.

111. Yazdi A.S., Röcken M., Ghoreschi K. Cutaneous immunology: Basics and new concepts. *Semin. Immunopathol.* 2016;38:3–10. doi: 10.1007/s00281-015-0545-x.

112. Zhang L., Wang W., Wang S. Effect of vaccine administration modality on immunogenicity and efficacy. *Expert Rev. Vaccines.* 2015;14:1509–1523. doi: 10.1586/14760584.2015.1081067.

113. Wangmo K., Laven R., Cliquet F., Wasniewski M., Yang A. Comparison of antibody titres between intradermal and intramuscular rabies vaccination using inactivated vaccine in cattle in Bhutan. *PLoS ONE.* 2019;14:e0209946. doi: 10.1371/journal.pone.0209946.

114. Wahl M., Hermodsson S. Intradermal, subcutaneous or intramuscular administration of hepatitis B vaccine: Side effects and antibody response. *Scand. J. Infec. Dis.* 1987;19:617–621. doi: 10.3109/00365548709117195.

115. Ols S., Yang L., Thompson E.A., Pushparaj P., Tran K., Liang F., Lin A., Eriksson B., Hedestam G.B.K., Wyatt R.T., et al. Route of vaccine administration alters antigen trafficking but not innate or adaptive immunity. *Cell Rep.* 2020;30:3964–3971. doi: 10.1016/j.celrep.2020.02.111.

116. Smith G.N., Griffiths B., Mollison D., Mollison P.L. Uptake of IgG after intramuscular and subcutaneous injection. *Lancet.* 1972;1:1208–1212. doi: 10.1016/S0140-6736(72)90926-9.

117. Hervé C., Laupèze B., Del Giudice G., Didierlaurent A.M., Tavares Da Silva F. The how's and what's of vaccine reactogenicity. *NPJ Vaccines.* 2019;4:39. doi: 10.1038/s41541-019-0132-6.

118. Hopf S., Garner-Spitzer E., Hofer M., Kundi M., Wiedermann U. Comparable immune responsiveness but increased reactogenicity after subcutaneous versus intramuscular administration of tick borne

encephalitis (TBE) vaccine. *Vaccine.* 2016;34:2027–2034. doi: 10.1016/j.vaccine.2015.12.057.

119. Frazer I.H., Jones B., Dimitrakakis M., Mackay I.R. Intramuscular versus low-dose intradermal hepatitis B vaccine. Assessment by humoral and cellular immune response to hepatitis B surface antigen. *Med. J. Aust.* 1987;146:242–245. doi: 10.5694/j.1326-5377.1987.tb120230.x.
120. McCluskie M.J., Weeratna R.D., Payette P.J., Davis H.L. Parenteral and mucosal prime-boost immunization strategies in mice with hepatitis B surface antigen and CpG DNA. *FEMS Immunol. Med. Microbiol.* 2002;32:179–185. doi: 10.1111/j.1574-695X.2002.tb00551.x.
121. Contreras M., Kasaija P.D., Merino O., de la Cruz-Hernandez N.I., Gortazar C., de la Fuente J. Oral vaccination with a formulation combining *Rhipicephalus microplus* subolesin with heat inactivated Mycobacterium bovis reduces tick infestations in cattle. *Front. Cell. Infect. Microbiol.* 2019;9:45. doi: 10.3389/fcimb.2019.00045.
122. Squier C.A. The permeability of oral mucosa. *Crit. Rev. Oral. Biol. Med.* 1991;2:13–32. doi: 10.1177/10454411910020010301.
123. Otczyk D.C., Cripps A.W. Mucosal immunization: Realistic alternative. *Hum. Vaccin.* 2010;6:978–1006. doi: 10.4161/hv.6.12.13142.
124. Gupta A., Chaphalkar S.R. Vaccine Adjuvants: The current necessity of life. *Shiraz E-Med. J.* 2015;16:e28061. doi: 10.17795/semj28061.
125. Uddowla S., Freytag L.C., Clements J.D. Effect of adjuvants and route of immunizations on the immune response to recombinant plague antigens. *Vaccine.* 2007;25:7984–7993. doi: 10.1016/j.vaccine.2007.09.030.
126. Radaelli E., Santagostino S.F., Sellers R.S., Brayton C.F. Immune relevant and immune deficient mice: Options and opportunities in translational research. *ILAR J.* 2018;59:211–246. doi: 10.1093/ilar/ily026.
127. Turner R.J., Held S.D., Hirst J.E., Billinghurst G., Wootton R.J. An immunological assessment of group-housed rabbits. *Lab. Anim.* 1997;31:362–372. doi: 10.1258/002367797780596194.
128. Tabor A.E., Ali A., Rehman G., Rocha Garcia G., Zangirolamo A.F., Malardo T., Jonsson N.N. Cattle tick *Rhipicephalus microplus* host interface: A Review of resistant and susceptible host responses. *Front. Cell. Infect. Microbiol.* 2017;7:506. doi: 10.3389/fcimb.2017.00506.

129. Bakheit M.A., Latif A.A. The innate resistance of Kenana cattle to tropical theileriosis (*Theileria annulata* infection) in the Sudan. *Ann. N. Y. Acad. Sci.* 2002;969:159–163. doi: 10.1111/j.1749-6632.2002.tb04370.x.

130. Glass E.J., Jensen K. Resistance and susceptibility to a protozoan parasite of cattle-gene expression differences in macrophages from different breeds of cattle. *Vet. Immunol. Immunopathol.* 2007;120:20–30. doi: 10.1016/j.vetimm.2007.07.013.

131. Piper E.K., Jonsson N.N., Gondro C., Lew-Tabor A.E., Moolhuijzen P., Vance M.E., Jackson L.A. Immunological profiles of Bos taurus and Bos indicus cattle infested with the cattle tick, *Rhipicephalus (Boophilus) microplus. Clin. Vaccine Immunol.* 2009;16:1074–1086. doi: 10.1128/CVI.00157-09.

132. Plowright W. The duration of immunity in cattle following inoculation of rinderpest cell culture vaccine. *Am. J. Hyg.* 1984;92:285–296. doi: 10.1017/S0022172400064524.

133. Gustafson C.E., Kim C., Weyand C.M., Goronzy J.J. Influence of immune aging on vaccine responses. *J. Allergy Clin. Immunol.* 2020;145:1309–1321. doi: 10.1016/j.jaci.2020.03.017.

134. Uspensky I., Ioffe-Uspensky I. The relationship between engorged female weight and egg number in ixodid ticks: A biological interpretation of linear regression parameters. *Acarologia.* 1999;40:9–17.

135. Pereira M. Daily mean number of eggs laid by the southern cattle tick (Acari: Ixodidae) compared with mean egg mass weight. *J. Econ. Entomol.* 1998;91:153–158. doi: 10.1093/jee/91.1.153.

136. Do Amaral M.A., Prata M.C., Daemon E., Furlong J. Biological parameters of cattle ticks fed on rabbits. *Rev. Bras. Parasitol. Vet.* 2012;21:22–27. doi: 10.1590/S1984-29612012000100005.

137. Bailey K.P. Notes on the rearing of Rhipicephalus appendiculatus and their infection with *Theileria parva* for experimental transmission. *Bull. Epizoot. Dis. Afr.* 1960;8:33–64.

138. Wang H., Paesen G.C., Nuttall P.A., Barbour A.G. Male ticks help their mates to feed. *Nature.* 1998;391:753–754. doi: 10.1038/35773.

139. Stjernberg L., Berglund J. Detecting ticks on light versus dark clothing. *Scand. J. Infec. Dis.* 2005;37:361–364. doi: 10.1080/00365540410021216.

140. Senbill H., Hazarika L.K., Baruah A., Borah D.K., Bhattacharyya B., Rahman S. Life cycle of the southern cattle tick, *Rhipicephalus* (*Boophilus*) *microplus* Canestrini 1888 (Acari: Ixodidae) under laboratory conditions. *Syst. Appl. Acarol.* 2018;23:1169–1179. doi: 10.11158/saa.23.6.12.

141. Scheerlinck J.-P.Y. The immune system of sheep and goats. *Encycl. Immunobiol.* 2016;1:526–531. doi: 10.1016/B978-0-12-374279-7.12017-X.

142. Castro J., Cunningham M., Dolan T., Dransfield R., Newson R., Young A. Effects on cattle of artificial infestations with the tick *Rhipicephalus appendiculatus*. *Parasitology.* 1985;90:21–33. doi: 10.1017/S0031182000048988.

143. Merino O., Alberdi P., Pérez de la Lastra J.M., de la Fuente J. Tick vaccines and the control of tick-borne pathogens. *Front. Cell Infect. Microbiol.* 2013;3:30. doi: 10.3389/fcimb.2013.00030.

144. Valle M.R., Guerrero F.D. Anti-tick vaccines in the omics era. *Front. Biosci.* 2018;10:122–136. doi: 10.2741/e812.

145. De la Fuente J., Rodríguez M., Redondo M., Montero C., García-García J.C., Méndez L., Serrano E., Valdés M., Enriquez A., Canales M., et al. Field studies and cost-effectiveness analysis of vaccination with Gavac against the cattle tick *Boophilus microplus*. *Vaccine.* 1998;16:366–373. doi: 10.1016/S0264-410X(97)00208-9.

146. De la Fuente J., Rodríguez M., Montero C., Redondo M., García-García J.C., Méndez L., Serrano E., Valdés M., Enríquez A., Canales M., et al. Vaccination against ticks (*Boophilus* spp.): The experience with the Bm86-based vaccine Gavac(TM) *Genet. Anal. Biomol. Eng.* 1999;15:143–148. doi: 10.1016/S1050-3862(99)00018-2.

147. De la Fuente J., Almazán C., Canales M., Pérez de la Lastra J.M., Kocan K.M., Willadsen P. A ten-year review of commercial vaccine performance for control of tick infestations on cattle. *Anim. Health Res. Rev.* 2007;8:23–28. doi: 10.1017/S1466252307001193.

148. George J., Pound J., Davey R. In: *Acaricides for Controlling Ticks on Cattle and the Problem of Acaricide Resistance.* Bowman A., Nuttall P., editors. Cambridge University Press; Cambridge, UK: 2008. pp. 408–423.

149. Fernandez-Ruvalcaba M., Preciado-De-La Torre F., Cruz-Vazquez C., Garcia-Vazquez Z. Anti-tick effects of *Melinis minutiflora* and

Andropogon gayanus grasses on plots experimentally infested with *Boophilus microplus* larvae. *Exp. Appl. Acarol.* 2004;32:293–299. doi: 10.1023/B:APPA.0000023233.63268.cc.

150. Ferguson H.M., Ng'habi K.R., Walder T., Kadungula D., Moore S.J., Lyimo I., Russell T.L., Urassa H., Mshinda H., Killeen G.F., et al. Establishment of a large semi-field system for experimental study of African malaria vector ecology and control in Tanzania. *Malar. J.* 2008;7:158. doi: 10.1186/1475-2875-7-158.
151. Ng'habi K.R., Mwasheshi D., Knols B.G., Ferguson H.M. Establishment of a self-propagating population of the African malaria vector *Anopheles arabiensis* under semi-field conditions. *Malar. J.* 2010;9:356. doi: 10.1186/1475-2875-9-356.
152. Thompson K.C., Roa E.J., Romero N.T. Anti-tick grasses as the basis for developing practical tropical tick control packages. *Trop. Anim. Health Prod.* 1978;10:179–182. doi: 10.1007/BF02235337.
153. Furlong J. Infesting power of Boophilus microplus larvae (acari: Ixodidae) in *Melinis minutiflora*, *Brachiaria decumbens* e *Brachiaria mutica* pastures. *Cienc. Rural.* 1998;28:635–640. doi: 10.1590/S0103-84781998000400016.
154. Designed by the Skyline Consults Ltd, Uganda. [(accessed on 20 January 2021)]. Available online: https://www.skylinearchitectsug.com/
155. Halloran M.E., Struchiner C.J. Study designs for dependent happenings. *Epidemiology.* 1991;2:331–338. doi: 10.1097/00001648-199109000-00004.
156. John T.J., Samuel R. Herd immunity and herd effect: New insights and definitions. *Eur. J. Epidemiol.* 2000;16:601–606. doi: 10.1023/A:1007626510002.
157. Francis M.J. A Veterinary vaccine development process map to assist in the development of new vaccines. *Vaccine.* 2020;38:4512–4515. doi: 10.1016/j.vaccine.2020.05.007.

CHAPTER 7

Major Scientific Hurdles in HIV Vaccine Development: Historical Perspective and Future Directions

Tiza Ng'uni[1], Caroline Chasara[1], and Zaza M. Ndhlovu[1,2]

[1]KwaZulu-Natal Research Institute for Tuberculosis and HIV (K-RITH), Nelson R. Mandela School of Medicine, University of KwaZulu-Natal, Durban, South Africa
[2]Ragon Institute of Massachusetts General Hospital, Massachusetts Institute of Technology, and Harvard University, Cambridge, MA, United States

ABSTRACT

Following the discovery of HIV as a causative agent of AIDS, the expectation was to rapidly develop a vaccine; but thirty years later, we still do not have a licensed vaccine. Progress has been hindered by the extensive genetic variability of HIV and our limited understanding of immune responses required to protect against HIV acquisition. Nonetheless, valuable knowledge accrued from numerous basic and translational science research studies and

Citation: Ng'uni, T., Chasara, C., & Ndhlovu, Z. M. (2020). Major Scientific Hurdles in HIV Vaccine Development: Historical Perspective and Future Directions. Frontiers in immunology, 11, 590780. https://doi.org/10.3389/fimmu.2020.590780

vaccine trials has provided insight into the structural biology of the virus, immunogen design and novel vaccine delivery systems that will likely constitute an effective vaccine. Furthermore, stakeholders now appreciate the daunting scientific challenges of developing an effective HIV vaccine, hence the increased advocacy for collaborative efforts among academic research scientists, governments, pharmaceutical industry, philanthropy, and regulatory entities. In this review, we highlight the history of HIV vaccine development efforts, highlighting major challenges and future directions.

Keywords: HIV, history of HIV-1 vaccines, efficacy trials, HIV prevention, HIV-1 vaccine design

INTRODUCTION

The HIV/AIDS epidemic remains a major global health challenge and continues to exert significant strain on healthcare resources in sub-Saharan Africa. According to the UNAIDS, globally, approximately 37.9 million people were living with HIV infection in 2018. In addition, there were 1.7 million new infections with approximately 770,000 AIDS-related deaths in the same year despite widespread rollout of antiretroviral therapy (ART) (1, 2). The global HIV incidence-to-prevalence ratio of 0.05 indicates that the number of HIV-infected people will continue to rise unless more effective preventive strategies are employed to reduce transmission (3). There is broad scientific consensus that the most effective approach to control and eventually end the HIV epidemic is to develop a preventive AIDS vaccine that is safe, effective, cost efficient, and easily accessible worldwide (4). Regrettably, despite over 30 years of rigorous HIV research and numerous vaccine trials, there is no licensed HIV vaccine currently on the market (5). The aim of this review is to discuss past and present approaches to vaccine development and clinical trials to date. The review also highlights current gaps in knowledge and proposes new directions and novel strategies toward developing an efficacious preventive HIV vaccine.

CHALLENGES IN HIV VACCINE DEVELOPMENT

The development of potent antiretroviral therapies, now delivered as a single pill once a day, has transformed HIV infection into a clinically manageable chronic disease. Globally, over 19 million people are now on life-long treatment, and test-and-treat strategies and oral pre-exposure prophylaxis (PrEP) could further reduce HIV transmissions. However, despite these remarkable advances, prolonged combined antiretroviral therapy (cART) mediated suppression of plasma viral loads to undetectable levels does not eradicate the virus, which often rapidly rebounds upon treatment interruption. In addition, while cART has decreased mortality and morbidity among people living with HIV (PLWH), long-term cART treatment is associated with increased occurrence of a range of serious non-AIDS events (SNAEs). These SNAEs include cardiovascular diseases, cancer, liver disease, long-term peripheral and central nervous system complications, renal and metabolic disorders, and osteoporosis (6). The many logistical limitations and cost challenges that come with providing life-long care to those living with HIV highlight the need for a preventive HIV vaccine (7). Desirable attributes of an HIV vaccine include elicitation of long-lasting all-round protection with a limited number of doses administered to the patient, the vaccine should also be affordable, easy to administer, and easy to store without the need of a cold chain. The HIV vaccine can either be preventive or therapeutic, which means it can either block HIV infection or could be used to treat HIV infected individuals (8).

Over the years, the greatest challenge in developing an effective HIV vaccine has been the high rate of mutation and recombination during viral replication (9). The enormous genetic diversity of HIV is mainly driven by the high rate of variability of the viral envelope (Env) glycoprotein, which ironically happens to be the main target of neutralizing antibodies (10). The HIV genome contains nine genes which encode 16 proteins including the major structural proteins Gag, Pol, and Env; accessory proteins Nef, Vif, Vpu, and Vpr; and regulatory proteins Tat and Rev. HIV diversity, which is mainly generated by the error prone viral reverse transcriptase, has various implications for disease progression and responses to ART (11). The high mutation rates of approximately 1–10 mutations per genome per replication cycle, extensive conformational adaptability, and massive glycan shielding of the Env enable the virus to evade the effects of neutralizing antibodies and other immune responses (12). Nonetheless, despite the high rate of variability, it has progressively been shown that polyvalent HIV vaccines can be developed and used to target conserved domains on the viral envelope (13). Most current efforts are aimed at inducing broadly neutralizing antibodies (bNAbs), which can neutralize the majority of HIV strains. The ability of bNAbs to neutralize a wide spectrum of HIV strains (broad cross-reactivity) is a major advantage (14). Moreover, the safety and remarkable antiviral activity of highly potent HIV specific bNAbs have been demonstrated in pre-clinical and clinical trials (15).

Beside high viral mutation and recombination rates, extraordinary worldwide genetic diversity is yet another hurdle to the development of a vaccine. HIV is composed of 4 groups: M (main), O (outlier), N (non-M/non-O), and P (pending). Group M is further subdivided into 9 subtypes/clades denoted by the letters A, B, C, D, F, G, H, J, and K. It has been shown that amino acid variations within subtypes can be as high as 30% with those between subtypes reaching as high as 42%. These amino acid variations are based on the subtypes and region of the genome being examined (16, 17). The difficulty of developing a universal vaccine is further compounded by the fact that 10–20% of HIV infected people in several parts of Africa, are infected with two or more viral variants (subtypes and recombinant forms) that circulate in these regions (18).

Other challenges that impact HIV vaccine development include an incomplete understanding of the correlates of immune protection, lack of appropriate animal models, and limited investments by the pharmaceutical industry (19, 20). In addition, most traditional immunogen delivery systems are unable to induce potent and long-lasting immunity against HIV and the traditional live attenuated or whole-inactivated virus methods, employed in the design of measles, mumps, and rubella vaccines, are not appropriate for HIV due to legitimate safety and regulatory concerns associated with the risk of permanent integration of proviral HIV DNA into the host genome (4, 21).

THE DEVELOPMENT OF EARLY HIV VACCINES

Following the isolation of HIV as a causative agent for AIDS in 1983–1984, numerous vaccine prototypes have to date failed to protect against HIV infection (22, 23). In 1984, it was speculated that an HIV vaccine would be developed and available for testing in approximately two years (24). In hindsight, it is clear that researchers underestimated the complexity of such a scientific undertaking. We now know that HIV is unlike any other viral disease for which effective vaccines were empirically developed (25, 26). Given the risk of irreversible HIV integration into the human genome, rational design of subunit vaccines is the only viable option. The introduction of recombinant DNA technologies in the mid-1980s presented the best approach to develop a safe and effective HIV vaccine. This idea was derived from the hepatitis B model for which the hepatitis B surface antigen was successfully cloned and expressed in yeast cells, thus allowing for a new recombinant hepatitis B vaccine to be manufactured and licensed in

1986 (24). Using the hepatitis B recombinant DNA model, HIV researchers developed a subunit vaccine based on genetically engineered antigens representing the outer envelope glycoproteins of HIV. The design was based on previous vaccine approaches that used virus subunits, synthetic peptides, and vaccinia-vector vaccines in animal models (27–29). Rapid developments in the molecular biology of HIV, such as the identification of the major viral structural proteins (30) and the cloning and sequencing of the HIV genome (31), benefited initial vaccine development efforts. However, the genetic variability of HIV still remained the biggest obstacle to the success of these vaccine development efforts (32).

HIV VACCINES BASED ON THE INDUCTION OF NEUTRALIZING ANTIBODIES

Initially, scientists believed that neutralizing antibodies would be adequate to protect against HIV infection (33) and many of the HIV vaccines in this category were designed to primarily target the envelope glycoproteins, gp120 or gp160 (24). The first experimental immunization of humans against HIV/AIDS was done in Zaire (now the Democratic Republic of Congo) by Zagury and colleagues in 1986 (34). A small group of Zairians were vaccinated with a vaccinia vector (a recombinant HIV-vaccinia virus) expressing gp160, an envelope glycoprotein. The purpose of the study was to assess whether vaccination could induce neutralizing antibodies and cytotoxic T lymphocyte (CTL) responses in HIV negative individuals. Administration of the recombinant vaccinia virus vaccine by scarification elicited weak humoral and cellular immune responses that were boosted by four additional immunizations to achieve anamnestic humoral responses and cellular responses, which persisted for more than a year (35). Following this pioneering in human experimental HIV/AIDS vaccine trial, over 250 clinical trials have been carried out, with the majority being early-phase trials (phase 1 or 2) (36). Approximately 140 of these trials were conducted in the United States with several others being carried out in African countries and Thailand (24). Major vaccine trials have been completed and timelines are shown in Table 1.

Table 1: Illustration of completed and documented HIV vaccine trials

Vaccine Trial	Year	Site	Target group	Vaccine	Immune response	Result	Reference
VaxSyn	1987	Canada (Clade B)	72 adults	Recombinant envelope glycoprotein subunit (rgp160) of HIV	Neutralizing antibodies were detected	No vaccine efficacy	(37)
HIVAC-1e	1988	USA (Clade B)	35 male adults	Recombinant vaccinia virus designed to express HIV gp160	Vaccine was unable to confer protection against HIV	No vaccine efficacy	(38)
Vax004	1998–2002	North America (Clade B)	5,417 MSM and 300 women	AIDSVAX B/B gp120 with alum	Vaccine was unable to confer protection against HIV	No vaccine efficacy	(39, 40)
Vax003	1999–2003	Thailand (Clade B/E)	2,545 men and women IDUs	AIDSVAX B/E gp120 with alum	Vaccine was unable to confer protection against HIV	No vaccine efficacy	(39, 40)
HVTN 505	2009–2013	United States (Clade B)	2,504 men or transgender women who have sex with men	Three vaccinations with DNA encoding HIV clade B *gag*, *pol* and *nef* as well as *env* from HIV clades A, B and C followed by an Ad5 vector-based vaccine encoding clade B *gag* and *pol* as well as *env* from clades A, B and C	Vaccine was unable to prevent infection or decrease viral load in vaccinated volunteers	No vaccine efficacy	(41, 42)
STEP/HVTN 502 trial	2004–2007	North America the Caribbean South America, and Australia (Clade B),,	3,000 MSM and heterosexual men and women	MRKAd5 HIV-1 *gag/pol/nef* trivalent vaccine	Vaccine was unable to confer protection against HIV	No vaccine efficacy	(43, 44)
Phambili/ HVTN 503 trial	2003–2007	South Africa (Clade C)	801 adults	rAd5 (*gag/pol/nef*)	Vaccine was unable to confer protection against HIV	No vaccine efficacy	(24)
RV144	2003–2009	Thailand (Clade B)	16,402 community-risk men and women	ALVAC-HIV (vCP1521) and AIDSVAX B/E vaccines	IgG antibody avidity for Env in vaccine recipients with low IgA	31.2% vaccine efficacy at 42 months	(45, 46)

HVTN 305	2012–2017	Thailand (Clade B/E)	162 women and men	ALVAC-HIV and AIDSVAX B/E		No vaccine efficacy	(47)
HVTN 306	2013–2020	Thailand (Clade B/E)	360 men and women aged 20–40 years	ALVAC-HIV and AIDSVAX B/E	Vaccine was unable to confer protection against HIV	No vaccine efficacy	(48)
HVTN 097	2012–2013	South Africa (Clade B/E)	100 black Africans (men and women) aged 18–40 years	ALVAC-HIV (vCP1521) and AIDSVAX B/E	Induction of $CD4^+$ T cells directed to HIV-1 Env	No vaccine efficacy	(49)
HVTN 100	2015–2018	South Africa (Clade C)	252 men and women	ALVAC-HIV (vCP2438) and bivalent subtype C gp120/MF59	$CD4^+$ T-cell responses and gp120 binding antibody responses	No vaccine efficacy	(50)
HVTN 702	2016–2020	South Africa (702Clade C)	5,400 men and women	ALVAC-HIV (vCP2438) and bivalent subtype C gp120/MF59	Vaccine was unable to confer protection against HIV	No vaccine efficacy	(51)

MSM, men who have sex with men; IDUs, IV drug users.

VaxSyn and HIVAC-1e Vaccines

The first HIV vaccine trial carried out in the US investigated a recombinant envelope glycoprotein (rgp160), VaxSyn, created in a baculovirus-insect cell system. This trial evaluated the safety and immunogenicity of a rgp160 candidate vaccine in 72 healthy, HIV-negative adults. The vaccine recipients were randomly assigned to one of four groups to receive intramuscular injections of 40 or 80 μg of rgp160, 10 μg of hepatitis B vaccine, or placebo in three doses (on days 0, 30, and 180) with an optional fourth dose on day 540. The placebo and hepatitis B vaccine groups served as control groups (37). Results showed that the vaccine was safe and well tolerated. It was observed that vaccines recipients receiving 40 or 80 μg of rgp160 displayed mostly weak serum antibody responses to HIV envelope proteins. However, antibody titers noticeably increased after the third dose and declined over an 18-month period. The administration of a fourth dose resulted in homologous neutralizing activity and enhanced complement-mediated antibody-dependent activity in some vaccine recipients (37). Other studies showed that the administration of VaxSyn at a dose of 640 μg resulted in increased immunogenicity and higher rates of homologous neutralizing antibody responses, despite the titer being low (52) but failed to elicit sufficient protective neutralizing antibodies (37). After the VaxSyn trial, numerous envelope proteins were evaluated in 35 phase I trials, between 1988 and 2003. Collectively, these vaccine constructs induced binding and neutralizing antibodies that were remarkably durable and also primed $CD4^+$ T cell responses but no apparent CTL responses (24).

In 1988, a second recombinant vaccinia virus designed to express HIV gp160 (HIVAC-1e) entered phase 1 clinical trials in the US. The study enrolled 35 HIV-negative males (31 of whom had prior smallpox immunisations and 4 of whom were vaccinia naive). Study participants were randomly allocated to receive either standard New York strain vaccinia virus or HIVAC-le. Results showed that vaccinia-naive subjects shed virus from the vaccination site for longer and at a higher titer than the vaccinia-primed individuals. It was also observed that *in-vitro* T-cell proliferative responses to one or more HIV antigen preparations developed in some vaccinia-primed subjects inoculated with HIVAC-1e. However, T-cell responses were short-lived, and no HIV-specific antibodies were detectable. Nonetheless, two vaccinia-naive subjects vaccinated with HIVAC-1e showed robust T-cell responses to homologous and heterologous whole virus stains and to the recombinant gp160 protein, which were detectable for over a year. HIV Env specific antibodies also developed in both subjects. Interestingly, despite HIVAC-1e inducing transient and robust T-cell responses, it failed to produce antibodies against HIV infection in other subjects (24, 38). This led researchers to posit that antibody responses could be improved by priming with a recombinant vaccinia vector expressing the HIV-1 envelope and later boosting with an envelope protein.

In 1991, a combined approach was used in a phase 1 vaccine trial conducted in the US. This trial primed with HIVAC-1e and later boosted with VaxSyn. The prime-boost approach, used in several clinical trials, significantly enhanced both humoral and cellular immune responses and induced neutralizing antibodies (24, 53, 54). Despite these promising results, the use of vaccinia virus vectors raised several concerns. For instance, there was notable decrease in the immunogenicity of the vector in individuals previously vaccinated against smallpox (55, 56). In addition, there were concerns that administering the replicating vaccinia to immunosuppressed individuals could result in severe disease (55, 56). These concerns led to the development of non-replicating poxvirus vectors in the early 1990s, based on two models, namely, a highly attenuated vaccinia virus (NYVAC) or an avian poxvirus, Canarypox (ALVAC), that is not able to replicate in mammalian cells (57, 58).

ALVAC-HIV Vector Vaccine

In 1993, an ALVAC-vector HIV vaccine, vCP125, expressing gp160 was tested alone or as a prime-boost combination with an adjuvanted gp160 subunit. The results revealed that the ALVAC-HIV vaccine significantly primed the neutralizing antibody response of the protein boost and induced CTL activity (59). Other ALVAC-vectors (vCP205, vCP300, vCP1433, vCP1452, and vCP1521) were developed not only to express the HIV envelope but also to express *gag* and other HIV genes to induce broader cell-mediated immune responses (60, 61). In 1999, the second ALVAC-1 vector HIV vaccine was tested in Ugandan infants born to HIV infected mothers in a randomized placebo-controlled double-blind phase 1 trial. The vaccine construct expressed multiple genes such as *gp120*, the anchor region of *gp41*, *gag*, and *protease* (62–64). Even though the vaccine design was based on clade B genes, the researchers justified testing it in Uganda where clades A and D dominate based on the widespread cross-clade cellular immune responses observed in pre-clinical studies (63, 64). The results showed that the vaccine was safe in infants. Notably, vCP1521 was the prime employed in the Thai RV144 trial (discussed below) (45, 61).

VAX003 and VAX004 Efficacy Trials

1994 saw the emergence of two possible vaccine candidates (consisting of formulations of bivalent recombinant gp120 and alum) that were used in efficacy trials. These vaccine concepts were advanced to efficacy trials because they conferred protection to chimpanzees following HIV challenge and were safe and immunogenic in phase 1/2 clinical trials in humans (65, 66). The first two efficacy trials were carried out, from 1998 to 2003, by VaxGen in North America (VAX004; ClinicalTrials.gov Identifier: NCT00002441) and Thailand (VAX003; ClinicalTrials.gov Identifier: NCT00006327) (67–69). Based on the knowledge gained regarding genetic variability of HIV strains and the ability to use various co-receptors, the two initial candidate HIV vaccines were redesigned as bivalent gp120 vaccines (AIDSVAX B/B) for the North American trial and (AIDSVAX B/E) for the Thailand trial (the AIDSVAX B/E gp120 boost was also used in the RV144 trial) (45, 70).

The two redesigned gp120 vaccines were derived from R5 and X4 strains (HIV strains using CCR5 and CXCR4 co-receptors, respectively) (71). The VAX004 efficacy trial recruited 5417 volunteers who were mainly men who have sex with men (MSM) in North America, and the VAX003 trial recruited 2545 volunteers comprising intravenous injection drug users in Bangkok, Thailand. Unfortunately, in 2003, data analysis revealed that the two vaccines did not prevent HIV acquisition and did not ameliorate disease (39, 40).

VACCINES DESIGNED TO STIMULATE T CELL IMMUNITY TO HIV

0Multiple failures in antibody-based vaccines such as the VaxGen gp120 trials prompted the HIV vaccine field to begin pursuing T cell-based vaccines. There was a growing body of evidence demonstrating that $CD8^+$ T cell responses play a key role in controlling HIV infection. Animal studies showed that depletion of $CD8^+$ T cells in acute infection led to loss of virus control (72–75). Human studies also showed that the emergence of HIV-specific $CD8^+$ T cell responses coincided with the decline of viral load to a set point (76). Discovery of $CD8^+$ T cell driven viral mutations was another piece of evidence highlighting the importance of $CD8^+$ T cell responses in immune mediated control of HIV infection. Moreover, cumulative data showing $CD8^+$ T cell responses are largely responsible for spontaneous control of viremia for prolonged periods in the absence of medication, provided the impetus for the vaccine field to seriously pursue the development of T cell-based HIV vaccines (72, 77, 78). The candidate model vaccine vectors used for T cell-based vaccines were live recombinant viral vectors, mainly pox and adenovirus vectors (particularly the replication-defective adenovirus 5 [Ad5]), as well as DNA vaccines (79–81).

T Cell-Based Phase 2b Efficacy Trials

The first large T cell-based vaccine trial, the STEP trial also called HVTN 502 trial (ClinicalTrials.gov Identifier: NCT00095576), was a phase 2b multicentre, double-blind, randomized, placebo-controlled "test-of-concept" study that tested the efficacy of the MRKAd5 HIV-1*gag*/*pol*/*nef* vaccine. In December 2004, 3,000 participants were enrolled from Australia, Brazil, Canada, the Dominican Republic, Haiti, Jamaica, Peru, Puerto Rico, and the US where HIV subtype B is prevalent. The aim of the STEP trial was to determine whether the vaccine could prevent HIV infection, decrease the viral load in HIV-infected individuals, or both. Participants were randomly assigned to either vaccine or placebo groups, in a 1:1 ratio, to receive 3 injections of MRKAd5 HIV-1 *gag/pol/nef* vaccine or placebo (43, 44). The MRKAd5 HIV-1*gag/pol/nef* vaccine was also tested in a phase 2b trial (the Phambili/HVTN 503 trial; ClinicalTrials.gov Identifier: NCT00413725) in 801 adult South Africans. The goal of this trial was to evaluate the effectiveness of the vaccine in preventing infection in Southern Africa, where HIV subtype C is predominant. Unfortunately, vaccination and enrolment into both the STEP and Phambili trials were terminated in September 2007 following preliminary assessment that demonstrated no efficacy (24, 82). Furthermore, multivariate analysis of baseline risk factors revealed that the vaccination resulted in increased risk of HIV infection in some volunteers. In 2008, the final results of the trial were published, showing that cell-mediated immune responses elicited by this vaccine did not prevent HIV infection or blunt peak viral load (44). Additionally, there were slightly more HIV infections in the vaccine group compared to the placebo group in uncircumcised men with Ad5-neutralizing antibodies (43).

HVTN 505 (ClinicalTrials.gov Identifier: NCT00865566) was the next T cell-based phase 2b randomized, placebo-controlled, efficacy trial initiated in 2009 to test a potential DNA-primed vaccine in 2504 men or transgender women who have sex with men using a prime-boost regimen. The regimen consisted of three vaccinations with DNA encoding HIV clade B *gag*, *pol*, and *nef* as well as *env* from HIV clades A, B and C followed by an Ad5 vector-based vaccine encoding clade B *gag* and *pol* and *env* from clades A, B, and C (41, 42, 83). The experimental group received a DNA-primed vaccine injection on days 0, 28, and 56 followed by an Ad5 vector-based vaccine injection on day 168. The placebo group received placebo injections on days 0, 28, 56, and 168. Unfortunately, the trial was prematurely terminated after 47 months because interim analysis showed that the vaccine was not able to prevent infection or decrease viral load in vaccinated volunteers. Moreover, there was a slight increase in breakthrough infections in vaccine recipients compared to placebo controls (24).

The failure of conventional approaches to generate $CD8^+$ T cell-based vaccines prompted the scientific community to search for new approaches to HIV vaccine design. In 2008, NIAID encouraged researchers to go back to basics. A meeting convened by NIAID endorsed the expansion of research agenda to answer both basic science and novel vaccine design questions (24).

A VACCINE AGAINST HIV-1 IS POSSIBLE

The multiple setbacks in the HIV vaccine field led to substantial discussions regarding the optimal path toward a vaccine. The unexpected success of the RV144 trial (ClinicalTrials.gov Identifier: NCT00223080) showing modest but significant vaccine-induced protection (31.2% by 42 months) against HIV acquisition provided renewed hope that an HIV vaccine is possible. The RV144 was a randomized, double-blind phase 3 efficacy trial that utilized a recombinant canarypox vector vaccine, ALVAC-HIV (vCP1521), expressing Env (clade E), group-specific antigen (Gag) (clade B), and protease (Pro) (clade B), and an alum-adjuvanted AIDSVAX B/E and a bivalent HIV glycoprotein 120 (gp120) subunit vaccine (45, 46, 84). Vaccine recipients were given four priming injections of ALVAC-HIV (vCP1521) at months 0, 1, 3, and 6 with two booster injections of AIDSVAX B/E administered at months 3 and 6 (70). Immune correlates analyses revealed that the regimen induced HIV-specific humoral and cellular immune responses which resulted in reduced risk of HIV infection (inversely correlated with risk of HIV infection). Vaccine-induced responses included IgG antibodies binding to the HIV Env variable loops 1 and 2 (V1V2) and antibody-dependent cellular cytotoxicity (ADCC) in vaccine recipients with low IgA (70, 85–87). The magnitude and polyfunctionality of Env-specific $CD4^+$ T cells were also later shown to play a role in reducing the risk of HIV infection (70, 85–87).

Follow-Up Studies Based on the RV144 Trial

To date, only the RV144 trial has shown modest efficacy of 31.2% 42 months after the final vaccination. Rapid decline in protective antibody levels in most vaccine recipients led to the proposition that late boosts would induce durable protective immune response (88). Therefore, late boost studies (RV305 and RV306) were designed to assess immune responses generated in newly boosted vaccine recipients compared to RV144 vaccine recipients (47, 48).

RV305 Phase 2 Trial

The RV305 trial (ClinicalTrials.gov Identifier: NCT01435135) was a randomized, placebo-controlled, double-blind study conducted to investigate whether late boosts with ALVAC-HIV (vCP1521) or AIDSVAX B/E, administered either alone or in combination could enhance immune correlates of protection. The trial re-enrolled 162 healthy, HIV-negative Thai RV144 vaccine recipients to receive 2 additional boosts given 6–8 years after

RV144 vaccination. Study participants were randomized into one of three groups to receive either vaccine or placebo. Group 1 received ALVAC-HIV and AIDSVAX B/E, group 2 received AIDSVAX B/E, and group 3 received ALVAC-HIV, or placebo, at weeks 0 and 24 (47). Results showed that plasma immunoglobulin G (IgG), IgA, and neutralizing antibody responses at week 2 were all significantly higher in groups 1 and 2 compared to the responses observed 2 weeks following the last RV144 vaccination. The boost also resulted in higher antibody titers against various Env antigens (such as gp120 and V1V2) which were above the levels observed at the peak RV144 vaccine time point in plasma and mucosal secretions (47, 89). While the antibody titers increased following the first boost, they did not increase following the second boost in RV305 vaccine recipients. Moreover, the administration of late boosts did not result in lasting antibody responses as they rapidly declined after boosting in all groups (47). Overall, it was concluded that administering late boosts to RV144 vaccine recipients (6–8 years following their last vaccination) was safe and well tolerated. In addition, it was observed that despite AIDSVAX B/E alone or in combination with ALVAC-HIV generating higher humoral and $CD4^+$ T cells responses in RV305 vaccine recipients, these responses were short-lived and subsequent boosts did not increase their magnitude (47).

RV306 Phase 2 Trial

The RV306 trial (ClinicalTrials.gov Identifier: NCT01435135) was a randomized, placebo-controlled, double-blind study conducted at three clinical sites in Thailand. This study was designed to determine whether boosting the RV144 regimen at either month 12, 15, or 18 following initial vaccination would improve the quality, magnitude or duration of humoral, cellular, and mucosal responses. In addition, it was designed to also establish the optimal boosting interval for further clinical development (48). The study enrolled 367 healthy, HIV-negative individuals and randomly allocated them to one of five groups to receive vaccine or placebo. All groups received the original RV144 vaccination regimen at months 0, 1, 3, and 6 as follows: ALVAC-HIV at months 0 and 1 followed by either ALVAC-HIV and AIDSVAX B/E or placebo at months 3 and 6. Group 1 received only the RV144 series and no additional boost, group 2 received additional ALVAC-HIV and AIDSVAX B/E or placebo at month 12, group 3 received AIDSVAX B/E alone or placebo at month 12, group 4a received ALVAC-HIV and AIDSVAX B/E or placebo at month 15, and group 4b received ALVAC-HIV and AIDSVAX B/E or placebo at month 18. No serious vaccine-related adverse events were reported across active groups. Furthermore, it was observed that groups with late boosts (groups 2, 3, 4a, and 4b) had increased peak plasma IgG-binding antibody levels against gp70 V1V2 relative to group 1 vaccine recipients with no late boost. It was also observed that boosting at month 12 (groups 2 and 3) did not increase gp120 responses compared with the peak responses after the RV144 priming regimen at month 6, but boosting at month 15 (group 4a) improved responses to gp120 A244gD–D11 and boosting at month 18 (group 4b) improved responses to both gp120 A244gD–D11 and gp120 MNgD–D11. In addition, boosting at month 18 versus month 15 resulted in a significantly higher plasma IgG response to gp120 antigens but not gp70 V1V2 antigens. It was further observed that CD4 functionality and polyfunctionality scores following stimulation with HIV-1 Env peptides (92TH023) increased with delayed boosting. Additionally, the results showed that groups with late boosts had increased functionality and polyfunctionality scores relative to vaccine recipients with no late boost. Collectively, these results implied that additional boosting of the RV144 regimen with longer intervals between the initial vaccination and late boost could improve vaccine efficacy (48).

The promising results of RV144 clinical trials prompted the need to assess its efficacy against other clades (3, 90). RV144 originally designed to protect against HIV clade CRF01_AE BE infection in Thailand was modified to target HIV clade C. Therefore, HVTN designed and conducted a series of clinical trials including, HVTN 097, HVTN 100 and HVTN 702 (3).

HVTN 097 Phase 1b Vaccine Trial

The first vaccine concept tested by HVTN based on the RV144 concept was the HVTN 097 trial (ClinicalTrials.gov Identifier: NCT02109354), which was a randomized, controlled, double-blind phase 1b study done in South Africa (49). The trial was designed to assess the safety and immunogenicity of the vaccine regimen in healthy, HIV-uninfected South African adults. The regimen consisted of two prime doses of the experimental canarypox HIV vaccine ALVAC-HIV (vCP1521) followed by two booster shots of the AIDSVAX B/E. Study participants were randomized into three groups, in a 3:1:1 ratio, to receive the vaccine combined with tetanus and hepatitis B immunizations, the vaccine only or placebo. The tetanus and hepatitis B immunizations were included to assess possible cross-correlates of immune responses to HIV vaccine, however no significant differences in HIV immune responses were observed indicating that subsequent results were solely due to immune responses to HIV (49). The prime-boost vaccine regimen induced mostly Env-specific $CD4^+$ T cell responses at significantly higher levels compared to RV144 vaccine recipients (RV144 = 36.4%; HVTN 097 = 51.9%). IgG antibodies recognizing the V1V2 region and the IgG3 binding antibody responses to both gp120 and V1V2 antigens were also significantly higher among HVTN 097 vaccine recipients relative to RV144 recipients. ADCC antibody responses were also higher in HVTN 097 than in RV144, 72.6% (53 of 73) and 58.5% (114 of 195), respectively. These favourable results provided compelling rationale for conducting larger clinical trials in South Africa (49).

HVTN 100 Phase 1/2 Preventative Vaccine Trial

The HVTN 100 (ClinicalTrials.gov Identifier: NCT02404311) phase 1/2 randomized, controlled, double-blind study, was also conducted in South Africa to evaluate the safety, tolerability, and immunogenicity of the new modified vaccine regimen for subsequent efficacy testing. It consisted of an ALVAC-HIV (vCP2438) vector, expressing HIV Env gp120 (clade C ZM96), Env gp41, Gag, and Pro (all clade B) as well as a MF59-adjuvanted bivalent subtype C gp120 protein vaccine. The MF59 adjuvant was used to boost neutralizing antibodies and T cell responses (50). Enrolment took place from February to May 2015. Vaccine recipients received ALVAC-HIV (vCP2438) vector intramuscular (IM) injections at months 0 and 1 followed by co-administrations of ALVAC-HIV (vCP2438) and MF59-adjuvanted bivalent subtype C gp120 at months 3, 6, and 12 (50). The vaccine induced greater frequency of IgG3 responses to Env gp120, significantly higher $CD4^+$ T-cell responses and gp120 binding antibody responses compared to the RV144 regimen. Importantly, the IgG response exceeded the expected 63% threshold required for 50% vaccine efficacy that was calculated using a V1V2 correlate of protection model. Therefore, the HVTN 100 vaccine regime was advanced to a phase 2b/3 efficacy trial (HVTN 702) (50). It is important to note that despite the frequency of IgG3 and V1V2 antibody responses in HVTN 100 exceeding levels that were modeled to be required for protection, the correlate of reduced risk in RV144 was not response rate but rather the level of the antibodies (HIV-specific antibody responses resulting in reduced risk of HIV acquisition) (70, 85–87). Therefore, this criterion used for advancement to the phase 2b/3 HVTN 702 trial was not consistent with the findings of the RV144 trial.

HVTN 702 Phase 2b/3 Efficacy Trial

The HVTN 702 trial (ClinicalTrials.gov Identifier: NCT02968849) was a randomized, controlled, double-blind study conducted at 14 sites in South Africa from 2016 to 2020. The main objective was to assess the efficacy, safety, and tolerability of the ALVAC-HIV (vCP2438) plus bivalent Subtype C gp120/MF59 prime-boost vaccine regimen. The study enrolled 5407 HIV-uninfected sexually active individuals (both men and women), aged 18–35 years, and randomly allocated them to the vaccine or placebo arm. Vaccine recipients received an IM injection of ALVAC-HIV (vCP2438) at months 0 and 1 followed by co-administrations of ALVAC-HIV (vCP2438) + Bivalent Subtype C gp120/MF59 at months 3, 6, and 12 (51). The vaccine regimen employed in the HVTN 702 was modified to improve the efficacy and durability of the immune responses compared to RV144. Firstly, ALVAC-HIV vaccine construct contained clade C HIV genetic inserts to match those predominantly found in South Africa whereas the RV144 ALVAC contained clade B and E genetic inserts (those predominantly found in Thailand). Secondly, the boost in the HVTN 702 trial was a clade C genetically engineered HIV gp120 protein that was co-formulated with the MF59 adjuvant, whereas RV144 utilized the adjuvant alum. Thirdly, vaccine candidates in the HVTN 702 trial received five injections administered over a 12-month period (0, 1, 3, 6, and 12), whereas the RV144 vaccine candidates received four injections over a 6-month period (0, 1, 3, and 6). It was believed that a fifth dose at month 12 would potentially result in an extended protective effect (45, 51). Unfortunately, the trial was stopped on 23 January 2020 following an interim analysis by an independent data and safety monitoring board (DSMB). The DSMB analysed data from 2694 vaccine recipients and 2689 placebo recipients and discovered that 129 HIV infections occurred among the vaccine recipients, and 123 HIV infections occurred among the placebo recipients (91, 92). These findings indicated that the vaccine was not effective in preventing HIV infection and the DSMB recommended discontinuation of further vaccinations but allowed follow-up to continue. This was a very disappointing and unexpected result given that prior studies showed the vaccine exhibited greater immunogenicity compared to RV144. The borderline statistical significance of the RV144 results of 31.2% protective effect has led others to question the veracity of the analyses (93). Clearly, more work is needed to understand the discrepancy between immunogenicity in the HVTN 100 study and the lack of efficacy in the HVTN 702 study; this will inevitably include revisiting the RV144 analyses.

CURRENT STATUS OF THE HIV VACCINE FIELD

During the past 30 years, only a few HIV vaccine regimens have been tested in phase 2b clinical trials (3, 24, 26, 36, 94). More recently, there has been strong advocacy for adaptive clinical trials aimed at accelerating vaccine development by rapid evaluation of vaccine candidates in small human studies and rapidly advancing promising candidates to efficacy trials (95–97). The new accelerated approach has resulted in more than 100 HIV vaccine concepts being clinically tested. Similar approaches have been adopted in the accelerated development of novel coronavirus disease 2019 (COVID-19), vaccines. Collaborative efforts such as the Pox-Protein Public Private Partnership (P5) which includes private industry, government agencies, the Bill and Melinda Gates foundation, and HVTN have contributed to accelerating vaccine trials through training and establishment of vaccine testing sites across the world. Additionally, partnerships between Janssen Pharmaceutical company, academic labs, and HVTN have championed the development and testing of mosaic-based vaccines. These private-public partnerships have led to the current rich pipeline of new vaccine concepts in preclinical trials and various stages of clinical trials. Although the vaccine field is trying to accelerate the extensive pipeline of vaccine concepts to efficacy trials, the decline of HIV incidence worldwide and the wider deployment of other HIV prevention tools such as Pre- exposure prophylaxis (PrEP) has complicated vaccine testing landscape by necessitating very large, more complex, and very expensive vaccine trial designs.

Passive Immunization Studies

For years, passive immunisation with protective antibodies has been used in the prevention and treatment of several bacterial and viral infections, subsequently influencing the current HIV vaccine field (98). The diverse mechanism of action of antibodies (through their interaction with the innate and adaptive arms of the immune system), coupled with their ability to bind and neutralize viruses, continues to make the antibody-based approach appealing to researchers (99, 100). The identification of various bNAbs with increased breadth and potency such as PG9, PG16, PGT121, PGT145, VRC01, VRC07, and 3BNC117 has provided an opportunity for their potential application in HIV vaccine research (101). Moreover, animal studies have demonstrated the protective and therapeutic properties of numerous bNAbs (102). Additionally, bNAbs have been shown to reduce viremia and delay viral rebound following ART interruption in HIV-infected individuals (103–105). However, it is unknown whether bNAbs are able to prevent HIV infection in humans (106). Hence, the need for the continued evaluation of the protective efficacy of passive immunization with bNAbs.

Early Passive Immunization Studies

The wealth of evidence from early passive immunization studies in animal models has revealed that the passive infusion of bNAbs resulted in protection from HIV infection (101). While several animal models have been used in passive immunisation studies, the most commonly used are Non-Human Primate (NHPs) and humanized mouse models (101). NHPs are typically infected with either simian immunodeficiency virus (SIV) or chimeric Simian/Human Immunodeficiency Virus (SHIVs), expressing the HIV Env in a SIV backbone (101). However, it has been shown that antibodies specific for the HIV Env protein are unable to neutralize SIV due to the difference in the HIV and SIV Env protein composition (102, 107). It is for this reason that SHIVs have been frequently used to infect NHPs, while humanized mouse models are directly infected with HIV in antibody protection studies (101, 102).

One of the earliest antibody protection studies in a SHIV challenge model, used polyclonal HIV IgG derived from HIV-infected chimpanzees. It was shown that the passive transfer of HIV IgG to pig-tailed macaques protected them from SHIV (based on the HIV DH12 strain) challenge (108). Another study showed that PGT121, a potent bNAb, protected monkeys from SHIV-SF162P3 challenge at serum concentrations that were lower than those previously observed (109). Several antibody protection studies evaluating the passive transfer of bNAbs in NHP models have demonstrated their ability to confer robust protection from HIV infection, even at low concentrations (97, 109–114). Furthermore, proof-of-principle studies of first-generation antibodies such as b12, that targets the CD4 binding site, in SHIV challenged monkeys have provided insight into the mechanism and durability of antibody protection (101, 115, 116). Similarly, antibody protection studies in mouse models have highlighted the protective efficacy of neutralizing antibodies (NAbs), such as b12 (117, 118) and BAT123 (119, 120). In addition, several studies have demonstrated that immunodeficient mice transplanted with human hematopoietic stem cells (hu-HSC) or bone marrow/liver/thymus (BLT) and passively immunized with bNAbs such as 2G12 (121), VRCO1 (122, 123), PG16 (124), and PG126 (125) were protected against HIV infection.

Overall, the passive infusion of SHIV challenged monkeys and HIV challenged humanized mice with bNAbs, particularly the potent second-generation antibodies, has provided evidence of their ability to effectively protect against viral infection (101). While Fc receptor binding for antibody protection has proven to be important (126, 127), the specific mechanisms by which protection is rendered are not fully understood (101, 115). Nonetheless, phase 1 human studies previously conducted to evaluate the protective efficacy of bNAbs such as 3BNC117 and VRC01 have demonstrated short-term viral control (104, 128). The use of bNAb-based vaccines in human has generated tremendous interest and clinical trials are being conducted to investigate their ability to prevent HIV infection. For instance, antibody mediated prevention (AMP) studies are being conducted to test whether VRC01 can prevent HIV infection in men who have sex with men as well as heterosexual women (106).

Antibody Mediated Prevention Studies

Over the years, researchers have been studying and developing bNAbs as potential HIV vaccine candidates. Subsequently, the use of bNAb-based vaccines in human trials has generated tremendous interest and clinical trials such as the Antibody Mediated Prevention (AMP) studies (HVTN 703/HPTN 081 and HVTN 704/HPTN 085) are being conducted to test whether VRC01, a potent bNAb designed to target the $CD4^+$ binding site of the HIV-1 envelope glycoprotein, can prevent HIV infection in men who have sex with men as well as heterosexual women (106).

HVTN703/HPTN 081 Phase 2b Study

HVTN 703/HPTN 081 (ClinicalTrials.gov Identifier: NCT02568215) is a phase 2b randomized, controlled, double-blind study currently underway in sub-Saharan Africa. The study commenced in May 2016 and is expected to be completed by December 2020. This test-of-concept trial seeks to assess the safety, tolerability and efficacy of VRC01 in preventing HIV infection in healthy sexually active HIV-uninfected women. This AMP study has enrolled about 1900 HIV-uninfected sexually active women, aged 18-50 years, from several countries. Study participants, randomly allocated to one of three groups, in a 1:1:1 ratio, receive an intravenous (IV) infusion of 10 mg/kg VRC01 (low dose), 30 mg/kg VRC01 (high dose)m or placebo every 8 weeks (106, 129, 130).

HVTN704/HPTN 085 Phase 2b Study

The HVTN 704/HPTN 085 (ClinicalTrials.gov Identifier: NCT02716675) is another AMP study seeking to evaluate the safety, tolerability, and efficacy of VRC01 in preventing HIV-1 infection in healthy men and transgender (TG) men who have sex with men (MSM). The study commenced in March 2016 and the expected study completion date is February 2021. This study has enrolled 2701 HIV-uninfected men and transgender MSM in Brazil, Peru, Switzerland, and the United States (106, 130, 131). Participants in this study, like the HVTN 703/HPTN 081, were randomly allocated to one of three groups, in a 1:1:1 ratio, to receive a total of 10 IV infusion of 10 mg/kg VRC01 (low dose), 30 mg/kg VRC01 (high dose) or placebo every 8 weeks. The ultimate goal of the AMP trials is to identify and understand the characteristics of VRC01, such as optimal antibody concentration and effector functions, that correlate with protection against HIV infection (130).

CURRENT HIV VACCINE EFFICACY TRIALS

Currently, some of the ongoing phase 2b efficacy trials include HVTN 705/HPX2008 (Imbokodo study), HVTN 706/HPX3002 and PrepVacc. Janssen Pharmaceutical in collaboration with HVTN and academic labs are testing vaccine regimens that are designed to cover different types of HIV found across the world.

HVTN 705/HPX2008 (Imbokodo Study) Phase 2b Efficacy Trial

One such vaccine currently in efficacy trials is the HVTN 705/HPX2008 (Imbokodo study) (ClinicalTrials.gov Identifier: NCT03060629). This proof of concept study is a multicentre randomized, controlled, double-blind phase 2b/3 efficacy trial currently being conducted at 24 sites in 5 sub-Saharan African countries. This study commenced in November 2017 and is expected to be completed by May 2022. It aims to evaluate the efficacy, safety, and tolerability of a prime-boost regimen for the prevention of HIV infection. This study has enrolled 2600 healthy HIV-uninfected sexually active women, aged 18-35 years. Study participants, randomly assigned to either the experimental group or placebo group in a 1:1 ratio, received either the vaccine or placebo. The regimen consists of a tetravalent adenovirus vector vaccine, Ad26.Mos4.HIV (consisting of Ad26.Mos1.Gag-Pol, Ad26.Mos2.Gag-Pol, Ad26.Mos1.Env, and Ad26.Mos2S.Env clade C), and aluminium-phosphate adjuvanted clade C gp140 and Mosaic gp140 HIV protein vaccine.

Vaccine recipients receive intramuscular (IM) injections of Ad26.Mos4.HIV at months 0 and 3 followed by IM injections of Ad26.Mos4.HIV and aluminum-phosphate adjuvanted clade C gp140 at months 6 and 12 whereas those in the placebo group will receive intramuscular injections of placebo. The primary endpoints will include; assessment of vaccine efficacy, number of participants with reactogenicity signs or symptoms, and adverse events (AEs). The secondary endpoints will include, immunogenicity, immune response biomarkers as correlates of risk of subsequent HIV acquisition, and genomic sequences of viral isolates from vaccine and placebo recipients (132, 133).

HVTN 706/HPX3002/Mosaico Phase 3 Efficacy Trial

Another mosaic-based vaccine concept currently in clinical trials is the HVTN 706/HPX3002 (ClinicalTrials.gov Identifier: NCT03964415), or Mosaico trail. This multicentre, randomized, controlled, double-blind phase 3 efficacy trial is currently underway in Europe, North America, and South America. It commenced in October 2019 and is expected to be completed by September 2023. This study seeks to assess the safety and efficacy of the Ad26.Mos4.HIV and adjuvanted clade C gp140 and Mosaic gp140 protein vaccine prime-boost vaccine regimen in healthy, HIV-uninfected MSM and transgender people. This study has enrolled approximately 3800 participants, aged 18-60 years, and randomly allocated to receive either the vaccine or placebo as outlined in the HVTN 705/HPX2008. The primary endpoint is to assess vaccine efficacy. Secondary endpoints are, to assess the number of participants with solicited and unsolicited local and systemic adverse events (AEs), medically-attended adverse events (MAAEs) and SAEs, frequency and magnitude of HIV Env-specific humoral and cellular immune responses, antibody titers for Ad26, risky sexual behaviour, and Pre-Exposure Prophylaxis (PrEP) intake. Preliminary results reported at the International AIDS conference in Mexico city (IAS 2019), showed evidence of vaccine induced immune responses to different HIV strains circulating worldwide (133, 134).

PrepVacc Phase 2b Trial

Finally, the PrepVacc (ClinicalTrials.gov Identifier: NCT04066881) is another multicentre, randomised, controlled, double-blind phase 2b vaccine study currently underway in Mozambique, South Africa, Tanzania, and Uganda. The study seeks to evaluate the effectiveness/efficacy of a combination of two HIV vaccine regimens (DNA/AIDSVAX and DNA/CN54gp140+ MVA/CN54gp140) with PrEP (PrEPVacc). The study commenced in January 2020 and is projected to complete in March 2023. A total of 1668 healthy HIV-uninfected adults (18-40 years) are expected to be enrolled in the study and an equal number (278) will be randomised into one of six groups (i.e. Group A, B, C, D, E, and G) to receive either the vaccine regime or placebo with PrEP. The primary endpoints will include; evaluation of HIV infection in vaccine recipients and assessment of AEs associated with receiving either vaccine or PrEP regimes which may lead to the regimes being terminated. The ultimate goal is to determine if the vaccine leads to a decrease in HIV prevalence with adequate public health significance to justify implementation of the combination vaccine regimen (133, 135). A complete list of ongoing vaccine trials is illustrated in Table 2.

Table 2: Illustration of ongoing HIV vaccine trials

Vaccine Trial	Year	Site	Target group	Vaccine	Immune response	Result	References
HVTN 703	2016-2020	Sub-Saharan Africa (Clade C)	1900 women	VRC01 broadly neutralizing monoclonal antibody	—	Pending	(106, 130)
HVTN 704	2016-2020	Brazil Peru Switzerland, United States (Clade B),,	2701 men and transgender persons	VRC01 broadly neutralizing monoclonal antibody	—	Pending	(106, 130)
HVTN 705	2017-2022	Sub-Saharan Africa (Clade C)	2600 women	Ad26.Mos4.HIV and adjuvanted clade C gp140 and Mosaic gp140 protein vaccine	—	Pending	(132, 133)
HVTN 706	2019-2023	Europe North America and South America (Clade C),,	3800 MSM and transgender persons	Ad26.Mos4.HIV and adjuvanted clade C gp140 and Mosaic gp140 protein vaccine	—	Pending	(133, 134)
PrepVacc	2020-2023	Mozambique South Africa, Tanzania, and Uganda (Clade C),	1 668 Adults	DNA/AIDS-VAX and DNA/CN54gp140 + MVA/CN54gp140) with PrEP.	—	Pending	(133, 135)

MSM, men who have sex with men; IDUs, intravenous drug users.

NEW HIV VACCINE CONCEPTS AND TECHNOLOGIES

Vaccines have been used for centuries to prevent and treat various diseases thereby saving millions of lives. Importantly, widespread vaccinations led to the successful eradication of smallpox and significant reduction in other infectious diseases such as polio and measles (136, 137). While modelling research show that a vaccine is essential to conclusively end the HIV epidemic (94), traditional vaccine formulations such as live attenuated and inactivated pathogens and subunit vaccines which offer robust protection against many deadly diseases, are not suitable for HIV vaccines (138). Moreover, cumulative evidence suggests that subunit vaccine designs do not elicit immune responses to levels required for protection, hence the need to utilize novel approaches such as targeted stimulation of broadly neutralizing antibody (bNAb) producing B cell precursors, novel viral vectors and combinatorial vaccine approaches.

Broadly Neutralizing Antibody (bNAbs) Vaccine Design

The most significant advance in the HIV vaccine field over the past decade has been the identification of bNAbs. Initially, the lack of evidence of antibody mediated suppression of HIV infection *in vivo* led to the belief that the human body was not capable of generating antibodies that can neutralize HIV. The discovery of human sera from HIV infected individuals that could neutralize a broad range of lab-adapted HIV-1 strains led to the rapid isolation and characterization of bNAbs (139–142). From then on, numerous bNAbs have been isolated, some of which can neutralize up to 99% of all known HIV-1 isolates (143). It is now well established that bNAbs responses can be generated during natural infection, but are quite rare and tend to occur much later in chronic infection (144, 145).

Isolation of bNAbs from HIV infected individuals motivated the design of immunogens capable of inducing bNAbs through vaccination. Over the years, bNAbs research has generated deeper insight into the virology and humoral immunity to HIV-1 infection. The wealth of knowledge gained has led to several HIV immunogen design approaches including germ-line targeting immunogens to molecular structural stabilization of envelope trimers such as eOD-GT8 (146, 147), the use of soluble trimers that mimic the native Env spike such as the BG505 SOSI.664 (148), epitope targeted immunogen designs with minimal epitope fragments meant to focus the response on the right region of the Env spike, and minimize off target responses.

It has been shown that all bNAbs typically target the HIV envelope (Env) spike protein (149–151). However, HIV produces numerous non-functional Env proteins which divert antibody immune responses by displaying immunodominant epitopes, resulting in higher titers of non-neutralizing antibody responses (148, 152–154). Immunodominant epitopes tend to be easily accessible whereas most conserved epitopes (vulnerable sites) such as the CD4 binding site (CD4bs) have poor accessibility that limit bNAb recognition (154). One strategy used to increase the immunogenicity of immunodominant epitopes is epitope masking (155). Epitope masking is aimed at directing the immune response to sites of neutralization vulnerability by selectively allowing access to broadly neutralizing epitopes while masking immunodominant (non-neutralizing) regions (156). For instance, addition of glycans to mask immunodominant epitopes has been shown to reduce non-neutralizing antibody access (157, 158).

Another strategy to decrease the immunogenicity of non-neutralizing antibody (non-NAb) epitopes is the occlusion of immunodominant glycan holes. Serological studies conducted in an effort to get insight into the breadth of vaccine-elicited NAbs revealed that immunodominant strain-specific glycan holes on HIV Env contributed to the limited breadth of these monoclonal antibodies (mAbs) (157, 159–162). In addition, it has been shown that the addition of *N*-glycosylation sites to the V3 region or the glycan hole epitope at position 241/289 of the BG505 trimer suppressed the immunogenicity of its non-NAb epitopes while, in some instances, diverting the NAb responses to neoepitopes (163–165). Collectively, these novel strategies provide a platform for the optimization of the epitope targeted immunogen design approaches (97). Furthermore, advances in high throughput recombinant antibody technology has created the possibility of using bNAbs for prevention or treatment of HIV-1 infection as described above (166, 167).

While evidence has shown that cows immunized with BG505 SOSIP trimers rapidly developed broad and potent serum HIV specific NAb responses (168), large-scale bNAb research has not yet achieved the development of a vaccine that induces bNAb responses in other animal models or in humans (169, 170). Deep understanding of host and viral factors necessary and sufficient for the generation of bNAbs is imperative. The extensive Env diversity and the large glycan shield that cover the envelope trimer surface remain a major immunogen design challenge. Moreover, the extremely high somatic hypermutation required for bNAb function makes it very difficult to induce bNAbs by traditional vaccination protocols. To overcome the requirement for high mutation levels, novel approaches that involve priming the initial bNAb precursor B cells followed by sequential immunization aimed at driving the evolutionally intermediates are currently being evaluated in humans (171).

CMV-Based Vector Vaccines

Clearly, the limited immunogenicity of HIV is still a challenge. Most vaccines induce, weak, narrow, and short-lived immunity. Innovative approaches to overcome this limitation include the use of new viral vectors such as cytomegalovirus (CMV) or Ad26 viral vectors. CMV vectors have emerged as a type of virial vector with unique properties of inducing massive atypical immune responses that are capable of conferring sterilizing protection in animal studies (172). Special attributes of CMV vectored vaccines include the ability to maintain persistent immune stimulation and not being prone to attenuation by pre-existing immune responses. Also, CMV vector-induced HLA-E restricted $CD8^+$ T cell responses have the potential to provide vaccine efficacy in all individuals regardless of MHC-class I genotypes (173, 174).

CMV vectors represent a promising strategy in HIV vaccines because of the possibility of genetically reprogramming the vector to induce massive numbers of unusual $CD8^+$ T cell responses that can confer sterilizing immunity. The concept of genetic programming of the CMV vector to induce protective immune responses by vaccination was operationalized by Louis Picker and colleagues. In a series of studies, they used non-human primate models to demonstrate that CMV vectors can be genetically calibrated to induce MHC class II restricted as well as non-polymorphic MHC-E restricted $CD8^+$ T cell responses. Animals vaccinated with genetically engineered Rhesus CMV (RhCMV/) vectored SIV vaccines elicited very high frequencies of effector memory $CD8^+$ T cells that persisted in tissue sites and conferred stringent control of SIV/SHIV in 50% of the vaccinated animals without any apparent antibody responses (175–177).

Although CMV vectors show great promise, safety concerns regarding persistence and potential pathogenicity dampens enthusiasm to use such vaccines in humans. It is also not clear if humans can generate unconventional immune responses reported in animal studies. A study of HIV elite controllers showed that HLA-II restricted $CD8^+$ T cell responses exist in humans but are very rare (178), and highly unlikely to be elicited by vaccination. On the other hand, recent studies showed that HLA-E restricted responses are much more common in humans than previously appreciated, which opens up the possibility of inducing such responses by vaccination (179). To improve the safety of CMV vectors, the Picker group has managed to genetically modify CMV vectors to significantly reduce the capacity for the vector to widely disseminate while retaining its ability to superinfect, elicit, and maintain protective $CD8^+$ T cell responses (180). These safety improvements render support for testing CMV vectors in humans.

Combinatorial HIV Vaccine Design

This review has highlighted the inadequacy of standard vaccine approaches to elicit protective immunity against HIV. Therefore, it is imperative that future vaccines adopt multipronged approaches capable of eliciting more than one arm of the immune system. Examples of such approaches for HIV vaccines include the combination of HIV vaccines and non-vaccine approaches such as PrEP, microbicides or other non-vaccine HIV prevention methods. Non-human primate (NHP) studies have provided key proof-of-concept data (181, 182). The first NHP study of combined biomedical preventions (CBP), combined DNA prime recombinant adenovirus boost T cell-based vaccine with a vaginal microbicide gel (with a suboptimal concentration of an HIV-1 nucleocapsid zinc finger inhibitor) and administered the regime to rhesus macaques (183). CBP delayed simian-human immunodeficiency virus (SHIV) infection in the break through animals and resulted in reduced viral load compared to individual treatments, highlighting strong synergy between the two approaches (183). Another study evaluated the combined effect of a tenofovir-containing vaginal gel co-administered with a protein-based HIV vaccine (containing proteins from clades B and C, adjuvanted with MF59) designed to induce T- and B-cell immunity. The vaccine alone did not offer protection against repeated $SHIV_{162P3}$ challenges, and 1% tenofovir alone showed an efficacy of 46%. But a combination of the vaccine with 1% tenofovir increased protective efficacy to 81% (182). Together, these studies demonstrate the potential of combinatorial approaches to HIV vaccine development and underscore the need for concerted efforts in pursuing such approaches.

CONCLUDING REMARKS

Although considerable progress has been made in understanding the structural and molecular biology of the virus, no HIV vaccine candidate has progressed to licensure, in spite of decades of very expensive research. Nonetheless, there is reason for cautious optimism that at least one of the many vaccine candidates, highlighted in this review, currently in pre-clinical or in efficacy trials will succeed. But, given the history of HIV vaccines, it is not far-fetched to think that success for the current vaccine candidates, if any, is most likely to be modest and will require several iterations to achieve significant protection. The glaring knowledge gap regarding the nature of immunity required for protection warrants basic science research. Given that an effective vaccine will need to stimulate more than one arm of the immune system, we recommend that combinatorial vaccine approaches capable of inducing innate and adaptive immune responses should actively be pursued. Finally, the huge cost of basic research, manufacture and regulatory processes required to take a vaccine candidate from the bench all the way to market, can only be achieved through collaborative efforts from all stake holders. Basic research scientists, vaccine trialists, governments and philanthropists, regulatory bodies, and pharmaceutical companies must continue to pull together to overcome this global challenge.

AUTHOR CONTRIBUTIONS

Conceptualization: ZN formulated the idea and provided oversight in planning and organizing the literature search. Writing (original draft preparation) was done by TNg'u and CC with the supervision of ZN. Writing (review and editing): TNg'u, CC, and ZN edited the manuscript. All authors contributed to the article and approved the submitted version.

ACKNOWLEDGMENTS

We thank Prof. Thumbi Ndung'u and Dr. Daniel Muema for their careful and critical reading of the manuscript.

REFERENCES

1. Ditse Z, Mkhize NN, Yin M, Keefer M, Montefiori DC, Tomaras GD, et al. Effect of HIV Envelope Vaccination on the Subsequent Antibody Response to HIV Infection. *mSphere* (2020) 5:1–13. 10.1128/mSphere.00738-19
2. *Global HIV and AIDS statistics.* Available at: https://www.avert.org/global-hiv-and-aids-statistics (Accessed 24 April 2020).
3. Zhao LP, Fiore-Gartland A, Carpp LN, Cohen KW, Rouphael N, Fleurs L, et al. Landscapes of binding antibody and T-cell responses to pox-protein HIV vaccines in Thais and South Africans. *PLoS One* (2020) 15:e0226803. 10.1371/journal.pone.0226803
4. Berkley SF, Koff WC. Scientific and policy challenges to development of an AIDS vaccine. *Lancet* (2007) 370:94–101. 10.1016/S0140-6736(07)61054-X
5. Ahmed Y, Tian M, Gao Y. Development of an anti-HIV vaccine eliciting broadly neutralizing antibodies. *AIDS Res Ther* (2017) 14:50. 10.1186/s12981-017-0178-3
6. Chawla A, Wang C, Patton C, Murray M, Punekar Y, de Ruiter A, et al. Review of Long-Term Toxicity of Antiretroviral Treatment Regimens and Implications for an Aging Population. *Infect Dis Ther* (2018) 7:183–95. 10.1007/s40121-018-0201-6
7. Moretti S, Cafaro A, Tripiciano A, Picconi O, Butto S, Ensoli F, et al. HIV therapeutic vaccines aimed at intensifying combination antiretroviral therapy. *Expert Rev Vaccines* (2020) 19:71–84. 10.1080/14760584.2020.1712199
8. Calarota SA, Dai A, Trocio JN, Weiner DB, Lori F, Lisziewicz J. IL-15 as memory T-cell adjuvant for topical HIV-1 DermaVir vaccine. *Vaccine* (2008) 26:5188–95. 10.1016/j.vaccine.2008.03.067
9. Gao F, Weaver EA, Lu Z, Li Y, Liao HX, Ma B, et al. Antigenicity and immunogenicity of a synthetic human immunodeficiency virus type 1 group m consensus envelope glycoprotein. *J Virol* (2005) 79:1154–63. 10.1128/JVI.79.2.1154-1163.2005
10. Rudometov AP, Chikaev AN, Rudometova NB, Antonets DV, Lomzov AA, Kaplina ON, et al. Artificial Anti-HIV-1 Immunogen Comprising Epitopes of Broadly Neutralizing Antibodies 2F5, 10E8, and a Peptide Mimic of VRC01 Discontinuous Epitope. *Vaccines (Basel)* (2019) 7:1–18. 10.3390/vaccines7030083

11. Li G, De Clercq E. HIV Genome-Wide Protein Associations: a Review of 30 Years of Research. *Microbiol Mol Biol Rev* (2016) 80:679–731. 10.1128/MMBR.00065-15
12. Rathore U, Purwar M, Vignesh VS, Das R, Kumar AA, Bhattacharyya S, et al. Bacterially expressed HIV-1 gp120 outer-domain fragment immunogens with improved stability and affinity for CD4-binding site neutralizing antibodies. *J Biol Chem* (2018) 293:15002–20. 10.1074/jbc.RA118.005006
13. Marusic C, Vitale A, Pedrazzini E, Donini M, Frigerio L, Bock R, et al. Plant-based strategies aimed at expressing HIV antigens and neutralizing antibodies at high levels. Nef as a case study. *Transgenic Res* (2009) 18:499–512. 10.1007/s11248-009-9244-5
14. Del Moral-Sanchez I, Sliepen K. Strategies for inducing effective neutralizing antibody responses against HIV-1. *Expert Rev Vaccines* (2019) 18:1127–43. 10.1080/14760584.2019.1690458
15. Gruell H, Klein F. Antibody-mediated prevention and treatment of HIV-1 infection. *Retrovirology* (2018) 15:73. 10.1186/s12977-018-0455-9
16. Korber B, Gaschen B, Yusim K, Thakallapally R, Kesmir C, Detours V. Evolutionary and immunological implications of contemporary HIV-1 variation. *Br Med Bull* (2001) 58:19–42. 10.1093/bmb/58.1.19
17. Hemelaar J. The origin and diversity of the HIV-1 pandemic. *Trends Mol Med* (2012) 18:182–92. 10.1016/j.molmed.2011.12.001
18. Powell RL, Urbanski MM, Burda S, Kinge T, Nyambi PN. High frequency of HIV-1 dual infections among HIV-positive individuals in Cameroon, West Central Africa. *J Acquir Immune Defic Syndr* (2009) 50:84–92. 10.1097/QAI.0b013e31818d5a40
19. Barouch DH. Challenges in the development of an HIV-1 vaccine. *Nature* (2008) 455:613–9. 10.1038/nature07352
20. Plotkin SA. Increasing Complexity of Vaccine Development. *J Infect Dis* (2015) 212(Suppl 1):S12–16. 10.1093/infdis/jiu568
21. Rubens M, Ramamoorthy V, Saxena A, Shehadeh N, Appunni S. HIV Vaccine: Recent Advances, Current Roadblocks, and Future Directions. *J Immunol Res* (2015) 2015:560347. 10.1155/2015/560347
22. Coffin J, Haase A, Levy JA, Montagnier L, Oroszlan S, Teich N, et al. What to call the AIDS virus? *Nature* (1986) 321:10. 10.1038/321010a0

23. Gallo RC, Montagnier L. The chronology of AIDS research. *Nature* (1987) 326:435–6. 10.1038/326435a0
24. Esparza J. A brief history of the global effort to develop a preventive HIV vaccine. *Vaccine* (2013) 31:3502–18. 10.1016/j.vaccine.2013.05.018
25. Dowdle W. The search for an AIDS vaccine. *Public Health Rep* (1986) 101:232–3. [PMC free article] [PubMed] [Google Scholar]
26. Ensoli B, Cafaro A, Monini P, Marcotullio S, Ensoli F. Challenges in HIV Vaccine Research for Treatment and Prevention. *Front Immunol* (2014) 5:417. 10.3389/fimmu.2014.00417
27. Fischinger PJ, Robey WG, Koprowski H, Gallo RC, Bolognesi DP. Current status and strategies for vaccines against diseases induced by human T-cell lymphotropic retroviruses (HTLV-I, -II, -III). *Cancer Res* (1985) 45:4694s–9s.
28. Fauci AS. Current issues in developing a strategy for dealing with the acquired immunodeficiency syndrome. *Proc Natl Acad Sci USA* (1986) 83:9278–83. 10.1073/pnas.83.24.9278
29. McAleer WJ, Buynak EB, Maigetter RZ, Wampler DE, Miller WJ, Hilleman MR. Human hepatitis B vaccine from recombinant yeast. *Nature* (1984) 307:178–80. 10.1038/307178a0
30. Robey WG, Safai B, Oroszlan S, Arthur LO, Gonda MA, Gallo RC, et al. Characterization of envelope and core structural gene products of HTLV-III with sera from AIDS patients. *Science* (1985) 228:593–5. 10.1126/science.2984774
31. Wain-Hobson S, Sonigo P, Danos O, Cole S, Alizon M. Nucleotide sequence of the AIDS virus, LAV. *Cell* (1985) 40:9–17. 10.1016/0092-8674(85)90303-4
32. Wong-Staal F, Shaw GM, Hahn BH, Salahuddin SZ, Popovic M, Markham P, et al. Genomic diversity of human T-lymphotropic virus type III (HTLV-III). *Science* (1985) 229:759–62. 10.1126/science.2992084
33. Plotkin SA. Correlates of protection induced by vaccination. *Clin Vaccine Immunol* (2010) 17:1055–65. 10.1128/CVI.00131-10
34. Zagury D, Leonard R, Fouchard M, Reveil B, Bernard J, Ittele D, et al. Immunization against AIDS in humans. *Nature* (1987) 326:249–50. 10.1038/326249a0

35. Zagury D, Bernard J, Cheynier R, Desportes I, Leonard R, Fouchard M, et al. A group specific anamnestic immune reaction against HIV-1 induced by a candidate vaccine against AIDS. *Nature* (1988) 332:728–31. 10.1038/332728a0

36. Wang HB, Mo QH, Yang Z. HIV vaccine research: the challenge and the way forward. *J Immunol Res* (2015) 2015:503978. 10.1155/2015/503978

37. Dolin R, Graham BS, Greenberg SB, Tacket CO, Belshe RB, Midthun K, et al. The safety and immunogenicity of a human immunodeficiency virus type 1 (HIV-1) recombinant gp160 candidate vaccine in humans. NIAID AIDS Vaccine Clinical Trials Network. *Ann Intern Med* (1991) 114:119–27. 10.7326/0003-4819-114-2-119

38. Cooney EL, Collier AC, Greenberg PD, Coombs RW, Zarling J, Arditti DE, et al. Safety of and immunological response to a recombinant vaccinia virus vaccine expressing HIV envelope glycoprotein. *Lancet* (1991) 337:567–72. 10.1016/0140-6736(91)91636-9

39. Flynn NM, Forthal DN, Harro CD, Judson FN, Mayer KH, Para MF. rgp, H.I.V.V.S.G. Placebo-controlled phase 3 trial of a recombinant glycoprotein 120 vaccine to prevent HIV-1 infection. *J Infect Dis* (2005) 191:654–65. 10.1086/428404

40. Pitisuttithum P, Gilbert P, Gurwith M, Heyward W, Martin M, van Griensven F, et al. Randomized, double-blind, placebo-controlled efficacy trial of a bivalent recombinant glycoprotein 120 HIV-1 vaccine among injection drug users in Bangkok, Thailand. *J Infect Dis* (2006) 194:1661–71. 10.1086/508748

41. Catanzaro AT, Roederer M, Koup RA, Bailer RT, Enama ME, Nason MC, et al. Phase I clinical evaluation of a six-plasmid multiclade HIV-1 DNA candidate vaccine. *Vaccine* (2007) 25:4085–92. 10.1016/j.vaccine.2007.02.050

42. McEnery R. HVTN 505 trial expanded to see if vaccine candidates can block HIV acquisition. *IAVI Rep* (2011) 15:17.

43. Duerr A, Huang Y, Buchbinder S, Coombs RW, Sanchez J, del Rio C, et al. Extended follow-up confirms early vaccine-enhanced risk of HIV acquisition and demonstrates waning effect over time among participants in a randomized trial of recombinant adenovirus HIV vaccine (Step Study). *J Infect Dis* (2012) 206:258–66. 10.1093/infdis/jis342

44. Buchbinder SP, Mehrotra DV, Duerr A, Fitzgerald DW, Mogg R, Li D, et al. Efficacy assessment of a cell-mediated immunity HIV-1 vaccine (the Step Study): a double-blind, randomised, placebo-controlled, test-of-concept trial. *Lancet* (2008) 372:1881–93. 10.1016/S0140-6736(08)61591-3
45. Rerks-Ngarm S, Pitisuttithum P, Nitayaphan S, Kaewkungwal J, Chiu J, Paris R, et al. Vaccination with ALVAC and AIDSVAX to prevent HIV-1 infection in Thailand. *N Engl J Med* (2009) 361:2209–20. 10.1056/NEJMoa0908492
46. Kijak GH, Tovanabutra S, Rerks-Ngarm S, Nitayaphan S, Eamsila C, Kunasol P, et al. Molecular evolution of the HIV-1 Thai epidemic between the time of RV144 immunogen selection to the execution of the vaccine efficacy trial. *J Virol* (2013) 87:7265–81. 10.1128/JVI.03070-12
47. Rerks-Ngarm S, Pitisuttithum P, Excler JL, Nitayaphan S, Kaewkungwal J, Premsri N, et al. Randomized, Double-Blind Evaluation of Late Boost Strategies for HIV-Uninfected Vaccine Recipients in the RV144 HIV Vaccine Efficacy Trial. *J Infect Dis* (2017) 215:1255–63. 10.1093/infdis/jix099
48. Pitisuttithum P, Nitayaphan S, Chariyalertsak S, Kaewkungwal J, Dawson P, Dhitavat J, et al. Late boosting of the RV144 regimen with AIDSVAX B/E and ALVAC-HIV in HIV-uninfected Thai volunteers: a double-blind, randomised controlled trial. *Lancet HIV* (2020) 7:e238–48. 10.1016/S2352-3018(19)30406-0
49. Gray GE, Huang Y, Grunenberg N, Laher F, Roux S, Andersen-Nissen E, et al. Immune correlates of the Thai RV144 HIV vaccine regimen in South Africa. *Sci Transl Med* (2019) 11:1–17. 10.1126/scitranslmed.aax1880
50. Bekker LG, Moodie Z, Grunenberg N, Laher F, Tomaras GD, Cohen KW, et al. Subtype C ALVAC-HIV and bivalent subtype C gp120/MF59 HIV-1 vaccine in low-risk, HIV-uninfected, South African adults: a phase 1/2 trial. *Lancet HIV* (2018) 5:e366–78. 10.1016/S2352-3018(18)30071-7
51. Global Advocacy for HIV Prevention. Available at: https://www.avac.org/trial/hvtn-702 (Accessed 30 April 2020).
52. Keefer MC, Graham BS, Belshe RB, Schwartz D, Corey L, Bolognesi DP, et al. Studies of high doses of a human immunodeficiency virus type 1 recombinant glycoprotein 160 candidate vaccine in HIV

type 1-seronegative humans. The AIDS Vaccine Clinical Trials Network. *AIDS Res Hum Retroviruses* (1994) 10:1713–23. 10.1089/aid.1994.10.1713

53. Cooney EL, McElrath MJ, Corey L, Hu SL, Collier AC, Arditti D, et al. Enhanced immunity to human immunodeficiency virus (HIV) envelope elicited by a combined vaccine regimen consisting of priming with a vaccinia recombinant expressing HIV envelope and boosting with gp160 protein. *Proc Natl Acad Sci U S A* (1993) 90:1882–6. 10.1073/pnas.90.5.1882
54. Graham BS, Matthews TJ, Belshe RB, Clements ML, Dolin R, Wright PF, et al. Augmentation of human immunodeficiency virus type 1 neutralizing antibody by priming with gp160 recombinant vaccinia and boosting with rgp160 in vaccinia-naive adults. The NIAID AIDS Vaccine Clinical Trials Network. *J Infect Dis* (1993) 167:533–7. 10.1093/infdis/167.3.533
55. Redfield RR, Wright DC, James WD, Jones TS, Brown C, Burke DS. Disseminated vaccinia in a military recruit with human immunodeficiency virus (HIV) disease. *N Engl J Med* (1987) 316:673–6. 10.1056/NEJM198703123161106
56. Picard O, Lebas J, Imbert JC, Bigel P, Zagury D. Complication of intramuscular/subcutaneous immune therapy in severely immune-compromised individuals. *J Acquir Immune Defic Syndr (1988)* (1991) 4:641–3.
57. Cox WI, Tartaglia J, Paoletti E. Induction of cytotoxic T lymphocytes by recombinant canarypox (ALVAC) and attenuated vaccinia (NYVAC) viruses expressing the HIV-1 envelope glycoprotein. *Virology* (1993) 195:845–50. 10.1006/viro.1993.1442
58. Perkus ME, Tartaglia J, Paoletti E. Poxvirus-based vaccine candidates for cancer, AIDS, and other infectious diseases. *J Leukoc Biol* (1995) 58:1–13. 10.1002/jlb.58.1.1
59. Pialoux G, Excler JL, Riviere Y, Gonzalez-Canali G, Feuillie V, Coulaud P, et al. A prime-boost approach to HIV preventive vaccine using a recombinant canarypox virus expressing glycoprotein 160 (MN) followed by a recombinant glycoprotein 160 (MN/LAI). The AGIS Group, and l'Agence Nationale de Recherche sur le SIDA. *AIDS Res Hum Retroviruses* (1995) 11:373–81. 10.1089/aid.1995.11.373

60. Marovich MA. ALVAC-HIV vaccines: clinical trial experience focusing on progress in vaccine development. *Expert Rev Vaccines* (2004) 3:S99–104. 10.1586/14760584.3.4.s99
61. Thongcharoen P, Suriyanon V, Paris RM, Khamboonruang C, de Souza MS, Ratto-Kim S, et al. A phase 1/2 comparative vaccine trial of the safety and immunogenicity of a CRF01_AE (subtype E) candidate vaccine: ALVAC-HIV (vCP1521) prime with oligomeric gp160 (92TH023/LAI-DID) or bivalent gp120 (CM235/SF2) boost. *J Acquir Immune Defic Syndr* (2007) 46:48–55. 10.1097/QAI.0b013e3181354bd7
62. Mugerwa RD, Kaleebu P, Mugyenyi P, Katongole-Mbidde E, Hom DL, Byaruhanga R, et al. First trial of the HIV-1 vaccine in Africa: Ugandan experience. *BMJ* (2002) 324:226–9. 10.1136/bmj.324.7331.226
63. Cao H, Mani I, Vincent R, Mugerwa R, Mugyenyi P, Kanki P, et al. Cellular immunity to human immunodeficiency virus type 1 (HIV-1) clades: relevance to HIV-1 vaccine trials in Uganda. *J Infect Dis* (2000) 182:1350–6. 10.1086/315868
64. Cao H, Kaleebu P, Hom D, Flores J, Agrawal D, Jones N, et al. Immunogenicity of a recombinant human immunodeficiency virus (HIV)-canarypox vaccine in HIV-seronegative Ugandan volunteers: results of the HIV Network for Prevention Trials 007 Vaccine Study. *J Infect Dis* (2003) 187:887–95. 10.1086/368020
65. Berman PW, Gregory TJ, Riddle L, Nakamura GR, Champe MA, Porter JP, et al. Protection of chimpanzees from infection by HIV-1 after vaccination with recombinant glycoprotein gp120 but not gp160. *Nature* (1990) 345:622–5. 10.1038/345622a0
66. Berman PW, Murthy KK, Wrin T, Vennari JC, Cobb EK, Eastman DJ, et al. Protection of MN-rgp120-immunized chimpanzees from heterologous infection with a primary isolate of human immunodeficiency virus type 1. *J Infect Dis* (1996) 173:52–9. 10.1093/infdis/173.1.52
67. Cohen J. Thailand weighs AIDS vaccine tests. *Science* (1995) 270:904–7. 10.1126/science.270.5238.904
68. Cohen J. Planned tests in Thailand spark debate. *Science* (1997) 276:1197. 10.1126/science.276.5316.1197
69. Balter M. Impending AIDS vaccine trial opens old wounds. *Science* (1998) 279:650. 10.1126/science.279.5351.650
70. Haynes BF, Gilbert PB, McElrath MJ, Zolla-Pazner S, Tomaras GD, Alam SM, et al. Immune-correlates analysis of an HIV-1 vaccine efficacy

trial. *N Engl J Med* (2012) 366:1275–86. 10.1056/NEJMoa1113425

71. Berman PW. Development of bivalent rgp120 vaccines to prevent HIV type 1 infection. *AIDS Res Hum Retroviruses* (1998) 14 Suppl 3:S277–289.
72. McMichael A, Hanke T. The quest for an AIDS vaccine: is the CD8+ T-cell approach feasible? *Nat Rev Immunol* (2002) 2:283–91. 10.1038/nri779
73. McMichael AJ, Callan M, Appay V, Hanke T, Ogg G, Rowland-Jones S. The dynamics of the cellular immune response to HIV infection: implications for vaccination. *Philos Trans R Soc Lond B Biol Sci* (2000) 355:1007–11. 10.1098/rstb.2000.0637
74. McMichael AJ, Borrow P, Tomaras GD, Goonetilleke N, Haynes BF. The immune response during acute HIV-1 infection: clues for vaccine development. *Nat Rev Immunol* (2010) 10:11–23. 10.1038/nri2674
75. Mudd PA, Martins MA, Ericsen AJ, Tully DC, Power KA, Bean AT, et al. Vaccine-induced CD8+ T cells control AIDS virus replication. *Nature* (2012) 491:129–33. 10.1038/nature11443
76. Koup RA, Roederer M, Lamoreaux L, Fischer J, Novik L, Nason MC, et al. Priming immunization with DNA augments immunogenicity of recombinant adenoviral vectors for both HIV-1 specific antibody and T-cell responses. *PLoS One* (2010) 5:e9015. 10.1371/journal.pone.0009015
77. Walker BD. Elite control of HIV Infection: implications for vaccines and treatment. *Top HIV Med* (2007) 15:134–6.
78. Watkins DI. The hope for an HIV vaccine based on induction of CD8+ T lymphocytes–a review. *Mem Inst Oswaldo Cruz* (2008) 103:119–29. 10.1590/s0074-02762008000200001
79. Schoenly KA, Weiner DB. Human immunodeficiency virus type 1 vaccine development: recent advances in the cytotoxic T-lymphocyte platform "spotty business". *J Virol* (2008) 82:3166–80. 10.1128/JVI.01634-07
80. Barouch DH. Novel adenovirus vector-based vaccines for HIV-1. *Curr Opin HIV AIDS* (2010) 5:386–90. 10.1097/COH.0b013e32833cfe4c
81. Wilks AB, Christian EC, Seaman MS, Sircar P, Carville A, Gomez CE, et al. Robust vaccine-elicited cellular immune responses in breast milk following systemic simian immunodeficiency virus DNA prime and live virus vector boost vaccination of lactating rhesus monkeys. *J*

Immunol (2010) 185:7097–106. 10.4049/jimmunol.1002751

82. Cohen J. AIDS research. Promising AIDS vaccine's failure leaves field reeling. *Science* (2007) 318:28–9. 10.1126/science.318.5847.28
83. Catanzaro AT, Koup RA, Roederer M, Bailer RT, Enama ME, Moodie Z, et al. Phase 1 safety and immunogenicity evaluation of a multiclade HIV-1 candidate vaccine delivered by a replication-defective recombinant adenovirus vector. *J Infect Dis* (2006) 194:1638–49. 10.1086/509258
84. Watanaveeradej V, Benenson MW, Souza MD, Sirisopana N, Nitayaphan S, Tontichaivanich C, et al. Molecular epidemiology of HIV Type 1 in preparation for a Phase III prime-boost vaccine trial in Thailand and a new approach to HIV Type 1 genotyping. *AIDS Res Hum Retroviruses* (2006) 22:801–7. 10.1089/aid.2006.22.801
85. Gottardo R, Bailer RT, Korber BT, Gnanakaran S, Phillips J, Shen X, et al. Plasma IgG to linear epitopes in the V2 and V3 regions of HIV-1 gp120 correlate with a reduced risk of infection in the RV144 vaccine efficacy trial. *PLoS One* (2013) 8:e75665. 10.1371/journal.pone.0075665
86. Yates NL, Liao HX, Fong Y, deCamp A, Vandergrift NA, Williams WT, et al. Vaccine-induced Env V1-V2 IgG3 correlates with lower HIV-1 infection risk and declines soon after vaccination. *Sci Transl Med* (2014) 6:228ra239. 10.1126/scitranslmed.3007730
87. Zolla-Pazner S, deCamp A, Gilbert PB, Williams C, Yates NL, Williams WT, et al. Vaccine-induced IgG antibodies to V1V2 regions of multiple HIV-1 subtypes correlate with decreased risk of HIV-1 infection. *PLoS One* (2014) 9:e87572. 10.1371/journal.pone.0087572
88. Fischinger S, Shin S, Boudreau CM, Ackerman M, Rerks-Ngarm S, Pitisuttithum P, et al. Protein-based, but not viral vector alone, HIV vaccine boosting drives an IgG1-biased polyfunctional humoral immune response. *JCI Insight* (2020) 5:1–11. 10.1172/jci.insight.135057
89. Easterhoff D, Moody MA, Fera D, Cheng H, Ackerman M, Wiehe K, et al. Boosting of HIV envelope CD4 binding site antibodies with long variable heavy third complementarity determining region in the randomized double blind RV305 HIV-1 vaccine trial. *PLoS Pathog* (2017) 13:e1006182. 10.1371/journal.ppat.1006182
90. Bekker LG, Gray GE. Hope for HIV control in southern Africa: The continued quest for a vaccine. *PLoS Med* (2017) 14:e1002241. 10.1371/journal.pmed.1002241

91. Cohen J. Combo of two HIV vaccines fails its big test. *Science* (2020) 367:611–2. 10.1126/science.367.6478.611

92. National Institute of Allergy and Infectious Disease *Experimental HIV Vaccine Regimen Ineffective in Preventing HIV.* Available at: https://www.niaid.nih.gov/news-events/experimental-hiv-vaccine-regimen-ineffective-preventing-hiv (Accessed 30 April 2020).

93. Desrosiers RC. Protection against HIV Acquisition in the RV144 Trial. *J Virol* (2017) 91:1–3. 10.1128/JVI.00905-17

94. Barouch DH. A step forward for HIV vaccines. *Lancet HIV* (2018) 5:e338–9. 10.1016/S2352-3018(18)30095-X

95. Corey L, Nabel GJ, Dieffenbach C, Gilbert P, Haynes BF, Johnston M, et al. HIV-1 vaccines and adaptive trial designs. *Sci Transl Med* (2011) 3:79ps13. 10.1126/scitranslmed.3001863

96. Gilbert PB, Huang Y, Janes HE. Modeling HIV vaccine trials of the future. *Curr Opin HIV AIDS* (2016) 11:620–7. 10.1097/COH.0000000000000314

97. Alter G, Barouch D. Immune Correlate-Guided HIV Vaccine Design. *Cell Host Microbe* (2018) 24:25–33. 10.1016/j.chom.2018.06.012

98. Graham BS, Ambrosino DM. History of passive antibody administration for prevention and treatment of infectious diseases. *Curr Opin HIV AIDS* (2015) 10:129–34. 10.1097/COH.0000000000000154

99. Perreau M, Banga R, Pantaleo G. Targeted Immune Interventions for an HIV-1 Cure. *Trends Mol Med* (2017) 23:945–61. 10.1016/j.molmed.2017.08.006

100. Lu LL, Suscovich TJ, Fortune SM, Alter G. Beyond binding: antibody effector functions in infectious diseases. *Nat Rev Immunol* (2018) 18:46–61. 10.1038/nri.2017.106

101. Wang P. Anti-HIV Passive Immunization in Animal Models. *J HIV Retrovirus* (2018) 4:1–5. 10.21767/2471-9676.100036 [CrossRef] [Google Scholar]

102. Pegu A, Hessell AJ, Mascola JR, Haigwood NL. Use of broadly neutralizing antibodies for HIV-1 prevention. *Immunol Rev* (2017) 275:296–312. 10.1111/imr.12511

103. Bar KJ, Sneller MC, Harrison LJ, Justement JS, Overton ET, Petrone ME, et al. Effect of HIV Antibody VRC01 on Viral Rebound after Treatment Interruption. *N Engl J Med* (2016) 375:2037–50. 10.1056/NEJMoa1608243

104. Caskey M, Klein F, Lorenzi JC, Seaman MS, West AP, Jr., Buckley N, et al. Viraemia suppressed in HIV-1-infected humans by broadly neutralizing antibody 3BNC117. *Nature* (2015) 522:487–91. 10.1038/nature14411

105. Lynch RM, Boritz E, Coates EE, DeZure A, Madden P, Costner P, et al. Virologic effects of broadly neutralizing antibody VRC01 administration during chronic HIV-1 infection. *Sci Transl Med* (2015) 7:319ra206. 10.1126/scitranslmed.aad5752

106. Morris L, Mkhize NN. Prospects for passive immunity to prevent HIV infection. *PLoS Med* (2017) 14:e1002436. 10.1371/journal.pmed.1002436

107. Lifson JD, Haigwood NL. Lessons in nonhuman primate models for AIDS vaccine research: from minefields to milestones. *Cold Spring Harb Perspect Med* (2012) 2:a007310. 10.1101/cshperspect.a007310

108. Shibata R, Igarashi T, Haigwood N, Buckler-White A, Ogert R, Ross W, et al. Neutralizing antibody directed against the HIV-1 envelope glycoprotein can completely block HIV-1/SIV chimeric virus infections of macaque monkeys. *Nat Med* (1999) 5:204–10. 10.1038/5568

109. Moldt B, Rakasz EG, Schultz N, Chan-Hui PY, Swiderek K, Weisgrau KL, et al. Highly potent HIV-specific antibody neutralization in vitro translates into effective protection against mucosal SHIV challenge in vivo. *Proc Natl Acad Sci U S A* (2012) 109:18921–5. 10.1073/pnas.1214785109

110. Gautam R, Nishimura Y, Pegu A, Nason MC, Klein F, Gazumyan A, et al. A single injection of anti-HIV-1 antibodies protects against repeated SHIV challenges. *Nature* (2016) 533:105–9. 10.1038/nature17677

111. Julg B, Liu PT, Wagh K, Fischer WM, Abbink P, Mercado NB, et al. Protection against a mixed SHIV challenge by a broadly neutralizing antibody cocktail. *Sci Transl Med* (2017) 9:1–9. 10.1126/scitranslmed.aao4235

112. Liu J, Ghneim K, Sok D, Bosche WJ, Li Y, Chipriano E, et al. Antibody-mediated protection against SHIV challenge includes systemic clearance of distal virus. *Science* (2016) 353:1045–9. 10.1126/science.aag0491

113. Parren PW, Marx PA, Hessell AJ, Luckay A, Harouse J, Cheng-Mayer C, et al. Antibody protects macaques against vaginal challenge with a pathogenic R5 simian/human immunodeficiency virus at serum levels

giving complete neutralization in vitro. *J Virol* (2001) 75:8340–7. 10.1128/jvi.75.17.8340-8347.2001

114. Veazey RS, Shattock RJ, Pope M, Kirijan JC, Jones J, Hu Q, et al. Prevention of virus transmission to macaque monkeys by a vaginally applied monoclonal antibody to HIV-1 gp120. *Nat Med* (2003) 9:343–6. 10.1038/nm833
115. Hessell AJ, Hangartner L, Hunter M, Havenith CE, Beurskens FJ, Bakker JM, et al. Fc receptor but not complement binding is important in antibody protection against HIV. *Nature* (2007) 449:101–4. 10.1038/nature06106
116. Hessell AJ, Poignard P, Hunter M, Hangartner L, Tehrani DM, Bleeker WK, et al. Effective, low-titer antibody protection against low-dose repeated mucosal SHIV challenge in macaques. *Nat Med* (2009) 15:951–4. 10.1038/nm.1974
117. Gauduin MC, Parren PW, Weir R, Barbas CF, Burton DR, Koup RA. Passive immunization with a human monoclonal antibody protects hu-PBL-SCID mice against challenge by primary isolates of HIV-1. *Nat Med* (1997) 3:1389–93. 10.1038/nm1297-1389
118. Parren PW, Ditzel HJ, Gulizia RJ, Binley JM, Barbas 3rd CF, Burton DR, et al. Protection against HIV-1 infection in hu-PBL-SCID mice by passive immunization with a neutralizing human monoclonal antibody against the gp120 CD4-binding site. *AIDS* (1995) 9:F1–6. 10.1097/00002030-199506000-00001
119. Gauduin MC, Safrit JT, Weir R, Fung MS, Koup RA. Pre- and postexposure protection against human immunodeficiency virus type 1 infection mediated by a monoclonal antibody. *J Infect Dis* (1995) 171:1203–9. 10.1093/infdis/171.5.1203
120. Safrit JT, Fung MS, Andrews CA, Braun DG, Sun WN, Chang TW, et al. hu-PBL-SCID mice can be protected from HIV-1 infection by passive transfer of monoclonal antibody to the principal neutralizing determinant of envelope gp120. *AIDS* (1993) 7:15–21. 10.1097/00002030-199301000-00002
121. Luo XM, Lei MY, Feidi RA, West AP, Jr., Balazs AB, Bjorkman PJ, et al. Dimeric 2G12 as a potent protection against HIV-1. *PLoS Pathog* (2010) 6:e1001225. 10.1371/journal.ppat.1001225
122. Sun M, Li Y, Yuan Z, Lu W, Kang G, Fan W, et al. VRC01 antibody protects against vaginal and rectal transmission of human

immunodeficiency virus 1 in hu-BLT mice. *Arch Virol* (2016) 161:2449–55. 10.1007/s00705-016-2942-4

123. Veselinovic M, Neff CP, Mulder LR, Akkina R. Topical gel formulation of broadly neutralizing anti-HIV-1 monoclonal antibody VRC01 confers protection against HIV-1 vaginal challenge in a humanized mouse model. *Virology* (2012) 432:505–10. 10.1016/j.virol.2012.06.025
124. Stoddart CA, Galkina SA, Joshi P, Kosikova G, Long BR, Maidji E, et al. Efficacy of broadly neutralizing monoclonal antibody PG16 in HIV-infected humanized mice. *Virology* (2014) 462-463:115–25. 10.1016/j.virol.2014.05.036
125. Deruaz M, Moldt B, Le KM, Power KA, Vrbanac VD, Tanno S, et al. Protection of Humanized Mice From Repeated Intravaginal HIV Challenge by Passive Immunization: A Model for Studying the Efficacy of Neutralizing Antibodies In Vivo. *J Infect Dis* (2016) 214:612–6. 10.1093/infdis/jiw203
126. Bournazos S, Klein F, Pietzsch J, Seaman MS, Nussenzweig MC, Ravetch JV. Broadly neutralizing anti-HIV-1 antibodies require Fc effector functions for in vivo activity. *Cell* (2014) 158:1243–53. 10.1016/j.cell.2014.08.023
127. Halper-Stromberg A, Lu CL, Klein F, Horwitz JA, Bournazos S, Nogueira L, et al. Broadly neutralizing antibodies and viral inducers decrease rebound from HIV-1 latent reservoirs in humanized mice. *Cell* (2014) 158:989–99. 10.1016/j.cell.2014.07.043
128. Wadman M. HIV trial under scrutiny. *Nature* (2013) 493:279–80. 10.1038/493279a
129. ClinicalTrials.gov *Evaluating the Safety and Efficacy of the VRC01 Antibody in Reducing Acquisition of HIV-1 Infection in Women.* Available at: https://clinicaltrials.gov/ct2/show/NCT02568215?term=NCT02568215&draw=2&rank=1 (Accessed 22 May 2020).
130. Gilbert PB, Juraska M, deCamp AC, Karuna S, Edupuganti S, Mgodi N, et al. Basis and Statistical Design of the Passive HIV-1 Antibody Mediated Prevention (AMP) Test-of-Concept Efficacy Trials. *Stat Commun Infect Dis* (2017) 9:1–45. 10.1515/scid-2016-0001
131. ClinicalTrials.gov *Evaluating the Safety and Efficacy of the VRC01 Antibody in Reducing Acquisition of HIV-1 Infection Among Men and Transgender Persons Who Have Sex With Men.* Available at: https://clinicaltrials.gov/ct2/show/NCT02716675?term=NCT02716675&draw=2&rank=1 (Accessed 22 May 2020).

132. ClinicalTrials.gov *A Study to Assess the Efficacy of a Heterologous Prime/Boost Vaccine Regimen of Ad26.Mos4.HIV and Aluminum Phosphate-Adjuvanted Clade C gp140 in Preventing Human Immunodeficiency Virus (HIV) -1 Infection in Women in Sub-Saharan Africa.* Available at: https://clinicaltrials.gov/ct2/show/NCT03060629?term=NCT03060629&draw=2&rank=1 (Accessed 22 May 2020).

133. Bekker LG, Tatoud R, Dabis F, Feinberg M, Kaleebu P, Marovich M, et al. The complex challenges of HIV vaccine development require renewed and expanded global commitment. *Lancet* (2020) 395:384–8. 10.1016/S0140-6736(19)32682-0

134. ClinicalTrials.gov *A Study of Heterologous Vaccine Regimen of Adenovirus Serotype 26 Mosaic4 Human Immunodeficiency Virus(Ad26. Mos4.HIV), Adjuvanted Clade C gp140 and Mosaic gp140 to Prevent HIV-1 Infection Among Cis-gender Men and Transgender Individuals Who Have Sex With Cis-gender Men and/or Transgender Individuals (MOSAICO).* Available at: https://clinicaltrials.gov/ct2/show/NCT03964415?term=NCT03964415&draw=2&rank=1 (Accessed 22 May 2020).

135. ClinicalTrials.gov *A Combination Efficacy Study in Africa of Two DNA-MVA-Env Protein or DNA-Env Protein HIV-1 Vaccine Regimens With PrEP (PrEPVacc).* Available at: https://clinicaltrials.gov/ct2/show/NCT04066881?term=NCT04066881&draw=2&rank=1 (Accessed 22 May 2020).

136. van Wijhe M, McDonald SA, de Melker HE, Postma MJ, Wallinga J. Effect of vaccination programmes on mortality burden among children and young adults in the Netherlands during the 20th century: a historical analysis. *Lancet Infect Dis* (2016) 16:592–8. 10.1016/S1473-3099(16)00027-X

137. Younger DS, Younger AP, Guttmacher S. Childhood Vaccination: Implications for Global and Domestic Public Health. *Neurol Clin* (2016) 34:1035–47. 10.1016/j.ncl.2016.05.004

138. Plotkin SA. Vaccines: the fourth century. *Clin Vaccine Immunol* (2009) 16:1709–19. 10.1128/CVI.00290-09

139. Mascola JR, Louwagie J, McCutchan FE, Fischer CL, Hegerich PA, Wagner KF, et al. Two antigenically distinct subtypes of human immunodeficiency virus type 1: viral genotype predicts neutralization serotype. *J Infect Dis* (1994) 169:48–54. 10.1093/infdis/169.1.48

140. Sawyer LS, Wrin MT, Crawford-Miksza L, Potts B, Wu Y, Weber PA, et al. Neutralization sensitivity of human immunodeficiency virus type 1 is determined in part by the cell in which the virus is propagated. *J Virol* (1994) 68:1342–9. 10.1128/JVI.68.3.1342-1349.1994

141. Fouts TR, Binley JM, Trkola A, Robinson JE, Moore JP. Neutralization of the human immunodeficiency virus type 1 primary isolate JR-FL by human monoclonal antibodies correlates with antibody binding to the oligomeric form of the envelope glycoprotein complex. *J Virol* (1997) 71:2779–85. 10.1128/JVI.71.4.2779-2785.1997

142. Binley JM, Arshad H, Fouts TR, Moore JP. An investigation of the high-avidity antibody response to glycoprotein 120 of human immunodeficiency virus type 1. *AIDS Res Hum Retroviruses* (1997) 13:1007–15. 10.1089/aid.1997.13.1007

143. Zhang L, Irimia A, He L, Landais E, Rantalainen K, Leaman DP, et al. An MPER antibody neutralizes HIV-1 using germline features shared among donors. *Nat Commun* (2019) 10:5389. 10.1038/s41467-019-12973-1

144. Murin CD, Wilson IA, Ward AB. Antibody responses to viral infections: a structural perspective across three different enveloped viruses. *Nat Microbiol* (2019) 4:734–47. 10.1038/s41564-019-0392-y

145. Subbaraman H, Schanz M, Trkola A. Broadly neutralizing antibodies: What is needed to move from a rare event in HIV-1 infection to vaccine efficacy? *Retrovirology* (2018) 15:52. 10.1186/s12977-018-0433-2

146. Jardine JG, Kulp DW, Havenar-Daughton C, Sarkar A, Briney B, Sok D, et al. HIV-1 broadly neutralizing antibody precursor B cells revealed by germline-targeting immunogen. *Science* (2016) 351:1458–63. 10.1126/science.aad9195

147. Havenar-Daughton C, Sarkar A, Kulp DW, Toy L, Hu X, Deresa I, et al. The human naive B cell repertoire contains distinct subclasses for a germline-targeting HIV-1 vaccine immunogen. *Sci Transl Med* (2018) 10:1–15. 10.1126/scitranslmed.aat0381

148. Sanders RW, Derking R, Cupo A, Julien JP, Yasmeen A, de Val N, et al. A next-generation cleaved, soluble HIV-1 Env trimer, BG505 SOSIP.664 gp140, expresses multiple epitopes for broadly neutralizing but not non-neutralizing antibodies. *PLoS Pathog* (2013) 9:e1003618. 10.1371/journal.ppat.1003618

149. Burton DR, Hangartner L. Broadly Neutralizing Antibodies to HIV and Their Role in Vaccine Design. *Annu Rev Immunol* (2016) 34:635–59. 10.1146/annurev-immunol-041015-055515

150. Wibmer CK, Moore PL, Morris L. HIV broadly neutralizing antibody targets. *Curr Opin HIV AIDS* (2015) 10:135–43. 10.1097/COH.0000000000000153

151. deTaeye SW, Moore JP, Sanders RW. HIV-1 Envelope Trimer Design and Immunization Strategies To Induce Broadly Neutralizing Antibodies. *Trends Immunol* (2016) 37:221–32. 10.1016/j.it.2016.01.007

152. Moore PL, Crooks ET, Porter L, Zhu P, Cayanan CS, Grise H, et al. Nature of nonfunctional envelope proteins on the surface of human immunodeficiency virus type 1. *J Virol* (2006) 80:2515–28. 10.1128/JVI.80.5.2515-2528.2006

153. Tomaras GD, Yates NL, Liu P, Qin L, Fouda GG, Chavez LL, et al. Initial B-cell responses to transmitted human immunodeficiency virus type 1: virion-binding immunoglobulin M (IgM) and IgG antibodies followed by plasma anti-gp41 antibodies with ineffective control of initial viremia. *J Virol* (2008) 82:12449–63. 10.1128/JVI.01708-08

154. Seabright GE, Doores KJ, Burton DR, Crispin M. Protein and Glycan Mimicry in HIV Vaccine Design. *J Mol Biol* (2019) 431:2223–47. 10.1016/j.jmb.2019.04.016

155. Gach JS, Mara KJV, LaBranche CC, van Gils MJ, McCoy LE, Klasse PJ, et al. Antibody Responses Elicited by Immunization with BG505 Trimer Immune Complexes. *J Virol* (2019) 93:1–13. 10.1128/JVI.01188-19

156. Andrabi R, Bhiman JN, Burton DR. Strategies for a multi-stage neutralizing antibody-based HIV vaccine. *Curr Opin Immunol* (2018) 53:143–51. 10.1016/j.coi.2018.04.025

157. Klasse PJ, Ketas TJ, Cottrell CA, Ozorowski G, Debnath G, Camara D, et al. Epitopes for neutralizing antibodies induced by HIV-1 envelope glycoprotein BG505 SOSIP trimers in rabbits and macaques. *PLoS Pathog* (2018) 14:e1006913. 10.1371/journal.ppat.1006913

158. Yu WH, Zhao P, Draghi M, Arevalo C, Karsten CB, Suscovich TJ, et al. Exploiting glycan topography for computational design of Env glycoprotein antigenicity. *PLoS Comput Biol* (2018) 14:e1006093. 10.1371/journal.pcbi.1006093

159. Bradley T, Fera D, Bhiman J, Eslamizar L, Lu X, Anasti K, et al. Structural Constraints of Vaccine-Induced Tier-2 Autologous HIV

Neutralizing Antibodies Targeting the Receptor-Binding Site. *Cell Rep* (2016) 14:43–54. 10.1016/j.celrep.2015.12.017

160. Crooks ET, Tong T, Chakrabarti B, Narayan K, Georgiev IS, Menis S, et al. Vaccine-Elicited Tier 2 HIV-1 Neutralizing Antibodies Bind to Quaternary Epitopes Involving Glycan-Deficient Patches Proximal to the CD4 Binding Site. *PLoS Pathog* (2015) 11:e1004932. 10.1371/journal.ppat.1004932
161. Zhou T, Doria-Rose NA, Cheng C, Stewart-Jones GBE, Chuang GY, Chambers M, et al. Quantification of the Impact of the HIV-1-Glycan Shield on Antibody Elicitation. *Cell Rep* (2017) 19:719–32. 10.1016/j.celrep.2017.04.013
162. McCoy LE, van Gils MJ, Ozorowski G, Messmer T, Briney B, Voss JE, et al. Holes in the Glycan Shield of the Native HIV Envelope Are a Target of Trimer-Elicited Neutralizing Antibodies. *Cell Rep* (2016) 16:2327–38. 10.1016/j.celrep.2016.07.074
163. Garrity RR, Rimmelzwaan G, Minassian A, Tsai WP, Lin G, de Jong JJ, et al. Refocusing neutralizing antibody response by targeted dampening of an immunodominant epitope. *J Immunol* (1997) 159:279–89.
164. Ringe RP, Ozorowski G, Rantalainen K, Struwe WB, Matthews K, Torres JL, et al. Reducing V3 Antigenicity and Immunogenicity on Soluble, Native-Like HIV-1 Env SOSIP Trimers. *J Virol* (2017) 91:1–17. 10.1128/JVI.00677-17
165. Ringe RP, Pugach P, Cottrell CA, LaBranche CC, Seabright GE, Ketas TJ, et al. Closing and Opening Holes in the Glycan Shield of HIV-1 Envelope Glycoprotein SOSIP Trimers Can Redirect the Neutralizing Antibody Response to the Newly Unmasked Epitopes. *J Virol* (2019) 93:1–26. 10.1128/JVI.01656-18
166. Borducchi EN, Liu J, Nkolola JP, Cadena AM, Yu WH, Fischinger S, et al. Publisher Correction: Antibody and TLR7 agonist delay viral rebound in SHIV-infected monkeys. *Nature* (2018) 564:E8. 10.1038/s41586-018-0721-y
167. Mayer KH, Seaton KE, Huang Y, Grunenberg N, Isaacs A, Allen M, et al. Safety, pharmacokinetics, and immunological activities of multiple intravenous or subcutaneous doses of an anti-HIV monoclonal antibody, VRC01, administered to HIV-uninfected adults: Results of a phase 1 randomized trial. *PLoS Med* (2017) 14:e1002435. 10.1371/journal.pmed.1002435

168. Sok D, Le KM, Vadnais M, Saye-Francisco KL, Jardine JG, Torres JL, et al. Rapid elicitation of broadly neutralizing antibodies to HIV by immunization in cows. *Nature* (2017) 548:108–11. 10.1038/nature23301

169. Sanders RW, van Gils MJ, Derking R, Sok D, Ketas TJ, Burger JA, et al. HIV-1 VACCINES. HIV-1 neutralizing antibodies induced by native-like envelope trimers. *Science* (2015) 349:aac4223. 10.1126/science.aac4223

170. Pauthner M, Havenar-Daughton C, Sok D, Nkolola JP, Bastidas R, Boopathy AV, et al. Elicitation of Robust Tier 2 Neutralizing Antibody Responses in Nonhuman Primates by HIV Envelope Trimer Immunization Using Optimized Approaches. *Immunity* (2017) 46:1073–1088.e1076. 10.1016/j.immuni.2017.05.007

171. Steichen JM, Lin YC, Havenar-Daughton C, Pecetta S, Ozorowski G, Willis JR, et al. A generalized HIV vaccine design strategy for priming of broadly neutralizing antibody responses. *Science* (2019) 366:1–22. 10.1126/science.aax4380

172. Hansen SG, Wu HL, Burwitz BJ, Hughes CM, Hammond KB, Ventura AB, et al. Broadly targeted CD8(+) T cell responses restricted by major histocompatibility complex E. *Science* (2016) 351:714–20. 10.1126/science.aac9475

173. Liu J, Jaijyan DK, Tang Q, Zhu H. Promising Cytomegalovirus-Based Vaccine Vector Induces Robust CD8(+) T-Cell Response. *Int J Mol Sci* (2019) 20:1–16. 10.3390/ijms20184457

174. Murray SE, Nesterenko PA, Vanarsdall AL, Munks MW, Smart SM, Veziroglu EM, et al. Fibroblast-adapted human CMV vaccines elicit predominantly conventional CD8 T cell responses in humans. *J Exp Med* (2017) 214:1889–99. 10.1084/jem.20161988

175. Hansen SG, Ford JC, Lewis MS, Ventura AB, Hughes CM, Coyne-Johnson L, et al. Profound early control of highly pathogenic SIV by an effector memory T-cell vaccine. *Nature* (2011) 473:523–7. 10.1038/nature10003

176. Hansen SG, Vieville C, Whizin N, Coyne-Johnson L, Siess DC, Drummond DD, et al. Effector memory T cell responses are associated with protection of rhesus monkeys from mucosal simian immunodeficiency virus challenge. *Nat Med* (2009) 15:293–9. 10.1038/nm.1935

177. Hansen SG, Sacha JB, Hughes CM, Ford JC, Burwitz BJ, Scholz I, et al. Cytomegalovirus vectors violate CD8+ T cell epitope recognition paradigms. *Science* (2013) 340:1237874. 10.1126/science.1237874
178. Nyanhete TE, Frisbee AL, Bradley T, Faison WJ, Robins E, Payne T, et al. HLA class II-Restricted CD8+ T cells in HIV-1 Virus Controllers. *Sci Rep* (2019) 9:10165. 10.1038/s41598-019-46462-8
179. Hannoun Z, Lin Z, Brackenridge S, Kuse N, Akahoshi T, Borthwick N, et al. Identification of novel HIV-1-derived HLA-E-binding peptides. *Immunol Lett* (2018) 202:65–72. 10.1016/j.imlet.2018.08.005
180. Hansen SG, Marshall EE, Malouli D, Ventura AB, Hughes CM, Ainslie E, et al. A live-attenuated RhCMV/SIV vaccine shows long-term efficacy against heterologous SIV challenge. *Sci Transl Med* (2019) 11:1–28. 10.1126/scitranslmed.aaw2607
181. Veazey RS. Animal models for microbicide safety and efficacy testing. *Curr Opin HIV AIDS* (2013) 8:295–303. 10.1097/COH.0b013e328361d096
182. McNicholl JM. Combining biomedical preventions for HIV: Vaccines with pre-exposure prophylaxis, microbicides or other HIV preventions. *Hum Vaccin Immunother* (2016) 12:3202–11. 10.1080/21645515.2016.1231258
183. Cheng-Mayer C, Huang Y, Gettie A, Tsai L, Ren W, Shakirzyanova M, et al. Delay of simian human immunodeficiency virus infection and control of viral replication in vaccinated macaques challenged in the presence of a topical microbicide. *AIDS* (2011) 25:1833–41. 10.1097/QAD.0b013e32834a1d94

CHAPTER 8

Using Adjuvants to Drive T Cell Responses for Next-Generation Infectious Disease Vaccines

Rekha R. Rapaka, Alan S. Cross, and Monica A. McArthur

Center for Vaccine Development and Global Health, University of Maryland School of Medicine, 685 West Baltimore Street, Baltimore, MD 21201, USA

ABSTRACT

Using adjuvants to drive features of T cell responses to vaccine antigens is an important technological challenge in the design of new and improved vaccines against infections. Properties such as T helper cell function, T cell memory, and CD8+ T cell cytotoxicity may play critical roles in optimal and long-lived immunity through vaccination. Directly manipulating specific immune activation or antigen delivery pathways with adjuvants may selectively augment desired T cell responses in vaccination and may

Citation: Rapaka, R.R.; Cross, A.S.; McArthur, M.A. Using Adjuvants to Drive T Cell Responses for Next-Generation Infectious Disease Vaccines. Vaccines 2021, 9, 820. https://doi.org/10.3390/ vaccines9080820

improve the effectiveness and durability of vaccine responses in humans. In this review we outline recently studied adjuvants in their potential for antigen presenting cell and T cell programming during vaccination, with an emphasis on what has been observed in studies in humans as available.

Keywords: T cell, adjuvant, differentiation, antigen presenting cell, CD8+ T cell, CD4+ T cell, Th, memory, TLR, APC

INTRODUCTION

Adjuvants are substances that enhance the immune response to a vaccine. Adjuvants may occur naturally as part of a vaccine, as seen in the pathogen associated molecular patterns (PAMPs) contained in live-attenuated vaccines or may specifically be added to a vaccine preparation to enhance immune responses. Additionally, certain antigen delivery platforms may have adjuvant functions. For example, antigen delivery platforms, such as adenoviral vectors have intrinsic PAMPs, and liposomes may be modified to impact antigen processing pathways, enhancing the immune response to the vaccine antigen. Currently, aluminum salts, followed by oil-in-water emulsions, are the most common adjuvants in clinical use [1,2]. Adjuvants are hypothesized to exert their effects through stimulation of the innate immune system, with innate immune cell activation and cytokine and chemokine production impacting cellular recruitment and activation at the site of vaccine administration. Adjuvants may stimulate the innate immune system through cellular pattern recognition receptors (PRR) and activation of signal transduction pathways in antigen presenting cells (APCs), such as tissue resident dendritic cells. APC programming by different adjuvants may impact features of APC maturation and migratory function towards draining lymph nodes and other secondary lymphoid organs. In the process of presentation of antigen by APCs, APCs impact T and B cell priming within lymph nodes, key processes in vaccine-elicited adaptive immune responses. Adjuvants may also impact the longevity of antigen in the body, assisting antigen in forming a depot and enhancing antigen presentation through an extended period of immune cell stimulation.

Next-generation vaccines involve precisely defined antigens, often involving short mRNA or DNA sequences, specific polysaccharide conformations, or modified proteins. Advances in technologies for developing vaccine antigens, including methods to stabilize protein conformations as used in the SARS-CoV-2 mRNA vaccines [3], or focusing antigen epitopes to enhance T cell responses or skew the T cell repertoire [4], are in development. Despite advances in antigens used in vaccines, including novel platforms, and the use of structural biology and systems biology to design highly specific responses to antigens [5], the optimization and utilization of adjuvants has proceeded more slowly. Improvements in our understandings of adjuvants may impact aspects of T cell differentiation upon vaccination, and potentially the development of long-term immunologic memory.

T cells hold a critical role in shaping adaptive antibody responses in the setting of vaccination and infection. T cells regulate properties of evolving immune responses, such as the development of immunologic memory, and have direct effector functions against infected cells. The functions of T cells in vaccine-elicited responses are important and likely go far beyond optimizing affinity matured class-switched antibody responses in existing vaccines. Advancements in adjuvant science may allow selective activation of specific innate immune sensors in triggering specific inflammatory pathways in APCs, driving selected pathways of immune responses that may be more optimal for specific vaccines. Greater understanding of how T cell responses can be shaped by adjuvants may significantly improve the efficacy and longevity of protection in next-generation vaccines.

INDUCTION OF CYTOTOXIC CD8+ T CELL RESPONSES BY ADJUVANTS

As demonstrated in a variety of different infection models, and particularly for pathogens with a prominent intracellular lifecycle, CD8+ T cells may play a central role in host defense against infection [6,7,8,9,10,11]. Differentiation of antigen-specific effector and memory CD8+ T cells could be utilized in vaccine-mediated immune protection, impacting the detection and clearance of infected cells. CD8+ T cells directly kill cells that are identified as infected or abnormal through recognition of major histocompatibility class I (MHC I) complexed to peptide at the host cell surface. During an infection, APCs play a critical role in priming antigen naïve CD8+ T cells, through the presentation of antigen, and through cellular and receptor-mediated co-stimulation or inhibition, leading to the proliferation and differentiation of antigen specific CD8+ T cells with specific migratory and cytotoxic functions [12]. Enhancement of intracellular antigen presentation with strategies that emphasize intracellular antigen phases, such as with use of saponins, with DNA or RNA based vaccination, or with viral vector-based antigen presentation, may improve CD8+ T cell responses [13,14,15].

Saponins, derived from the bark of tree *Quillaja saponaria* or from the leaves of *Quillaja brasiliensis*, are a family of amphipathic adjuvants with surfactant properties. The *Quillaja saponaria* saponins when isolated by size and weight were found to have fractions which were immunogenic and not associated with toxicity in murine models, such as QS-7 and QS-21 [16,17]. Given the favorable features and relative abundance of QS-21, many studies of saponins as adjuvants has focused on this particular fraction. Saponin QS-21, when formulated in cholesterol containing liposomes, has been shown to enhance human monocyte dendritic cell uptake of antigen through a receptor independent, cholesterol dependent mechanism [18]. Additionally, QS-21, was found to accumulate in and destabilize lysosomes, increasing in a dose-dependent manner the production of cellular activation inflammatory cytokines IL-6 and TNF-α [18]. It is hypothesized that this process of antigen presentation, with antigen escape from the lysosome and entry into the cytosolic space, and antigen presentation of peptides on MHC I by dendritic cells, enhances priming of cytotoxic T cells in lymphoid tissue.

The value of saponins in eliciting cytotoxic CD8+ T cells in humans and animal models is illustrated from studies in the field of anti-cancer vaccines. For example, the NY-ESO-1 protein antigen emulsified in the ISCOMATRIX adjuvant (described in [19], composed of purified fractions of saponin, phospholipid and cholesterol resulting in particles with a 40 nm cage-like structure) demonstrated an ability to mature human monocyte derived dendritic cells, and these dendritic cells demonstrated a capacity to cross present HLA-A2 restricted epitopes of NY-ESO-1 to CD8+ T cells [20]. In humans with melanoma, when ISCOMATRIX adjuvanted NY-ESO-1 protein was given as a vaccine, vaccinees developed high frequencies of long lasting specific CD8+T cell responses in vivo on booster immunization to a wide array of NY-ESO-1 peptides [21]. Cytotoxic CD8+T cell responses have been observed in murine models of malignancy with protein vaccines adjuvanted with ISCOMATRIX [22], including lysis of human cancer cell lines expressing tumor antigen [20]. Extrapolating from studies in the field of cancer, infections with a major intracellular component during the life cycle, as in viral infections, or infections with intracellular bacteria such as *Mycobacterium tuberculosis* or parasites, such as *Plasmodium falciparum*, may particularly have benefit from saponin-based adjuvants in vaccine design.

Saponins as adjuvants in infectious disease models have shown promise in promoting CD8+ T cell differentiation. In a model of bovine viral diarrhea virus (BVDV) vaccination in mice, mice vaccinated with BVDV antigen together with *Q. brasilensis* saponin had significantly increased production of CD8+ T cells producing the effector cytokine IFN-γ [23]. In mice immunized with Ebola virus glycoprotein, the Matrix-M adjuvant, derived from mixing 2 specific ISCOMATRIX types (as described in [24]), was demonstrated to significantly enhance production of antigen specific CD8+ T cells responses, compared to Ebola virus glycoprotein vaccine adjuvanted with alum, or with no adjuvant [25]. In this model, multifunctional CD8+ T cells, here CD8+ T cells simultaneously producing two or three of the cytokines, TNF-α, IL-2, or IFN-γ, were observed at increased frequencies if Ebola glycoprotein antigen was administered with Matrix-M, and occurred in a Matrix-M dose-dependent manner. In a murine model of Ebola infection, where CD8+ T cells are required for protection from infection [26], and where there is general precedent for multifunctional T cells in viral infections as limiting disease progression [10,27,28,29], this qualitative aspect of increased multifunctional CD8+ T cells observed with saponin adjuvant may be important for vaccine efficacy. Similarly, in a murine model of vaccination with the NVX-CoV2373 for SARS-CoV-2, the SARS-CoV-2 spike protein adjuvanted with Matrix-M led to the induction of increased CD8+ T cells producing cytokines, as well as multifunctional CD8+ T cells when compared to immunization of the spike protein without adjuvant [30]. Additionally, baboons immunized with NVX-CoV2373, if adjuvanted with Matrix-M, had increased production of IFN-γ cells in peripheral blood mononuclear cells stimulated with NVX-CoV2373. In humans, Matrix-M adjuvant has recently been evaluated with the Ebola virus glycoprotein for safety and immunogenicity [31], and has also been utilized in large scale trials in humans of the SARS-CoV-2 spike glycoprotein NVX-CoV-2373 vaccine during the COVID-19 during the pandemic [32,33]; studies of CD8+ T cell responses in humans for these vaccines are ongoing.

Enhancing production of cytotoxic CD8+ T cells may also be imparted by the delivery vehicle carrying the antigen, such as with liposomes. Liposomes consist of lipid bilayers encasing an aqueous core and may vary in size, charge, and lipid composition among other features [34]. Liposomes are notable for their ability to protect antigens from degradation, interact with cell membranes and optimize antigen internalization, and to carry a wide variety of antigens with lipophilic or hydrophilic properties. The biochemical composition of liposomes may impact their ability to enhance T cell priming. Vaccination with liposomes containing unsaturated fatty acids coupled to protein antigen in mice led to increased induction of antigen specific CD8+ T cells, in addition to CD4+ T cells; while if the liposomes only contained saturated fatty acids, only antigen specific CD4+ T cells were induced and CD8+ T cells were not significantly induced [35]. It was further demonstrated in this study by confocal microscopy that liposomes containing unsaturated fatty acids effectively targeted processing pathways leading to the cytosol, outside of lysosomes, which would more readily allow for peptide presentation onto MHC class I [35]. Hence, the specific composition of liposomes may direct the process of APC processing of vaccine antigen, impacting the processes of antigen presentation directly, as well as the process of APC maturation. Sterically stabilized liposomes (SSL), or liposomes containing large molecules, such as polyethylene glycol, coupled with protein antigen, have been shown in human dendritic cells to induce proliferation of CD8+ T cells at a rate of 300 times the rate of proliferation of CD4+ T cells [36]. Increased CD8+ T cell generation is hypothesized to occur as more antigen delivered by these liposomes was observed to be present in the cytosol compared to within lysosomes, with cytosolic antigen processed by transporter associated with antigen-processing 1 (TAP1) dependent pathways. Dendritic cells (DC) from TAP1 knock-out mice stimulated with protein antigen and complexed to SSL had significantly impaired antigen-specific CD8+T cell proliferation compared to DC from wild-type mice [36]. These studies collectively demonstrate that liposomes may enhance CD8+ T cell induction by impacting APC processing of antigen, and that the specific molecular composition of liposomes may be manipulated to selectively drive CD8+ T cell responses.

The first mRNA vaccines in use in humans for prevention of COVID-19, BNT162b2 (Pfizer) and mRNA-1273 (NIAID-Moderna) make use of an antigen delivery strategy similar to that of liposomes, specifically with lipid nanoparticles (NP) containing cholesterol, polyethylene glycol to enhance stability, and cationic lipids which bind and stabilize mRNA [37]. Lipid NP may range in size from 10–1000 nM, are designed to be endocytosed, and have a solid particulate nature. Lipid NP are optimized to carry nucleic acids, and in contrast to liposomes, do not necessarily have an aqueous core. mRNA based vaccines have good precedent for generation of antigen specific CD8+ T cells [38,39,40]. Both BNT162b2 and mRNA-1273 vaccines elicit SARS-CoV-2 spike glycoprotein antigen specific CD8+ T cells in humans. The mRNA vaccine BNT162b2 elicits high levels of IFN-γ producing CD8+ T cells in the peripheral blood specific for the spike glycoprotein receptor binding domain, which were observed after booster immunization; these frequencies were also notably higher than as observed in individuals who had developed and recovered from COVID-19 [41]. Interestingly, in this study, a higher dose of the vaccine, but without booster dosing, led to very low levels of CD8+ T cells responses, suggesting the critical nature of booster vaccination to enhance antigen specific CD8+ T cell frequencies in the blood. With mRNA-1273, spike glycoprotein specific CD8+ T cells were also observed in volunteers after booster immunization [42]. In a separate study where BNT162b2 was administered to volunteers, followed by booster at 22 days, several volunteers developed multifunctional CD8+ T cells with persistence of antigen-specific memory CD8+ T cells in the blood through at least 85 days of monitoring [43].

The live attenuated yellow fever vaccine and smallpox vaccinia vaccine induce high frequencies of multifunctional CD8+T cells in humans, with increased antigen load and pattern recognition receptor activation hypothesized to contribute to the effect [44,45,46,47]. Viral vector-based vaccines, where a virus is used as an antigen expression platform, with additional gene alterations to ensure reduced virulence and/or replication incompetence also have precedent for induction of T cell responses including CD8+ T cells. Several new viral vector-based vaccines have come into clinical use in humans recently, including for Ebola hemorrhagic fever and for COVID-19 [48,49,50,51]. The role of CD8+T cells in host defense may be central in prevention of infection; for example, in a small cohort of cynomolgus macaques vaccinated with recombinant adenovirus 5 encoding Ebola virus glycoprotein, 4 of 5 macaques were not protected if CD8+ T cells were specifically depleted in vivo prior to Ebola challenge [52]. The Ad26. COV2.S and ChAdOx1 COVID-19 vaccines both elicit spike glycoprotein antigen specific CD8+ T cells in humans [51,53].

Adjuvants with the capacity to stimulate intracellular compartment-associated toll-like receptor 9 (TLR9) or toll-like receptor 3 (TLR3) may also have promise in enhancing development of CD8+ T cells upon vaccination in humans. CpG, which are unmethylated cytosine and guanine oligonucleotide motifs, are commonly found in viruses and bacteria. There are four types of synthetic CpG adjuvants, with different immunostimulatory properties [54]. Recognition of CpG occurs at the cellular level with activation of TLR9 within APCs. It is important to note that differences in CpG ligands and interaction with TLR9 diverge between mice and humans, hence, the relevance of CpG as an adjuvant for a specific vaccine must be gathered in primate models prior to approaching studies in humans. Macaques immunized with a plasmid DNA vaccine containing the simian immunodeficiency virus *tat* gene and also containing synthetic

CpG sequences, developed high levels of Tat protein-specific CD8+ T cells [55]. Mechanistically, this effect was thought to be at least partially driven by the CpG sequences and their ability to stimulate the early innate immune response, as macaques vaccinated with the vector containing unmethylated CpG without *tat* insertion had significantly reduced simian/human immunodeficiency virus replication on challenge compared to macaques not receiving any vaccination. In patients with melanoma, serial vaccination with CpG oligonucleotides combined with melanoma antigen A protein along with incomplete Freund's adjuvant led to the induction of tumor antigen specific CD8+ T cell responses [56]. However, frequencies of melanoma A specific CD8+ T cells were reduced by 1-fold if CpG was not included in the vaccine. Importantly, the specific CpG structure may have differential effects on human APCs and their interactions with CD8+ T cells, as observed in studies with human PBMCs and the impact of different CpG on naïve vs. memory CD8+ T cells [57].

Polyinosinic:polycytidylic acid (poly(I:C)) is a TLR3 agonist and a synthetic double-stranded RNA mimicking viral nucleic acid, that has also been investigated for potential adjuvant effect in enhancing CD8+ T cell responses during vaccination. In a model of murine Ebola infection, a filovirus-like particle vaccine adjuvanted with poly (I:C) increased the efficacy of the vaccine in reducing filovirus levels on challenge and led to increased multifunctional CD8+ T cell responses to the vaccine, compared to vaccination without poly (I:C) [58]. In Langerhans cells derived from patients with human papillomavirus 16 (HPV16) infection and advanced cervical intraepithelial neoplasm, there were significantly enhanced inflammatory responses on exposure to HPV16 virus-like particles (HPV16 cVLP) if co-cultured with stabilized poly(I:C) [59]. Further, Langerhans cells stimulated with HPV16 cVLP and stabilized poly(I:C) on co-culture with autologous CD8+ T cells had significantly increased production of IFN-γ producing antigen-specific CD8+ T cells compared to culture without stabilized poly(I:C) [59]. Poly(I:C) has also been demonstrated to further enhance the immunostimulatory potential of live-attenuated vaccines, which already contain several intrinsic PAMPs. For example, in a murine model of live attenuated influenza vaccination, treatment with poly(I:C) after vaccination led to increased dendritic cell activation and increased CD8+ T cell responses within the lung [60]. These observations underscore the potential of CpG and poly(I:C) as adjuvants in enhancing multifunctional, antigen-specific CD8+ T cell responses at systemic sites, and at the mucosa.

More translational studies are needed to establish the benefit of poly (I:C) in promoting CD8+ T cell responses in humans.

Collectively, saponins, liposomes and lipid NP, viral vectors, and certain TLRs agonists may play critical roles in impacting antigen presenting cells to prime and differentiate antigen specific CD8+ T cells during vaccination (see Table 1). Future vaccines will further develop these and similar modalities in settings where CD8+ T cell dependent functions are critical in human defense against infection.

Table 1: Adjuvants enhancing CD8+ T cell differentiation in human/non-human primate studies

Adjuvant Enhancing CD8+ T Cell Differentiation in Human/ Non-Human Primate Studies	**Potential Mechanism(s)**	**Pathogen(s)/ Antigens for Which the Adjuvant Has Been Tested**	**Select References**
Saponins (e.g., QS-21, ISCOMATRIX *)	Enhances dendritic cell uptake of antigen through receptor-independent, cholesterol-dependent mechanism	Melanoma and prostate cancer protein antigens, bovine diarrhea virus, Ebola virus, SARS-CoV-2	[18,20,21,22,23,25,30]
	Accumulates in and destabilizes, lysosomes, increasing production of cellular activation inflammatory cytokines IL-6 and TNF-α		
Liposomes * mRNA/lipid nanoparticles *	Lysosome destabilization, enhanced cytosolic antigen	Ovalbumin, SARS-CoV-2	[35,36,41,42]
	Induction of CD8+ T cell proliferation		
	Enhance APC processing of vaccine antigen and influence APC maturation		
Live attenuated viruses	Increased antigen load	Yellow fever virus, smallpox virus	[44,45,46,47]
	Broad pattern recognition receptor activation		
	Antigen processing through MHC I		

Viral vectors *	Elicitation of antigen-specific CD8+ T cells, trafficking of antigen into pathways for MHC I expression	Ebola virus, SARS-CoV-2	[48,49,50,51,52,53]
Unmethylated CpG	Stimulation of innate immune system through TLR9 and stimulation of antigen-specific CD8+ T cells	SIV, melanoma antigen A	[55,56]

* An antigen delivery platform with adjuvant functions.

INDUCTION OF T FOLLICULAR HELPER CELL RESPONSES BY ADJUVANTS

The disruption of T follicular helper (Tfh) cell responses is associated with an impaired antibody response to infection or vaccination in numerous models [61], and enhancement of Tfh production in vaccination correlates with improved B cell memory in response to vaccination [62]. Tfh cells are involved in germinal center formation and processes of B cell affinity maturation and selection. Tfh selection of B cell clones producing somatically hypermutated, high-affinity antibodies is a central process in B cell memory development. Tfh are defined by their localization in germinal centers, are associated with cell surface markers CXCR5, ICOS, PD-1, and CD40L, and produce high quantities of cytokines IL-4 and IL-21. While these cells are not easily surveyed in humans given their localization during an immune response within central lymphoid tissues rather than the peripheral blood, circulating memory Tfh-like cells have been characterized in a variety of infectious disease models [63,64]. In humans, these cells display a T central memory cell phenotype (CCR7+/CD45RA-), express lymph node B cell zone homing molecule CXCR5, and are able to stimulate B cell plasma cell differentiation and immunoglobulin production in vitro [64].

Increased induction of Tfh has been associated with certain adjuvants in development, in particular adjuvants with capacity to activate the toll-like receptor 4 (TLR4) innate immune cell signaling pathway. The glucopyranosyl lipid adjuvant-stable emulsion (GLA-SE) adjuvant, developed by the Infectious Diseases Research Institute in Seattle, has been associated with increased Tfh induction in human vaccination. Glucopyranosyl Lipid A (GLA) is a synthetic toll-like receptor TLR4 agonist which binds to the MD2 portion of TLR4 in humans, and as GLA-SE is delivered as an oil (squalene) in water stable nano-emulsion (SE). In a clinical trial of volunteers receiving an experimental malaria peptide vaccine, there were significantly increased frequencies of Tfh induced when the peptide vaccine was adjuvanted with GLA-SE compared to when adjuvanted with alum [65]. Further, volunteers receiving the vaccine with the GLA-SE adjuvant compared to alum had a longer-lived antibody response, as well as a heightened extrafollicular antibody response to the peptide antigen; and the use of GLA-SE was also associated with induction of specific Tfh clonotypes in humans [65]. Here, the direct choice of adjuvant impacts the properties of adaptive CD4+ T cells responses in humans, with dominant Tfh clonotypes elicited, as well as long-lived antibody responses. Similarly, in mice immunized with recombinant protein, the GLA-SE adjuvant enhances production of Tfh compared to alum, SE alone, or GLA without SE, and the GLA-SE adjuvant is associated with increased memory B cell development on vaccination [66].

A similar adjuvant, GLA-LSQ, which contains the GLA adjuvant but emulsified in a liposomal adjuvant containing QS21 saponin (LSQ), enhances Tfh responses in the setting of peptide immunization in mice [67]. Here, Tfh were found in increased frequencies in draining lymph nodes in mice immunized with a peptide with GLA-LSQ adjuvant compared to an aluminum-based adjuvant. Given increased localization of Tfh to lymph nodes during a vaccine-elicited immune response, this supports the concept that use of GLA-LSQ adjuvant during antigen presentation enhances Tfh presence at sites impacting the process of antibody affinity maturation. Booster immunization with peptide antigen combined with GLA-LSQ adjuvant demonstrated even further increase in Tfh in draining lymph nodes, occurring in association with significant quantities of antigen specific plasmablasts, which were notably absent in the mice receiving the aluminum-based adjuvant [67].

TLR4 agonist monophosphoryl lipid A (MPLA), derived from bacterial lipopolysaccharide and similar to GLA in ability to activate TLR4, enhances Tfh responses when added to heat-inactivated rabies virus vaccine in a murine model of immunization [68]. Mice immunized with rabies vaccine with MPLA adjuvant had increased co-stimulatory molecule expression on dendritic cells and increased frequencies of Tfh in draining lymph nodes, compared to mice receiving the rabies vaccine without adjuvant. Immunization of mice with rabies vaccine with MPLA led to enhanced frequencies of plasma cells at 2 weeks as well as increased total IgG, and IgG isotype antibody titers. Delivery of antigen in lipid NP may enhance Tfh differentiation with the MPLA adjuvant. In mice immunized with protein antigen in lipid-based synthetic NP vesicles and adjuvanted with MPLA there was significantly enhanced Tfh frequencies induced compared to mice receiving soluble protein antigen with MPLA alone [69]. Further, mice immunized with a *Plasmodium vivax* circumsporozoite antigen, VMP001, formulated as a NP (VMP001-NP) adjuvanted with MPLA demonstrated germinal center formation near depots of VMP001-NPs in draining lymph nodes and had significantly higher IgG responses compared to VMP001-NPs adjuvanted with alum at 400 days after immunization [69].

The MF59 adjuvant is another adjuvant with evidence of enhancing Tfh responses during vaccination. MF59 is an oil-in-water emulsion containing the oil squalene, combined with surfactants and buffered with citric acid. In contrast to the mechanisms of GLA and MPLA based adjuvants, MF59 is believed to act by enhancing numbers of APC recruited to the site of vaccination, influencing total numbers of activated APC in draining lymph nodes [70]. Additionally, the adjuvant effect of MF59 may be dependent on MyD88 adaptor protein activation through a TLR independent signaling mechanism [71]. In mice immunized with influenza hemagglutinin antigen emulsified in MF59, there were significantly increased numbers of Tfh induced on vaccination compared to vaccination without MF59, and frequencies of Tfh correlated with frequencies of B cells within lymph node germinal centers [72]. Similar increases of lymph node germinal center B cells observed as late as 4 months after immunization were noted in a murine model of *Staphylococcus aureus* protein antigen vaccination when adjuvanted with MF59 [73]. However, in human volunteers immunized with the MF59 adjuvanted trivalent influenza vaccine, there was no significant difference in numbers of Tfh observed compared to trivalent influenza vaccine administered without MF59, while numbers of H1N1 specific Tfh significantly correlated with antibody responses at early and late timepoints after vaccination [74]. Interestingly, SARS-CoV-2 spike protein vaccine adjuvanted with MF59 led to the induction of Tfh in the blood in human volunteers at lower doses of vaccine and required booster immunization [75]. Additional studies are needed to determine the potential benefit of MF59 during vaccination in specifically enhancing Tfh responses in humans.

More recently, it has been demonstrated that combination strategies may enhance Tfh induction during vaccination. The synthetic TLR 7/8 agonist, 3M-052, enhances production of Tfh during protein subunit vaccination in macaques when presented in poly(lactic-co-glycolic acid) (PLGA) NP and co-administered with GLA compared to administration of protein subunit vaccination with alum [76]. Of note, in this study, the highest frequencies of long-lived antibody-secreting plasma cells in the bone marrow were identified in macaques that had received protein subunit vaccination with 3M-052 in PLGA NP with or without GLA, compared to vaccine receipt with GLA alone or with alum alone. Hence the combination of 3M-052 formulated in PLGA NP with GLA enhanced generation of Tfh in protein-based vaccination, and was associated with durable antibody-based immune responses.

These studies show promise that TLR4 agonism improves APC presentation of either inactivated virus, peptide, or recombinant protein antigen vaccines, increases T follicular helper cell generation, and enhances B cell and/or class-switched antibody responses (see Table 2). Additionally, MF59 and new synthetic adjuvants, such as 3M-052 targeting TLR7/8, may also play a role in enhancing Tfh induction in future vaccines. New strategies, such as bacterial enzymatic combinatorial chemistry on lipooligosaccharide (as reviewed in [77]) may yield new and diverse TLR4 agonist molecules that may have capacity for further increased Tfh differentiation. Future evaluation of this group of adjuvants to augment vaccination responses in humans should be considered in settings where antibody-based immunity and B cell memory in particular, are critical components of the protective immune response. In settings of immunosuppression where humoral immune responses are suboptimal, such as in the setting of HIV infection or in elderly patients, adjuvants that enhance Tfh responses may play an important role in augmenting B cell maturation and affinity matured, long-lived antibody responses.

Table 2: Adjuvants enhancing Tfh cell subset differentiation in human/non-human primate studies

Adjuvant Enhancing Tfh Cell Subset Differentiation in Human/Non-Human Primate Studies	Potential Mechanism(s)	Pathogen(s)/Antigens for Which the Adjuvant Has been Tested	Select References
GLA-SE	Activation of TLR4 and enhancement of antigen presentation	Malaria peptide antigens	[65]
3M-052 formulated with PLGA nanoparticles +GLA	Activation of TLR7 and TLR8; with TLR4 activation and nanoparticle-based presentation to APCs	HIV-1 envelope protein	[76]

Th1, Th2, AND Th17 IMMUNITY AND ADJUVANTS

Naïve CD4+ T cells, upon T cell receptor stimulation and guided by activation of master transcription factors, differentiate into T cells armed with different general types of effector functions. During T cell priming, APC derived cytokines and cell surface co-stimulatory or inhibitory molecules influence this process of T cell differentiation [78]. Hence, the interaction of vaccine or pathogen with APC may critically impact the direction of T-helper (Th) cell subset differentiation. Th1, Th2, and Th17 represent three major CD4+ Th subsets that may be preferentially induced by vaccination and further impacted by adjuvants. The relative importance of these respective Th subsets in host defense diverges significantly depending on the infection. Selective use of adjuvants to enhance optimal human immune defense pathways is a major goal for next generation vaccines.

Th1 Cell Induction and Adjuvants

Th1 cells are generally defined by activation of transcriptional factors T-bet and STAT5, and are characterized by high quantities of IFN-γ production upon T cell receptor (TCR) stimulation. Th1 immunity is associated with enhanced macrophage effector function against pathogens, as well as production of antibody isotypes with ability to enhance Fc-receptor mediated phagocytosis such as IgG1 in humans (the murine homolog of IgG2a). An adjuvant that has shown high ability to drive Th1 responses in vaccination is CpG. In a murine model of vaccination with whole cell lysates of *Leishmania* promastigotes, if the vaccine was administered with CpG there was 10-fold enhanced recruitment of CD4+ T cells producing IFN-γ to the skin upon intradermal *Leishmania* challenge compared to without CpG adjuvant [79], and in a macaque model of this infection the addition of CpG as an adjuvant to a heat-killed *Leishmania* vaccine enhanced protection from disease [80]. In an intranasal model of immunization of influenza in mice, addition of CpG significantly enhanced recruitment of CD4+ T cells in the lung producing IFN-γ, TNF-α, or multifunctional CD4+ T cells producing both cytokines [81]. When serving as an adjuvant to the Hepatitis B protein subunit vaccine in clinical use, CpG is associated with rapid development of high titer antibody responses in humans [82] and is associated with increased levels of seroprotective antibody titers and antibody durability in the setting of HIV infection [83]. The ability of this adjuvant to improve antibody responses in the setting of HIV suggests promise for enhancing antibody responses in other settings of immunodeficiency.

The Adjuvant System (AS) adjuvants, or defined combination adjuvants containing classical and recently identified immunomodulators are being utilized in current and next generation vaccine design. The AS01 and AS04 adjuvants, both licensed for use as adjuvants in humans, are combination adjuvants that have been demonstrated to enhance Th1 immune responses. The AS01 complex is a liposome-based adjuvant containing monophosphoryl A and the saponin QS21. In studies of the *Mycobacterium tuberculosis* modified protein antigen vaccine M72, when adjuvanted with AS01 and administered to volunteers, the vaccine was found to yield high levels of M72 specific CD4+ T cell multifunctional responses with expression of either TNF-α, IFN-γ, or IL-2, and high IFN-γ responses in serum after booster vaccination, compared to vaccine response when using other adjuvants [84]. A malaria vaccine candidate, circumsporozoite-based protein antigen RTS,S when adjuvanted with alum-based adjuvants failed to provide immunologic protection in a malaria controlled human challenge model [85], but when adjuvanted with AS01 provided ~50% vaccine efficacy under controlled human malaria challenge [86]. On analysis of the immune responses elicited by RTS,S adjuvanted with AS01, protected vaccinees had higher levels of antigen specific CD4+ T cells as well as higher levels of IFN-γ producing CD4+ T cells as observed two weeks after booster immunization and at later timepoints. From mechanistic studies, the increased production of Th1 type immune responses mediated by AS01 is suspected to be related to the production of early, innate IFN-γ responses by APCs such as macrophages [87]. Increased APC IFN-γ responses promote a Th1 environment locally, with increased IL-12 and IL-18 expression by APCs influencing downstream programming of T cells in the draining lymph node during APC interaction [87]. Hence the earliest interactions with antigen presenting cells with the adjuvant- vaccine complex may influence the Th-type adaptive immune response observed with this vaccine (see Table 3).

Table 3: Adjuvants enhancing Th1 cell subset differentiation in human/non-human primate studies

Adjuvant Enhancing Th1 Cell Subset Differentiation in Human/Non-Human Primate Studies	Potential Mechanism(s)	Pathogen(s)/Antigens for Which the Adjuvant Has Been Tested	Select References
Unmethylated CpG	Activation of TLR9 Enhanced recruitment of CD4+ T cells	*Leishmania* influenza, hepatitis B	[79,80,81,83]
Adjuvant system AS01 and AS04	Induction of antigen-specific CD4+ T cells Production of early, innate IFN-γ responses by APCs	*Mycobacterium tuberculosis*, malaria, human papillomavirus, influenza	[84,86,87,89,90]
mRNA/lipid nanoparticles *	High level intracellular antigen expression, promoting MHC II antigen presentation	SARS-CoV-2	[41,91]

* An antigen delivery platform with adjuvant functions.

The AS04 complex consists of MPLA adsorbed onto aluminum hydroxide is also associated with enhanced Th1 responses, and is currently in trials in humans for human papillomavirus (HPV) and hepatitis B vaccination [88]. Human dendritic cells stimulated with the HPV vaccine Cervarix, which contains AS04, induced TNF-α and IL-6 production and increased dendritic cell co-stimulatory molecule expression, with the effect found to be specific to the MPLA component [89]. In a murine model of influenza vaccination, virus like particles expressing influenza M2 ectodomain protein (M2eVLP), when adjuvanted with AS04 had superior protective benefit upon lethal influenza challenge after vaccination relative to M2eVLP adjuvanted with alum or MPLA alone or in the absence of any adjuvant [90]. It was noted that M2eVLP vaccination with AS04 also led to higher production of CD4+T cells in the lungs expressing IFN-γ and pore forming protein granzyme B on challenge relative to other groups. These data suggest that the Th1 response induced by the vaccine and recalled with challenge may relate to the mechanism of enhanced immune protection. Additionally, this study demonstrates that features of the combination AS04 are distinct from the separate components as adjuvants. Collectively, these studies underscore the value of AS01 and AS04, and similar adjuvants, in enhancing Th1 responses.

The mRNA COVID-19 vaccines BNT162b2 and mRNA-1273 are associated with a pronounced Th1 response compared to Th2 response in humans. Induction of CD4+ T cells producing either IFN-γ or TNF-α with mRNA-1273 was increased when a higher dose of vaccine was used compared to a lower dose, and notably the magnitude of Th1 response was preserved in adults over the age of 71 compared to adults ages 56–70 [91]. The Th2 response, as assessed by Th production of IL-4 or IL-13 was notably lower and unaffected by vaccine dose or age group. The BNT162b2 vaccine demonstrated a similar polarity towards driving Th1 responses in humans, with minimal IL-4 producing Th responses elicited in subjects at any vaccine dose tested [41]. Given the challenge of inducing reliable T cell responses during vaccination in older adults, these results suggest that the mRNA-lipid NP vaccines may be promising in eliciting strong Th1 responses generally, including in older adults.

Th2 Cell Induction and Adjuvants

Differentiation of Th2 cells is associated with activation of the master transcriptional factor GATA-3, and with Th production of cytokine IL-4, as well as IL-5, IL-9, IL-13, and IL-25. Th2 immunity with an antibody-predominant response is believed to be essential in human defense against parasitic infections, such as in resistance to gastrointestinal worm infections [92,93,94]. There is less known regarding potential Th2 driving adjuvants, compared to Th1 driving adjuvants, perhaps given the success of alum in driving Th2 responses while suppressing Th1 responses (see Table 4). Despite the duration of use of alum as an adjuvant in humans, the mechanism of alum as an adjuvant is unclear. alum's adjuvant effect is hypothesized to be related to initiating an antigen depot with prolonged recruitment of immune cells, including eosinophils, to the injection site; pattern recognition and activation of the cytosolic NLRP3 inflammasome; and elicitation of endogenous danger signals such as uric acid and IL-33 [95,96]. In a vaccine against Schistosomiasis, protein antigen glutathione *S*-transferase (28GST) derived from *S. mansoni* administered with aluminum hydroxide significantly reduced baboon adult worm quantities and was also associated with reduced infection driven inflammation in the liver [97]. In adult volunteers vaccinated with recombinant *S. hematobium* 28GST adjuvanted with aluminum hydroxide, high levels of 28GST specific IgG1 production were elicited, and PBMCs collected from volunteers after booster dosing and stimulated with 28GST in vitro demonstrated high production of IL-5 and IL-13 consistent with a Th2 dominant response [98]. Similarly, the vaccine for human hookworm, *Necator Americanus* antigen Glutatione-S-Transferase 1 (*Na*-GST-1) was assessed for immunogenicity in humans comparing adjuvants aluminum hydroxide (Alhydrogel) vs. Alhydrogel with glucopyranosyl lipid adjuvant. IgG1 and IgG3 isotypes were predominantly elicited, and there was no increase in antibody titers or rate of response with addition of glycopyranosyl lipid adjuvant over the three dose vaccine series [99]. Other adjuvants, such as cholera toxin, MF59, or polysaccharide based adjuvants, such as glucohexaose analogue beta-glu6, may hold promise in Th2 skewing of the Th response, though several recent studies highlight these groups of adjuvants as having broad capacity to enhance immune responses beyond a Th2 predominant pattern [100,101,102,103,104].

Table 4: Adjuvants enhancing Th2 cell subset differentiation in human/non-human primate studies

Adjuvant Enhancing Th2 Cell Subset Differentiation in Human/Non-Human Primate Studies.	Potential Mechanism(s)	Pathogen(s)/Antigens for Which the Adjuvant Has Been Tested	Select References
Alum	Antigen deposition with prolonged recruitment of immune cells (i.e., eosinophils) to injection site	*Schistomsoma mansoni, S. hematobium, Necator Americanus*	[93,95,97,98,99]
	Pattern recognition and activation of the cytosolic inflammasome		
	Elicitation of endogenous danger signals (i.e., uric acid, IL-33)		

Th17 Cell Induction by Adjuvants

Th17 responses play a critical role in host defense against a variety of bacterial and fungal infections at mucosal sites, as evidenced through disease models and studies in mice and humans with Th17 deficiency [105,106,107,108,109,110]. Beyond host defense mechanisms, Th17 responses are implicated in mechanisms of autoimmune disease [111]. Master transcriptional factor RORγT is essential for Th17 development and function, and Th17 are characterized by production of cytokine IL-17 (also known as IL-17A), and potentially the antimicrobial peptide associated cytokine IL-22 [112,113]. Differentiation of Th17 occurs in the setting of increased quantities of IL-23, IL-6, and IL-1β within the cytokine milieu, suggesting that this environment could potentially be recapitulated by adjuvants to enhance Th17 responses. However, given the relatively recent identification of this subset less is known regarding potential adjuvants driving Th17 differentiation. The cationic liposome adjuvant CAF01 has recently been investigated for its potential to drive Th17 responses. In a murine model of vaccination with tuberculosis fusion protein antigen H28, when adjuvanted with CAF01, it was noted that splenocytes on restimulation with antigen 13 weeks after immunization produced high levels of IL-17A as well as IFN-γ [114]. Furthermore, the IL-17A responses were essentially absent when the H28 was instead adjuvanted with either Alhydrogel, Quil A, or Montanide. The Th17 elicited by the vaccine adjuvanted with CAF01 were found to be stable as a cell population for nearly 2 years, and was capable of activation and recruitment to the lung on *M. tuberculosis* challenge [114]. This study is suggestive of the potential of CAF01 as an adjuvant in driving Th17 memory responses alongside Th1 responses against protein antigens. In mucosal vaccination against tuberculosis in mice with ESAT6 peptide antigen, addition of the *E. coli* type II heat labile enterotoxin as an adjuvant led to the induction of high levels of Th17 cells [115]. In this model, vaccine efficacy was lost in the absence of IL-17 signaling and Th17 cells induced by vaccination were important for the generation of inducible bronchus associated lymphoid tissue (iBALT) in a CXCL13 dependent manner [115]. iBALT induction was hypothesized to play a role in retention of antigen specific T cells within the lungs for rapid effector function upon *M. tuberculosis* exposure. Of note, IL-17 signaling is not required for primary immunity against *M. tuberculosis*, but was shown to have a critical role in mucosal vaccine based immune protection against *M. tuberculosis* elicited with *E. coli* heat labile enterotoxin.

Cholera toxin (CT) has also been shown to enhance Th17 responses in a murine model of anthrax vaccination. When mice were intranasally vaccinated with irradiated *Bacillus anthracis* spores, only if CT was included as an adjuvant with the immunization was there protection against lethal challenge, and blockade of IL-17A reduced survival [116]. On evaluation of the T cells in mice vaccinated with *B. anthracis* spores adjuvanted with CT, there was significantly enhanced generation of CD4+ T cells producing IL-17A compared to IFN-γ or IL-5.

More recently, studies with primary human peripheral blood mononuclear cells (PBMC) have shown that cholera toxin, non-toxic multiple mutated CT (mmCT), and double mutant *E. coli* derived heat labile toxin (dmLT) may be a potential adjuvant in enhancing Th17 differentiation in humans. When each of these modified proteins were introduced into PBMC culture, and T cells activated with Staphylococcal enterotoxin B, significantly enhanced levels of IL-17A were observed [117]. Th17 were specifically identified in the co-culture system through intracellular cytokine staining of CD4+ T cells, and blockade of the IL-1 receptor in PBMC cultures limited IL-17A production, suggesting the importance of IL-1 signaling in Th17 differentiation. Monocytes were identified as upregulating IL-1β production on stimulation with either CT, mmCT, or dmLT, suggesting a potential cellular target for these adjuvants [117].

In a macaque model of intranasal vaccination with Pneumococcal surface protein A (PspA) complexed with cationic cholesteryl group-bearing pullulan nanogel (cCHP nanogel), there was significantly enhanced levels of CD4+ T cells producing IL-17 in the blood, as well as enhanced CD4+T cell production of IL-4 and increased serum antigen-specific IgG1, compared to macaques receiving PspA alone [118]. Mice intranasally vaccinated with PspA presented in cCHP nanogel similarly generate CD4+ T cells that produce IL-17 as well as IL-4 responses and were able to resist a lethal challenge with *Streptococcus pneumoniae* compared to intranasal vaccination with PspA alone [119,120]. Collectively, this data suggests that mucosal vaccination with cCHP enhances Th17, as well as Th2 responses.

More recent work has suggested that activation of the lectin receptor dectin-1 may enhance Th17 differentiation by APCs. Human dendritic cells treated with influenza hemagglutinin antigen bound to an agonist anti-dectin-1 monoclonal antibody induced differentiation of naïve CD4+ T cells towards hemagglutinin antigen specific Th17 cells [121]. Separate studies have shown that dectin-1 agonists zymosan and curdlan polarize human dendritic cells towards induction of Th17 immune responses [122,123]. Additional work is needed to substantiate the role of dectin-1 receptor agonists in eliciting Th17 responses in the setting of vaccination in humans.

Insights into potential benefit of CAF01 liposomes, specific bacterial heat labile enterotoxins, and cCHP nanogels in driving Th17 differentiation suggest future paths for adjuvant design for extracellular mucosal infections (see Table 5). Given the prominence of Th17 in host defense pathways associated with a variety of bacterial pneumonias, mucosal viral infections, such as HIV, as well as fungal infections, such as invasive candidiasis, Th17 driving adjuvants may hold particular benefit for vaccine strategies in these settings.

Table 5: Adjuvants enhancing Th17 cell subset differentiation in human/non-human primate studies

Adjuvant Enhancing Th17 Cell Subset Differentiation in Human/Non-Human Primate Studies	Potential Mechanism(s)	Pathogen(s)/Antigens for Which the Adjuvant Has Been Tested	Select References
Heat labile bacterial toxins and derivatives (e.g., dmLT)	Monocyte upregulation of IL-1β production	*Mycobacterium tuberculosis, Staphylococcus aureus, Bacillus anthracis*	[115,117]
	Induction of iBALT through Th17 and elicitation of CXCR13		
cCHP nanogel *	Particulate mucosal adherent immunization, enhancing local APC activation	*Streptococcus pneumoniae*	[118,119,120]

* An antigen delivery platform with adjuvant functions.

FUTURE DIRECTIONS

The utilization of different adjuvants to drive desired pathways in T cell differentiation is a critical goal for developing effective and long-lasting immunity to infection. Our understandings of the immune pathways activated by different adjuvants, as well as of critical immune pathways in host defense against different infections are informing future directions in adjuvant science and vaccine design. With our continually evolving knowledge of mechanisms of APC pattern recognition and APC activation and signaling, advances in adjuvant design will follow. Progress will also be dependent on implementation of more translational studies of vaccines, with greater emphasis on assessment of immunologic responses to adjuvants, as well as mechanisms of adjuvants, in non-human primates and humans. Scalability and absence of toxicities are also key features in next-generation adjuvants driving T cell differentiation. Additionally, structural manipulation and optimization of adjuvants, such as with the use of bacterial enzymatic combinatorial chemistry on lipooligosaccharide, may hold promise for future adjuvant design. Strategies involving combination adjuvants as well as adjuvants with more specific molecular targets may also help to impact T cell differentiation. The response to an infectious disease in real-time, as illustrated with the COVID-19 pandemic, brought new classes of vaccine and adjuvant into broad use in humans safely and rapidly. Studies of properties of T cell responses from the large populations that have received recently developed adenoviral, protein, and mRNA lipid NP COVID-19 vaccines may pave new directions for adjuvants and delivery platforms for other infectious diseases. In this promising time in vaccinology, advancing the development of adjuvants driving T cell responses in humans may help to address our greatest challenges in infectious disease.

AUTHOR CONTRIBUTIONS

Conceptualization, R.R.R. and M.A.M.; writing—original draft preparation, R.R.R.; writing—review and editing, R.R.R., A.S.C., M.A.M.; funding acquisition, R.R.R. All authors have read and agreed to the published version of the manuscript.

REFERENCES

1. Di Pasquale A., Preiss S., Silva F.M.D.O.E., Garçon N. Vaccine Adjuvants: From 1920 to 2015 and Beyond. *Vaccines.* 2015;3:320–343. doi: 10.3390/vaccines3020320.
2. Del Giudice G., Rappuoli R., Didierlaurent A.M. Correlates of adjuvanticity: A review on adjuvants in licensed vaccines. *Semin. Immunol.* 2018;39:14–21. doi: 10.1016/j.smim.2018.05.001.
3. Wrapp D., Wang N., Corbett K.S., Goldsmith J.A., Hsieh C.-L., Abiona O., Graham B.S., McLellan J.S. Cryo-EM structure of the 2019-nCoV spike in the prefusion conformation. *Science.* 2020;367:1260–1263. doi: 10.1126/science.abb2507.
4. Calvo-Calle J.M., Hammer J., Sinigaglia F., Clavijo P., Moya-Castro Z.R., Nardin E.H. Binding of malaria T cell epitopes to DR and DQ molecules in vitro correlates with immunogenicity in vivo: Identification of a universal T cell epitope in the Plasmodium falciparum circumsporozoite protein. *J. Immunol.* 1997;159:1362–1373.
5. Berzofsky J.A., Ahlers J.D., Belyakov I.M. Strategies for designing and optimizing new generation vaccines. *Nat. Rev. Immunol.* 2001;1:209–219. doi: 10.1038/35105075.
6. Yap K., Braciale T., Ada G. Role of T-cell function in recovery from murine influenza infection. *Cell. Immunol.* 1979;43:341–351. doi: 10.1016/0008-8749(79)90178-3.
7. Schmitz J.E., Kuroda M.J., Santra S., Sasseville V.G., Simon M.A., Lifton M.A., Racz P., Tenner-Racz K., Dalesandro M., Scallon B.J., et al. Control of Viremia in Simian Immunodeficiency Virus Infection by CD8+ Lymphocytes. *Science.* 1999;283:857–860. doi: 10.1126/science.283.5403.857.
8. Schmidt M.E., Varga S.M. The CD8 T Cell Response to Respiratory Virus Infections. *Front. Immunol.* 2018;9:678. doi: 10.3389/fimmu.2018.00678.
9. Woodworth J.S.M., Behar S.M. Mycobacterium tuberculosis-Specific CD8+ T Cells and Their Role in Immunity. *Crit. Rev. Immunol.* 2006;26:317–352. doi: 10.1615/CritRevImmunol.v26.i4.30.
10. Betts M.R. HIV nonprogressors preferentially maintain highly functional HIV-specific CD8+ T cells. *Blood.* 2006;107:4781–4789. doi: 10.1182/blood-2005-12-4818.

11. Harty J.T., Tvinnereim A.R., White D.W. CD8+ T Cell Effector Mechanisms in Resistance to Infection. *Annu. Rev. Immunol.* 2000;18:275–308. doi: 10.1146/annurev.immunol.18.1.275.
12. Zhang N., Bevan M.J. CD8+ T Cells: Foot Soldiers of the Immune System. *Immunity.* 2011;35:161–168. doi: 10.1016/j.immuni.2011.07.010.
13. Beura L.K., Jameson S., Masopust D. Is a Human CD8 T-Cell Vaccine Possible, and if So, What Would It Take? *Cold Spring Harb. Perspect. Biol.* 2017;10:a028910. doi: 10.1101/cshperspect.a028910.
14. Koup R.A., Douek D.C. Vaccine Design for CD8 T Lymphocyte Responses. *Cold Spring Harb. Perspect. Med.* 2011;1:a007252. doi: 10.1101/cshperspect.a007252.
15. Lazzaro S., Giovani C., Mangiavacchi S., Magini D., Maione D., Baudner B., Geall A.J., De Gregorio E., D'Oro U., Buonsanti C. CD8 T-cell priming upon mRNA vaccination is restricted to bone-marrow-derived antigen-presenting cells and may involve antigen transfer from myocytes. *Immunology.* 2015;146:312–326. doi: 10.1111/imm.12505.
16. Kensil C.R., Patel U., Lennick M., Marciani D. Separation and characterization of saponins with adjuvant activity from Quillaja saponaria Molina cortex. *J. Immunol.* 1991;146:431–437.
17. Ragupathi G., Gardner J.R., Livingston P., Gin D.Y. Natural and synthetic saponin adjuvant QS-21 for vaccines against cancer. *Expert Rev. Vaccines.* 2011;10:463–470. doi: 10.1586/erv.11.18.
18. Welsby I., Detienne S., N'Kuli F., Thomas S., Wouters S., Bechtold V., De Wit D., Gineste R., Reinheckel T., Elouahabi A., et al. Lysosome-Dependent Activation of Human Dendritic Cells by the Vaccine Adjuvant QS-21. *Front. Immunol.* 2017;7:526. doi: 10.3389/fimmu.2016.00663.
19. Morelli A.B., Becher D., Koernig S., Silva A., Drane D., Maraskovsky E. ISCOMATRIX: A novel adjuvant for use in prophylactic and therapeutic vaccines against infectious diseases. *J. Med. Microbiol.* 2012;61:935–943. doi: 10.1099/jmm.0.040857-0.
20. Maraskovsky E., Sjölander S., Drane D.P., Schnurr M., Le T.T.T., Mateo L., Luft T., Masterman K.-A., Tai T.-Y., Chen Q., et al. NY-ESO-1 Protein Formulated in ISCOMATRIX Adjuvant Is a Potent Anticancer Vaccine Inducing Both Humoral and CD8+ T-Cell-Mediated Immunity and Protection against NY-ESO-1+ Tumors. *Clin.*

Cancer Res. 2004;10:2879–2890. doi: 10.1158/1078-0432.CCR-03-0245.

21. Chen J.-L., Dawoodji A., Tarlton A., Gnjatic S., Tajar A., Karydis I., Browning J., Pratap S., Verfaille C., Venhaus R.R., et al. NY-ESO-1 specific antibody and cellular responses in melanoma patients primed with NY-ESO-1 protein in ISCOMATRIX and boosted with recombinant NY-ESO-1 fowlpox virus. *Int. J. Cancer.* 2015;136:E590–E601. doi: 10.1002/ijc.29118.
22. Barr A.M., Silva A., Prato S., Belz G., Maraskovsky E., Morelli A.B. Therapeutic ISCOMATRIX™ adjuvant vaccine elicits effective anti-tumor immunity in the TRAMP-C1 mouse model of prostate cancer. *Cancer Immunol. Immunother.* 2020;69:1959–1972. doi: 10.1007/s00262-020-02597-6.
23. Cibulski S., Rivera-Patron M., Suárez N., Pirez M., Rossi S., Yendo A.C., de Costa F., Gosmann G., Fett-Neto A., Roehe P.M., et al. Leaf saponins of Quillaja brasiliensis enhance long-term specific immune responses and promote dose-sparing effect in BVDV experimental vaccines. *Vaccine.* 2018;36:55–65. doi: 10.1016/j.vaccine.2017.11.030.
24. Bengtsson K.L., Morein B., Osterhaus A.D. ISCOM technology-based Matrix M™ adjuvant: Success in future vaccines relies on formulation. *Expert Rev. Vaccines.* 2011;10:401–403. doi: 10.1586/erv.11.25.
25. Bengtsson K.L., Song H., Stertman L., Liu Y., Flyer D.C., Massare M.J., Xu R.-H., Zhou B., Lu H., Kwilas S.A., et al. Matrix-M adjuvant enhances antibody, cellular and protective immune responses of a Zaire Ebola/Makona virus glycoprotein (GP) nanoparticle vaccine in mice. *Vaccine.* 2016;34:1927–1935. doi: 10.1016/j.vaccine.2016.02.033.
26. Warfield K.L., Olinger G., Deal E.M., Swenson D.L., Bailey M., Negley D.L., Hart M.K., Bavari S. Induction of Humoral and CD8+ T Cell Responses Are Required for Protection against Lethal Ebola Virus Infection. *J. Immunol.* 2005;175:1184–1191. doi: 10.4049/jimmunol.175.2.1184.
27. Duvall M.G., Precopio M.L., Ambrozak D.A., Jaye A., McMichael A.J., Whittle H.C., Roederer M., Rowland-Jones S.L., Koup R.A. Polyfunctional T cell responses are a hallmark of HIV-2 infection. *Eur. J. Immunol.* 2008;38:350–363. doi: 10.1002/eji.200737768.
28. Snyder L.D., Chan C., Kwon D., Yi J.S., Martissa J.A., Copeland C.A.F., Osborne R.J., Sparks S.D., Palmer S., Weinhold K.J. Polyfunctional T-Cell Signatures to Predict Protection from Cytomegalovirus after

Lung Transplantation. *Am. J. Respir. Crit. Care Med.* 2016;193:78–85. doi: 10.1164/rccm.201504-0733OC.

29. Harari A., Dutoit V., Cellerai C., Bart P., Du Pasquier R., Pantaleo G. Functional signatures of protective antiviral T-cell immunity in human virus infections. *Immunol. Rev.* 2006;211:236–254. doi: 10.1111/j.0105-2896.2006.00395.x.
30. Tian J.-H., Patel N., Haupt R., Zhou H., Weston S., Hammond H., Logue J., Portnoff A.D., Norton J., Guebre-Xabier M., et al. SARS-CoV-2 spike glycoprotein vaccine candidate NVX-CoV2373 immunogenicity in baboons and protection in mice. *Nat. Commun.* 2021;12:1–14. doi: 10.1038/s41467-020-20653-8.
31. Fries L., Cho I., Krähling V., Fehling S.K., Strecker T., Becker S., Hooper J., Kwilas S., Agrawal S., Wen J., et al. Randomized, Blinded, Dose-Ranging Trial of an Ebola Virus Glycoprotein Nanoparticle Vaccine With Matrix-M Adjuvant in Healthy Adults. *J. Infect. Dis.* 2019;222:572–582. doi: 10.1093/infdis/jiz518.
32. Keech C., Albert G., Cho I., Robertson A., Reed P., Neal S., Plested J.S., Zhu M., Cloney-Clark S., Zhou H., et al. Phase 1–2 Trial of a SARS-CoV-2 Recombinant Spike Protein Nanoparticle Vaccine. *N. Engl. J. Med.* 2020;383:2320–2332. doi: 10.1056/NEJMoa2026920.
33. Shinde V., Bhikha S., Hoosain Z., Archary M., Bhorat Q., Fairlie L., Lalloo U., Masilela M.S., Moodley D., Hanley S., et al. Efficacy of NVX-CoV2373 Covid-19 Vaccine against the B.1.351 Variant. *N. Engl. J. Med.* 2021;384:1899–1909. doi: 10.1056/NEJMoa2103055.
34. Schwendener R.A. Liposomes as vaccine delivery systems: A review of the recent advances. *Ther. Adv. Vaccines.* 2014;2:159–182. doi: 10.1177/2051013614541440.
35. Taneichi M., Ishida H., Kajino K., Ogasawara K., Tanaka Y., Kasai M., Mori M., Nishida M., Yamamura H., Mizuguchi J., et al. Antigen chemically coupled to the surface of liposomes are cross-presented to CD8+ T cells and induce potent antitumor immunity. *J. Immunol.* 2006;177:2324–2330. doi: 10.4049/jimmunol.177.4.2324.
36. Ignatius R., Mahnke K., Rivera M., Hong K., Isdell F., Steinman R.M., Pope M., Stamatatos L. Presentation of proteins encapsulated in sterically stabilized liposomes by dendritic cells initiates CD8(+) T-cell responses in vivo. *Blood.* 2000;96:3505–3513. doi: 10.1182/blood.V96.10.3505.h8003505_3505_3513.

37. Let's talk about lipid nanoparticles. *Nat. Rev. Mater.* 2021;6:99. doi: 10.1038/s41578-021-00281-4.
38. Pardi N., Hogan M., Porter F.W., Weissman D. mRNA vaccines—A new era in vaccinology. *Nat. Rev. Drug Discov.* 2018;17:261–279. doi: 10.1038/nrd.2017.243.
39. Bialkowski L., Van Weijnen A., Van Der Jeught K., Renmans D., Daszkiewicz L., Heirman C., Stange G., Breckpot K., Aerts J.L., Thielemans K. Intralymphatic mRNA vaccine induces CD8 T-cell responses that inhibit the growth of mucosally located tumours. *Sci. Rep.* 2016;6:22509. doi: 10.1038/srep22509.
40. Van Gulck E.E., Vlieghe E.E., Vekemans M., Van Tendeloo V., Van De Velde A., Smits E., Anguille S.S., Cools N.N., Goossens H., Mertens L.L., et al. mRNA-based dendritic cell vaccination induces potent antiviral T-cell responses in HIV-1-infected patients. *AIDS.* 2012;26:F1–F12. doi: 10.1097/QAD.0b013e32834f33e8.
41. Sahin U., Muik A., Derhovanessian E., Vogler I., Kranz L.M., Vormehr M., Baum A., Pascal K., Quandt J., Maurus D., et al. COVID-19 vaccine BNT162b1 elicits human antibody and TH1 T-cell responses. *Nature.* 2020;586:594–599. doi: 10.1038/s41586-020-2814-7.
42. Jackson L.A., Anderson E.J., Rouphael N.G., Roberts P.C., Makhene M., Coler R.N., McCullough M.P., Chappell J.D., Denison M.R., Stevens L.J., et al. An mRNA Vaccine against SARS-CoV-2—Preliminary Report. *N. Engl. J. Med.* 2020;383:1920–1931. doi: 10.1056/NEJMoa2022483.
43. Sahin U., Muik A., Vogler I., Derhovanessian E., Kranz L.M., Vormehr M., Quandt J., Bidmon N., Ulges A., Baum A., et al. BNT162b2 vaccine induces neutralizing antibodies and poly-specific T cells in humans. *Nat. Cell Biol.* 2021:1–10. doi: 10.1038/s41586-021-03653-6.
44. Demkowicz W., Littaua R., Wang J., Ennis F. Human cytotoxic T-cell memory: Long-lived responses to vaccinia virus. *J. Virol.* 1996;70:2627–2631. doi: 10.1128/jvi.70.4.2627-2631.1996.
45. Ennis F.A., Cruz J., Demkowicz J.W.E., Rothman A.L., McClain D.J. Primary Induction of Human CD8+Cytotoxic T Lymphocytes and Interferon-γ–Producing T Cells after Smallpox Vaccination. *J. Infect. Dis.* 2002;185:1657–1659. doi: 10.1086/340517.
46. Ahmed R., Akondy R. Insights into human CD8 + T-cell memory using the yellow fever and smallpox vaccines. *Immunol. Cell Biol.* 2011;89:340–345. doi: 10.1038/icb.2010.155.

47. Akondy R., Johnson P.L.F., Nakaya H., Edupuganti S., Mulligan M.J., Lawson B., Miller J.D., Pulendran B., Antia R., Ahmed R. Initial viral load determines the magnitude of the human CD8 T cell response to yellow fever vaccination. *Proc. Natl. Acad. Sci. USA*. 2015;112:3050–3055. doi: 10.1073/pnas.1500475112.

48. Tomori O., Kolawole M.O. Ebola virus disease: Current vaccine solutions. *Curr. Opin. Immunol.* 2021;71:27–33. doi: 10.1016/j.coi.2021.03.008.

49. Folegatti P.M., Ewer K.J., Aley P.K., Angus B., Becker S., Belij-Rammerstorfer S., Bellamy D., Bibi S., Bittaye M., Clutterbuck E.A., et al. Safety and immunogenicity of the ChAdOx1 nCoV-19 vaccine against SARS-CoV-2: A preliminary report of a phase 1/2, single-blind, randomised controlled trial. *Lancet.* 2020;396:467–478. doi: 10.1016/S0140-6736(20)31604-4.

50. Solforosi L., Kuipers H., Jongeneelen M., Huber S.K.R., van der Lubbe J.E., Dekking L., Czapska-Casey D.N., Gil A.I., Baert M.R., Drijver J., et al. Immunogenicity and efficacy of one and two doses of Ad26.COV2.S COVID vaccine in adult and aged NHP. *J. Exp. Med.* 2021;218 doi: 10.1084/jem.20202756.

51. Sadoff J., Le Gars M., Shukarev G., Heerwegh D., Truyers C., de Groot A.M., Stoop J., Tete S., Van Damme W., Leroux-Roels I., et al. Interim Results of a Phase 1–2a Trial of Ad26.COV2.S Covid-19 Vaccine. *N. Engl. J. Med.* 2021;384:1824–1835. doi: 10.1056/NEJMoa2034201.

52. Sullivan N.J., Hensley L., Asiedu C., Geisbert T.W., Stanley D., Johnson J., Honko A., Olinger G., Bailey M.Q., Geisbert J.B., et al. CD8+ cellular immunity mediates rAd5 vaccine protection against Ebola virus infection of nonhuman primates. *Nat. Med.* 2011;17:1128–1131. doi: 10.1038/nm.2447.

53. Ewer K.J., Barrett J.R., Belij-Rammerstorfer S., Sharpe H., Makinson R., Morter R., Flaxman A., Wright D., Bellamy D., The Oxford COVID Vaccine Trial Group et al. T cell and antibody responses induced by a single dose of ChAdOx1 nCoV-19 (AZD1222) vaccine in a phase 1/2 clinical trial. *Nat. Med.* 2021;27:270–278. doi: 10.1038/s41591-020-01194-5.

54. Bode C., Zhao G., Steinhagen F., Kinjo T., Klinman D.M. CpG DNA as a vaccine adjuvant. *Expert Rev. Vaccines.* 2011;10:499–511. doi: 10.1586/erv.10.174.

55. Cafaro A., Titti F., Fracasso C., Maggiorella M., Baroncelli S., Caputo A., Goletti D., Borsetti A., Pace M., Fanales-Belasio E., et al. Vaccination with DNA containing tat coding sequences and unmethylated CpG motifs protects cynomolgus monkeys upon infection with simian/human immunodeficiency virus (SHIV89.6P) *Vaccine.* 2001;19:2862–2877. doi: 10.1016/S0264-410X(01)00002-0.
56. Speiser D., Liénard D., Rufer N., Rubio-Godoy V., Rimoldi D., Lejeune F., Krieg A.M., Cerottini J.-C., Romero P. Rapid and strong human CD8+ T cell responses to vaccination with peptide, IFA, and CpG oligodeoxynucleotide 7909. *J. Clin. Investig.* 2005;115:739–746. doi: 10.1172/JCI23373.
57. Rothenfusser S., Hornung V., Ayyoub M., Britsch S., Towarowski A., Krug A., Sarris A., Lubenow N., Speiser D., Endres S., et al. CpG-A and CpG-B oligonucleotides differentially enhance human peptide–specific primary and memory CD8+ T-cell responses in vitro. *Blood.* 2004;103:2162–2169. doi: 10.1182/blood-2003-04-1091.
58. Martins K.A.O., Steffens J.T., Van Tongeren S.A., Wells J.B., Bergeron A.A., Dickson S.P., Dye J.M., Salazar A.M., Bavari S. Toll-Like Receptor Agonist Augments Virus-Like Particle-Mediated Protection from Ebola Virus with Transient Immune Activation. *PLoS ONE.* 2014;9:e89735. doi: 10.1371/journal.pone.0089735.
59. Da Silva D.M., Woodham A.W., Skeate J.G., Rijkee L.K., Taylor J.R., Brand H.E., Muderspach L.I., Roman L.D., Yessaian A.A., Pham H.Q., et al. Langerhans cells from women with cervical precancerous lesions become functionally responsive against human papillomavirus after activation with stabilized Poly-I:C. *Clin. Immunol.* 2015;161:197–208. doi: 10.1016/j.clim.2015.09.003.
60. Girón J.V.P., Belicha-Villanueva A., Hassan E., Gómez-Medina S., Cruz J.L.G., Lüdtke A., Ruibal P., Albrecht R., Garcia-Sastre A., Muñoz-Fontela C. Mucosal Polyinosinic-Polycytidylic Acid Improves Protection Elicited by Replicating Influenza Vaccines via Enhanced Dendritic Cell Function and T Cell Immunity. *J. Immunol.* 2014;193:1324–1332. doi: 10.4049/jimmunol.1400222.
61. Pallikkuth S., Parmigiani A., Silva S.Y., George V.K., Fischl M., Pahwa R., Pahwa S. Impaired peripheral blood T-follicular helper cell function in HIV-infected nonresponders to the 2009 H1N1/09 vaccine. *Blood.* 2012;120:985–993. doi: 10.1182/blood-2011-12-396648.

62. Cárdeno A., Magnusson M., Quiding-Järbrink M., Lundgren A. Activated T follicular helper-like cells are released into blood after oral vaccination and correlate with vaccine specific mucosal B-cell memory. *Sci. Rep.* 2018;8:1–15. doi: 10.1038/s41598-018-20740-3.
63. Morita R., Schmitt N., Bentebibel S.-E., Ranganathan R., Bourdery L., Zurawski G., Foucat E., Dullaers M., Oh S., Sabzghabaei N., et al. Human Blood CXCR5+CD4+ T Cells Are Counterparts of T Follicular Cells and Contain Specific Subsets that Differentially Support Antibody Secretion. *Immunity.* 2011;34:108–121. doi: 10.1016/j.immuni.2010.12.012.
64. Chevalier N., Jarrossay D., Ho E., Avery D.T., Ma C., Yu D., Sallusto F., Tangye P.S., Mackay C. CXCR5 Expressing Human Central Memory CD4 T Cells and Their Relevance for Humoral Immune Responses. *J. Immunol.* 2011;186:5556–5568. doi: 10.4049/jimmunol.1002828.
65. Hill D., Pierson W., Bolland D.J., Mkindi C., Carr E., Wang J., Houard S., Wingett S.W., Audran R., Wallin E.F., et al. The adjuvant GLA-SE promotes human Tfh cell expansion and emergence of public TCRβ clonotypes. *J. Exp. Med.* 2019;216:1857–1873. doi: 10.1084/jem.20190301.
66. Desbien A.L., Cauwelaert N.D., Reed S.J., Bailor H.R., Liang H., Carter D., Duthie M., Fox C.B., Orr M.T., Reed S.G. IL-18 and Subcapsular Lymph Node Macrophages are Essential for Enhanced B Cell Responses with TLR4 Agonist Adjuvants. *J. Immunol.* 2016;197:4351–4359. doi: 10.4049/jimmunol.1600993.
67. Radtke A.J., Anderson C.F., Riteau N., Rausch K., Scaria P., Kelnhofer E.R., Howard R.F., Sher A., Germain R.N., Duffy P. Adjuvant and carrier protein-dependent T-cell priming promotes a robust antibody response against the Plasmodium falciparum Pfs25 vaccine candidate. *Sci. Rep.* 2017;7:40312. doi: 10.1038/srep40312.
68. Chen C., Zhang C., Li R., Wang Z., Yuan Y., Li H., Fu Z., Zhou M., Zhao L. Monophosphoryl-Lipid A (MPLA) is an Efficacious Adjuvant for Inactivated Rabies Vaccines. *Viruses.* 2019;11:1118. doi: 10.3390/v11121118.
69. Moon J., Suh H., Li A.V., Ockenhouse C.F., Yadava A., Irvine D.J. Enhancing humoral responses to a malaria antigen with nanoparticle vaccines that expand Tfh cells and promote germinal center induction.

Proc. Natl. Acad. Sci. USA. 2012;109:1080–1085. doi: 10.1073/pnas.1112648109.

70. O'Hagan D., Ott G., De Gregorio E., Seubert A. The mechanism of action of MF59—An innately attractive adjuvant formulation. *Vaccine.* 2012;30:4341–4348. doi: 10.1016/j.vaccine.2011.09.061.
71. Seubert A., Calabro S., Santini L., Galli B., Genovese A., Valentini S., Aprea S., Colaprico A., D'Oro U., Giuliani M.M., et al. Adjuvanticity of the oil-in-water emulsion MF59 is independent of Nlrp3 inflammasome but requires the adaptor protein MyD88. *Proc. Natl. Acad. Sci. USA.* 2011;108:11169–11174. doi: 10.1073/pnas.1107941108.
72. Gavillet B.M., Eberhardt C.S., Auderset F., Castellino F., Seubert A., Tregoning J., Lambert P.-H., De Gregorio E., Del Giudice G., Siegrist C.-A. MF59 Mediates Its B Cell Adjuvanticity by Promoting T Follicular Helper Cells and Thus Germinal Center Responses in Adult and Early Life. *J. Immunol.* 2015;194:4836–4845. doi: 10.4049/jimmunol.1402071.
73. Lofano G., Mancini F., Salvatore G., Cantisani R., Monaci E., Carrisi C., Tavarini S., Sammicheli C., Paccani S.R., Soldaini E., et al. Oil-in-Water Emulsion MF59 Increases Germinal Center B Cell Differentiation and Persistence in Response to Vaccination. *J. Immunol.* 2015;195:1617–1627. doi: 10.4049/jimmunol.1402604.
74. Spensieri F., Siena E., Borgogni E., Zedda L., Cantisani R., Chiappini N., Schiavetti F., Rosa D., Castellino F., Montomoli E., et al. Early Rise of Blood T Follicular Helper Cell Subsets and Baseline Immunity as Predictors of Persisting Late Functional Antibody Responses to Vaccination in Humans. *PLoS ONE.* 2016;11:e0157066. doi: 10.1371/journal.pone.0157066.
75. Chappell K.J., Mordant F.L., Li Z., Wijesundara D.K., Ellenberg P., Lackenby J., Cheung S.T.M., Modhiran N., Avumegah M.S., Henderson C.L., et al. Safety and immunogenicity of an MF59-adjuvanted spike glycoprotein-clamp vaccine for SARS-CoV-2: A randomised, double-blind, placebo-controlled, phase 1 trial. *Lancet Infect. Dis.* 2021;10 doi: 10.1016/s1473-3099(21)00200-0.
76. Kasturi S.P., Rasheed M.A.U., Havenar-Daughton C., Pham M., Legere T., Sher Z.J., Kovalenkov Y., Gumber S., Huang J.Y., Gottardo R., et al. 3M-052, a synthetic TLR-7/8 agonist, induces durable HIV-1 envelope–specific plasma cells and humoral immunity in nonhuman

primates. *Sci. Immunol.* 2020;5:eabb1025. doi: 10.1126/sciimmunol.abb1025.

77. Gregg K.A., Harberts E., Gardner F.M., Pelletier M.R., Cayatte C., Yu L., McCarthy M.P., Marshall J.D., Ernst R.K. Rationally Designed TLR4 Ligands for Vaccine Adjuvant Discovery. *mBio.* 2017;8:e00492-17. doi: 10.1128/mBio.00492-17.
78. Ruterbusch M., Pruner K.B., Shehata L., Pepper M. In Vivo CD4+ T Cell Differentiation and Function: Revisiting the Th1/Th2 Paradigm. *Annu. Rev. Immunol.* 2020;38:705–725. doi: 10.1146/annurev-immunol-103019-085803.
79. Mendez S., Tabbara K., Belkaid Y., Bertholet S., Verthelyi D., Klinman D., Seder R.A., Sacks D.L. Coinjection with CpG-Containing Immunostimulatory Oligodeoxynucleotides Reduces the Pathogenicity of a Live Vaccine against Cutaneous Leishmaniasis but Maintains Its Potency and Durability. *Infect. Immun.* 2003;71:5121–5129. doi: 10.1128/IAI.71.9.5121-5129.2003.
80. Verthelyi D., Kenney R.T., Seder R.A., Gam A.A., Friedag B., Klinman D.M. CpG Oligodeoxynucleotides as Vaccine Adjuvants in Primates. *J. Immunol.* 2002;168:1659–1663. doi: 10.4049/jimmunol.168.4.1659.
81. Vogel A.J., Brown D.M. Single-Dose CpG Immunization Protects Against a Heterosubtypic Challenge and Generates Antigen-Specific Memory T Cells. *Front. Immunol.* 2015;6:327. doi: 10.3389/fimmu.2015.00327.
82. Halperin S.A., Ward B., Cooper C., Predy G., Diaz-Mitoma F., Dionne M., Embree J., McGeer A., Zickler P., Moltz K.-H., et al. Comparison of safety and immunogenicity of two doses of investigational hepatitis B virus surface antigen co-administered with an immunostimulatory phosphorothioate oligodeoxyribonucleotide and three doses of a licensed hepatitis B vaccine in healthy adults 18–55 years of age. *Vaccine.* 2012;30:2556–2563. doi: 10.1016/j.vaccine.2012.01.087.
83. Cooper C.L., Angel J.B., Seguin I., Davis H.L., Cameron B. CPG 7909 Adjuvant plus Hepatitis B Virus Vaccination in HIV-Infected Adults Achieves Long-Term Seroprotection for Up to 5 Years. *Clin. Infect. Dis.* 2008;46:1310–1314. doi: 10.1086/533467.
84. Leroux-Roels I., Forgus S., De Boever F., Clement F., Demoitié M.-A., Mettens P., Moris P., Ledent E., Leroux-Roels G., Ofori-Anyinam O. Improved CD4+ T cell responses to Mycobacterium tuberculosis in PPD-negative adults by M72/AS01 as compared to the M72/AS02

and Mtb72F/AS02 tuberculosis candidate vaccine formulations: A randomized trial. *Vaccine.* 2013;31:2196–2206. doi: 10.1016/j.vaccine.2012.05.035.

85. Gordon D.M., McGovern T., Krzych U., Cohen J.C., Schneider I., Lachance R., Heppner D.G., Yuan G., Hollingdale M., Slaoui M., et al. Safety, Immunogenicity, and Efficacy of a Recombinantly Produced Plasmodium falciparum Circumsporozoite Protein-Hepatitis B Surface Antigen Subunit Vaccine. *J. Infect. Dis.* 1995;171:1576–1585. doi: 10.1093/infdis/171.6.1576.
86. Kester K.E., Cummings J.F., Ofori-Anyinam O., Ockenhouse C.F., Krzych U., Moris P., Schwenk R., Nielsen R.A., Debebe Z., Pinelis E., et al. Randomized, Double-Blind, Phase 2a Trial of Falciparum Malaria Vaccines RTS,S/AS01B and RTS,S/AS02A in Malaria-Naive Adults: Safety, Efficacy, and Immunologic Associates of Protection. *J. Infect. Dis.* 2009;200:337–346. doi: 10.1086/600120.
87. Coccia M., Collignon C., Hervé C., Chalon A., Welsby I., Detienne S., Van Helden M.J., Dutta S., Genito C., Waters N.C., et al. Cellular and molecular synergy in AS01-adjuvanted vaccines results in an early IFNγ response promoting vaccine immunogenicity. *npj Vaccines.* 2017;2:1–14. doi: 10.1038/s41541-017-0027-3.
88. Garçon N., Di Pasquale A. From discovery to licensure, the Adjuvant System story. *Hum. Vaccines Immunother.* 2017;13:19–33. doi: 10.1080/21645515.2016.1225635.
89. Didierlaurent A.M., Morel S., Lockman L., Giannini S.L., Bisteau M., Carlsen H., Kielland A., Vosters O., Vanderheyde N., Schiavetti F., et al. AS04, an Aluminum Salt- and TLR4 Agonist-Based Adjuvant System, Induces a Transient Localized Innate Immune Response Leading to Enhanced Adaptive Immunity. *J. Immunol.* 2009;183:6186–6197. doi: 10.4049/jimmunol.0901474.
90. Lee Y.-N., Kim M.-C., Lee Y.-T., Hwang H.S., Cho M.-K., Lee J.S., Ko E.-J., Kwon Y.-M., Kang S.-M. AS04-adjuvanted virus-like particles containing multiple M2 extracellular domains of influenza virus confer improved protection. *Vaccine.* 2014;32:4578–4585. doi: 10.1016/j.vaccine.2014.06.040.
91. Anderson E.J., Rouphael N.G., Widge A.T., Jackson L.A., Roberts P.C., Makhene M., Chappell J.D., Denison M.R., Stevens L.J., Pruijssers

A.J., et al. Safety and Immunogenicity of SARS-CoV-2 mRNA-1273 Vaccine in Older Adults. *N. Engl. J. Med.* 2020;383:2427–2438. doi: 10.1056/NEJMoa2028436.

92. Grencis R.K. Immunity to Helminths: Resistance, Regulation, and Susceptibility to Gastrointestinal Nematodes. *Annu. Rev. Immunol.* 2015;33:201–225. doi: 10.1146/annurev-immunol-032713-120218.
93. Turner J., Faulkner H., Kamgno J., Cormont F., Van Snick J., Else K., Grencis R.K., Behnke J.M., Boussinesq M., Bradley J. Th2 Cytokines Are Associated with Reduced Worm Burdens in a Human Intestinal Helminth Infection. *J. Infect. Dis.* 2003;188:1768–1775. doi: 10.1086/379370.
94. Anthony R.M., Rutitzky L.I., Jr J.F.U., Stadecker M.J., Gause W.C. Protective immune mechanisms in helminth infection. *Nat. Rev. Immunol.* 2007;7:975–987. doi: 10.1038/nri2199.
95. Ii W.A.R., Okragly A.J., Patel C.N., Benschop R.J. IL-33 released by alum is responsible for early cytokine production and has adjuvant properties. *Sci. Rep.* 2015;5:srep13146. doi: 10.1038/srep13146.
96. Marrack P., McKee A.S., Munks M.W. Towards an understanding of the adjuvant action of aluminium. *Nat. Rev. Immunol.* 2009;9:287–293. doi: 10.1038/nri2510.
97. Boulanger D., Reid G., Sturrock R., Wolowczuk I., Balloul J., Grézel D., Pierce R., Otieno M., Guerret S., Grimaud J., et al. Immunization of mice and baboons with the recombinant Sm28GST affects both worm viability and fecundity after experimental infection with Schistosoma mansoni. *Parasite Immunol.* 1991;13:473–490. doi: 10.1111/j.1365-3024.1991.tb00545.x.
98. Riveau G., Deplanque D., Remoué F., Schacht A.-M., Vodougnon H., Capron M., Thiry M., Martial J., Libersa C., Capron A. Safety and Immunogenicity of rSh28GST Antigen in Humans: Phase 1 Randomized Clinical Study of a Vaccine Candidate against Urinary Schistosomiasis. *PLoS Negl. Trop. Dis.* 2012;6:e1704. doi: 10.1371/journal.pntd.0001704.
99. Diemert D.J., Freire J., Valente V., Fraga C.G., Talles F., Grahek S., Campbell D., Jariwala A., Periago M.V., Enk M., et al. Safety and immunogenicity of the Na-GST-1 hookworm vaccine in Brazilian and American adults. *PLoS Negl. Trop. Dis.* 2017;11:e0005574. doi: 10.1371/journal.pntd.0005574.

100. Mattsson J., Schön K., Ekman L., Fahlén-Yrlid L., Yrlid U., Lycke N.Y. Cholera toxin adjuvant promotes a balanced Th1/Th2/Th17 response independently of IL-12 and IL-17 by acting on Gsα in CD11b+ DCs. *Mucosal Immunol.* 2014;8:815–827. doi: 10.1038/mi.2014.111.

101. Marinaro M., Staats H., Hiroi T., Jackson R.J., Coste M., Boyaka P., Okahashi N., Yamamoto M., Kiyono H., Bluethmann H., et al. Mucosal adjuvant effect of cholera toxin in mice results from induction of T helper 2 (Th2) cells and IL-4. *J. Immunol.* 1995;155:4621–4629.

102. Li P., Wang F. Polysaccharides: Candidates of promising vaccine adjuvants. *Drug Discov. Ther.* 2015;9:88–93. doi: 10.5582/ddt.2015.01025.

103. Dong S.-F., Chen J.-M., Zhang W., Sun S.-H., Wang J., Gu J.-X., Boraschi D., Qu D. Specific immune response to HBsAg is enhanced by β-glucan oligosaccharide containing an α-(1→3)-linked bond and biased towards M2/Th2. *Int. Immunopharmacol.* 2007;7:725–733. doi: 10.1016/j.intimp.2007.01.004.

104. Knudsen N.P., Olsen A., Buonsanti C., Follmann F., Zhang Y., Coler R.N., Fox C.B., Meinke A., D'oro U., Casini D., et al. Different human vaccine adjuvants promote distinct antigen-independent immunological signatures tailored to different pathogens. *Sci. Rep.* 2016;6:19570. doi: 10.1038/srep19570.

105. Ye P., Garvey P.B., Zhang P., Nelson S., Bagby G., Summer W.R., Schwarzenberger P., Shellito J.E., Kolls J.K. Interleukin-17 and Lung Host Defense againstKlebsiella pneumoniaeInfection. *Am. J. Respir. Cell Mol. Biol.* 2001;25:335–340. doi: 10.1165/ajrcmb.25.3.4424.

106. Happel K.I., Dubin P.J., Zheng M., Ghilardi N., Lockhart C., Quinton L., Odden A.R., Shellito J.E., Bagby G.J., Nelson S., et al. Divergent roles of IL-23 and IL-12 in host defense against Klebsiella pneumoniae. *J. Exp. Med.* 2005;202:761–769. doi: 10.1084/jem.20050193.

107. Aujla S.J., Chan Y.R., Zheng M., Fei M., Askew D.J., Pociask D., Reinhart T., McAllister F., Edeal J., Gaus K., et al. IL-22 mediates mucosal host defense against Gram-negative bacterial pneumonia. *Nat. Med.* 2008;14:275–281. doi: 10.1038/nm1710.

108. Huang W., Na L., Fidel P.L., Schwarzenberger P. Requirement of Interleukin-17A for Systemic Anti–Candida albicansHost Defense in Mice. *J. Infect. Dis.* 2004;190:624–631. doi: 10.1086/422329.

109. Khader S., Bell G.K., Pearl J., Fountain J.J., Rangel-Moreno J., Cilley G., Shen F., Eaton S.M., Gaffen S.L., Swain S.L., et al. IL-23 and IL-17 in the establishment of protective pulmonary CD4+ T cell responses after vaccination and during Mycobacterium tuberculosis challenge. *Nat. Immunol.* 2007;8:369–377. doi: 10.1038/ni1449.

110. Milner J.D., Brenchley J.M., Laurence A., Freeman A.F., Hill B.J., Elias K., Kanno Y., Spalding C., Elloumi H.Z., Paulson M.L., et al. Impaired TH17 cell differentiation in subjects with autosomal dominant hyper-IgE syndrome. *Nat. Cell Biol.* 2008;452:773–776. doi: 10.1038/nature06764.

111. Yasuda K., Takeuchi Y., Hirota K. The pathogenicity of Th17 cells in autoimmune diseases. *Semin. Immunopathol.* 2019;41:283–297. doi: 10.1007/s00281-019-00733-8.

112. McDonald D.R. TH17 deficiency in human disease. *J. Allergy Clin. Immunol.* 2012;129:1429–1435. doi: 10.1016/j.jaci.2012.03.034.

113. Liang S.C., Tan X.-Y., Luxenberg D.P., Karim R., Dunussi-Joannopoulos K., Collins M., Fouser L.A. Interleukin (IL)-22 and IL-17 are coexpressed by Th17 cells and cooperatively enhance expression of antimicrobial peptides. *J. Exp. Med.* 2006;203:2271–2279. doi: 10.1084/jem.20061308.

114. Lindenstrøm T., Woodworth J., Dietrich J., Aagaard C., Andersen P., Agger E.M. Vaccine-Induced Th17 Cells Are Maintained Long-Term Postvaccination as a Distinct and Phenotypically Stable Memory Subset. *Infect. Immun.* 2012;80:3533–3544. doi: 10.1128/IAI.00550-12.

115. Gopal R., Rangel-Moreno J., Slight S., Lin Y., Nawar H.F., Junecko B.A.F., Reinhart T., Kolls J., Randall T.D., Connell T.D., et al. Interleukin-17-dependent CXCL13 mediates mucosal vaccine–induced immunity against tuberculosis. *Mucosal Immunol.* 2013;6:972–984. doi: 10.1038/mi.2012.135.

116. Datta S.K., Sabet M., Nguyen K.P.L., Valdez P.A., González-Navajas J.M., Islam S., Mihajlov I., Fierer J., Insel P.A., Webster N.J., et al. Mucosal adjuvant activity of cholera toxin requires Th17 cells and protects against inhalation anthrax. *Proc. Natl. Acad. Sci. USA.* 2010;107:10638–10643. doi: 10.1073/pnas.1002348107.

117. Larena M., Holmgren J., Lebens M., Terrinoni M., Lundgren A. Cholera Toxin, and the Related Nontoxic Adjuvants mmCT and dmLT, Promote Human Th17 Responses via Cyclic AMP–Protein

Kinase A and Inflammasome-Dependent IL-1 Signaling. *J. Immunol.* 2015;194:3829–3839. doi: 10.4049/jimmunol.1401633.

118. Fukuyama Y., Yuki Y., Katakai Y., Harada N., Takahashi H., Takeda S., Mejima M., Joo S., Kurokawa S., Sawada S., et al. Nanogel-based pneumococcal surface protein A nasal vaccine induces microRNA-associated Th17 cell responses with neutralizing antibodies against Streptococcus pneumoniae in macaques. *Mucosal Immunol.* 2015;8:1144–1153. doi: 10.1038/mi.2015.5.
119. Yuki Y., Nochi T., Kong I.G., Takahashi H., Sawada S.-I., Akiyoshi K., Kiyono H. Nanogel-based antigen-delivery system for nasal vaccines. *Biotechnol. Genet. Eng. Rev.* 2013;29:61–72. doi: 10.1080/02648725.2013.801226.
120. Kong I.G., Sato A., Yuki Y., Nochi T., Takahashi H., Sawada S., Mejima M., Kurokawa S., Okada K., Sato S., et al. Nanogel-Based PspA Intranasal Vaccine Prevents Invasive Disease and Nasal Colonization by Streptococcus pneumoniae. *Infect. Immun.* 2013;81:1625–1634. doi: 10.1128/IAI.00240-13.
121. Duluc D., Joo H., Ni L., Yin W., Upchurch K., Li D., Xue Y., Klucar P., Zurawski S., Zurawski G., et al. Induction and Activation of Human Th17 by Targeting Antigens to Dendritic Cells via Dectin-1. *J. Immunol.* 2014;192:5776–5788. doi: 10.4049/jimmunol.1301661.
122. Wei W.-C., Su Y.-H., Chen S.-S., Sheu J.-H., Yang N.-S. GM-CSF plays a key role in zymosan-stimulated human dendritic cells for activation of Th1 and Th17 cells. *Cytokine.* 2011;55:79–89. doi: 10.1016/j.cyto.2011.03.017.
123. Higashi T., Hashimoto K., Takagi R., Mizuno Y., Okazaki Y., Tanaka Y., Matsushita S. Curdlan Induces DC-Mediated Th17 Polarization via Jagged1 Activation in Human Dendritic Cells. *Allergol. Int.* 2010;59:161–166. doi: 10.2332/allergolint.09-OA-0103.

CHAPTER 9

From COVID-19 to Cancer mRNA Vaccines: Moving From Bench to Clinic in the Vaccine Landscape

Chiranjib Chakraborty[1, 2], Ashish Ranjan Sharma[2], Manojit Bhattacharya[3], and Sang-Soo Lee [2]

[1] Department of Biotechnology, School of Life Science and Biotechnology, Adamas University, Kolkata, India

[2] Institute for Skeletal Aging & Orthopedic Surgery, Hallym University-Chuncheon Sacred Heart Hospital, Chuncheon, Gangwon-do, South Korea

[3] Department of Zoology, Fakir Mohan University, Odisha, India

ABSTRACT

Recently, mRNA vaccines have become a significant type of therapeutic and have created new fields in the biopharmaceutical industry. mRNA vaccines are promising next-generation vaccines that have introduced a new age in vaccinology. The recent approval of two COVID-19 mRNA vaccines

Citation: Chakraborty C, Sharma AR, Bhattacharya M and Lee SS (2021) From COVID-19 to Cancer mRNA Vaccines: Moving From Bench to Clinic in the Vaccine Landscape. Front. Immunol. 12:679344. doi: 10.3389/fimmu.2021.679344

(mRNA-1273 and BNT162b2) has accelerated mRNA vaccine technology and boosted the pharmaceutical and biotechnology industry. These mRNA vaccines will help to tackle COVID-19 pandemic through immunization, offering considerable hope for future mRNA vaccines. Human trials with data both from mRNA cancer vaccines and mRNA infectious disease vaccines have provided encouraging results, inspiring the pharmaceutical and biotechnology industries to focus on this area of research. In this article, we discuss current mRNA vaccines broadly in two parts. In the first part, mRNA vaccines in general and COVID-19 mRNA vaccines are discussed. We presented the mRNA vaccine structure in general, the different delivery systems, the immune response, and the recent clinical trials for mRNA vaccines (both for cancer mRNA vaccines and different infectious diseases mRNA vaccines). In the second part, different COVID-19 mRNA vaccines are explained. Finally, we illustrated a snapshot of the different leading mRNA vaccine developers, challenges, and future prospects of mRNA vaccines.

Keywords: mRNA vaccines, mRNA-1273, BNT162b2, mRNA vaccine developers, COVID-19

INTRODUCTION

SARS-CoV-2 has rapidly created a worldwide pandemic, leading to significant health challenges and economic burdens for every country while also causing severe morbidity and mortality. To date, no approved proper treatments or therapeutic choice is available for this virus. Thousands of clinical studies have been registered to discover effective treatments. Simultaneously, this situation has created an urgent need for vaccine development. Vaccines are the most promising solution to fight against the pandemic. Several vaccine candidates are being developed to reduce morbidity and mortality and stop the pandemic. In total, 321 vaccine candidates have been documented from the global R&D setting for the development of COVID-19 vaccines (1). In the developmental phase of the COVID-19 vaccine, a broad range of vaccine approaches is being used, including traditional approaches and next-generation approaches. Several traditional approaches have been used to develop COVID-19 vaccines, such as live coronavirus vaccines, inactivated virus vaccines, and subunit vaccines (2). Similarly, next-generation vaccines for COVID-19 can be divided into protein- or peptide-based vaccines and nucleic acid-based vaccines (3). Nucleic acid-based vaccines are categorized into DNA vaccines and RNA vaccines. Previous studies have shown that conventional vaccine strategies, such as live attenuated vaccines, inactivated vaccines or subunit vaccines, may protect against a range of infectious diseases in the long term (4). In the case of the COVID-19 vaccine, vaccine candidates must be developed more rapidly and in large quantities. Additionally, vaccine candidates must be more effective in the fight against the pandemic. Therefore, mRNA-based vaccines are a more promising choice compared to conventional vaccine strategies because mRNA vaccine candidates have the capacity for rapid development with high effectiveness. These vaccine candidates also have the potential for low-cost manufacturing and safer administration. Therefore, mRNA vaccines have revolutionized the vaccinology field by addressing all of the current challenges (5, 6).

The mRNA vaccine development approach is developing quickly (**Figure 1**). Significant research investment in this field has allowed mRNA to become a potential candidate in the immunization landscape. Several major technological innovations have been developed in this area, and pre-clinical research data have been developed and accumulated during the last several years (5, 7). The first successful experiment was published in 1990. In this research, Wolff et al. successfully injected mRNA reporter genes into mouse skeletal muscle cells, and protein production was observed, documenting the first attempt at mRNA *in vivo* expression. This experiment demonstrated a successful method for mRNA vaccine development (8). Subsequently, several studies were performed on mRNA-based therapeutic development (**Figure 2**). Vasopressin mRNA was injected into a rat model to understand the uptake, transport, and expression of this mRNA (9). Several other significant innovations were performed that addressed problems in mRNA vaccine development. One of the important milestones was the assimilation of pseudouridine into mRNA, which provides biological stability and increased translational capacity (10). Another important discovery was optimizing the mRNA coding sequences. In this work, Thess et al. performed sequence engineering of erythropoietin (EPO) mRNA (11). However, codon optimization is not required for mammalian viruses and tumour antigens.

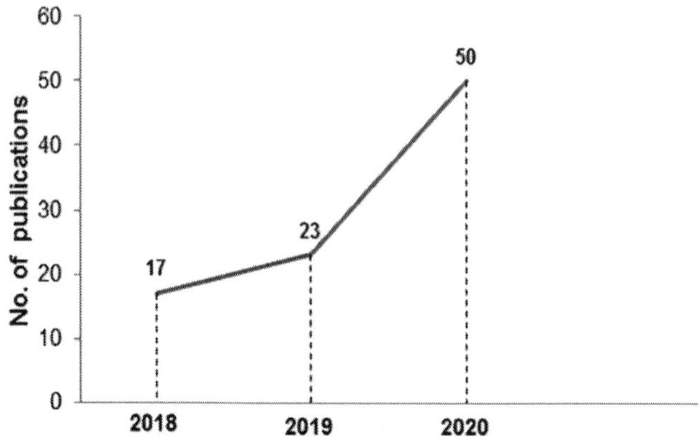

Figure 1: The number of publications in PubMed in the mRNA vaccine area from 2018 to 2020. The PubMed search was performed using the "mRNA vaccine" keyword on 10th Jan 2021.

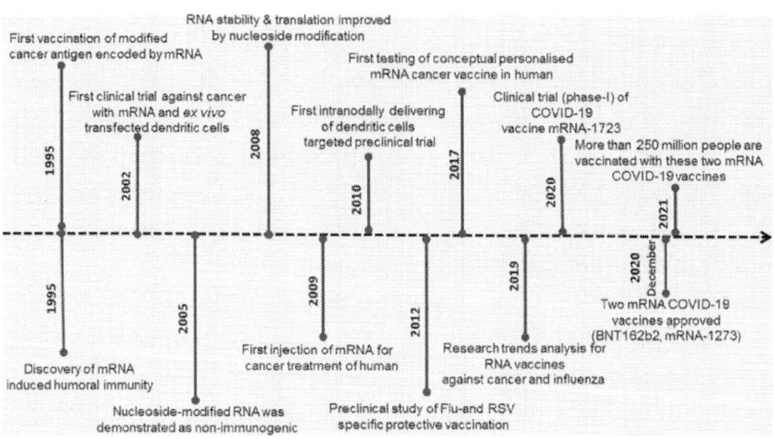

Figure 2: Timeline of the research breakthrough and progression of mRNA vaccine.

Simultaneously, high-performance liquid chromatography (HPLC) purification of mRNA was performed to identify the contaminants, leading to the generation of pure and therapeutic grade mRNA (12). The first mRNA vaccine entered into a Phase I trial with 13 study subjects in 2011, using a prostate-specific antigen RNA-based vaccine. In this study, mRNA-based prostate-specific antigens were transferred into dendritic cells, which were able to induce *in vitro* T cell-mediated antitumour immune responses (10, 11). Subsequently, several mRNA vaccines have been registered in clinical trials in recent years. Along with the other mRNA vaccine, two COVID-19 mRNA vaccines were developed very rapidly to fight against the pandemic.

In this review, we discuss all forms of mRNA vaccines from COVID-19 to cancer. First, we discuss the preparation of mRNA vaccines in general. Second, we discuss the different delivery systems for mRNA vaccines. Third, we discuss the immune response landscape of mRNA vaccines in general. Fourth, we discuss recent clinical trials of mRNA vaccines. Fifth, we discuss different COVID-19 mRNA vaccines. Finally, we illustrate the different leading mRNA vaccine developers as well as the future prospects of mRNA vaccines.

IN VITRO SYNTHESIS OF MRNA AND ENGINEERING SEQUENCES FOR MRNA VACCINE DEVELOPMENT

In general, mRNA vaccines code the antigen of interest, which contains 5′ and 3′ untranslated regions (UTRs). However, two types of mRNA vaccine constructs are available: nonreplicating mRNA (NRM) vaccine constructs and self-amplifying mRNA (SAM) vaccine constructs. In both cases, there was a universal 5′ cap, 5′ untranslated regions (UTRs), an open reading frame (ORF), 3′ untranslated regions (UTRs) and a 3′ poly(A) tail (Figure 3) (5, 7, 13). The ideal structure is described below.

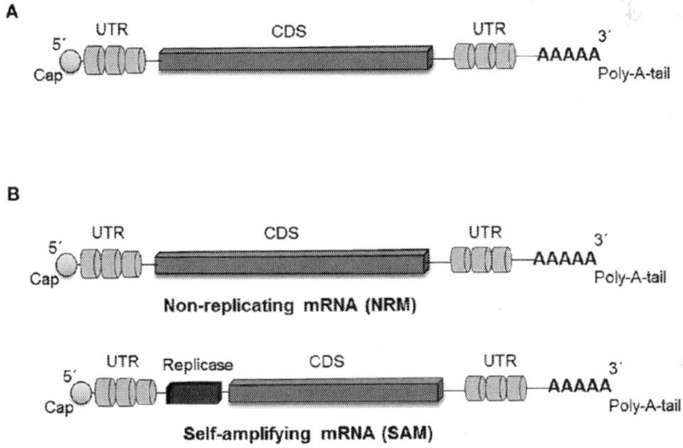

Figure 3: Critical parts of mRNA vaccine construct (A) Critical parts of mRNA vaccine constructs in general (B) Critical parts of two specific types of mRNA vaccine constructs [Non-replicating mRNA (NRM) vaccine and Self-amplifying mRNA (SAM) vaccine].

Ideal Structure of mRNA Vaccine Constructs

5′Cap Structure

CAP structure is an indispensable part of the eukaryotic mRNA. All eukaryotic mRNA has a cap structure, containing an evolutionarily conserved N7-methylated guanosine associated with the first nucleotide of the RNA (14). In the mRNA, the m7GpppN structure is added at the 5′ end as an mRNA cap with numerous functions (15). The cap protects the mRNA from quick degradation. Also, it helps the binding of the initiation factor eIF. There are three types of CAP structures: Cap 0, Cap 1, and Cap2 (16). Cap 0 [m7G(5')pppN1pN2p] can recruit eIF and helps to prevent mRNA degradation. Cap 0 also helps to stimulate interferon (IFN) mediated responses (17). Cap 1 [m7G(5')pppN1mpNp] is usually created by the methylation of the 2′-hydroxyl group of the Cap 0. Cap 1 is commonly found in cytoplasmic viruses. Cap 2 [m7G(5')pppN1mpN2mp] can be generated with an additional 2′-O-methylation of the Cap 1 (6, 14). However, the cap 2 function is still unclear though it has been known that approximately half of all mRNAs possess cap 2. Another cap has also been observed, i.e., m6Am cap. It is reported that cap 0 intermediate and other cap structures are available in eukaryotic mRNAs (18, 19). The cap structure is located on the 5′ end and impacts even protein production (6). In eukaryotes, there are two types of caps: Cap 0 and Cap 1 structures. Following the natural mRNA sequence, adding a regular cap [7-methylguanosine (m7G) cap] is required with the 5′ end of the mRNA sequence, which is an m7GpppN structure (7, 20). Therefore, an analog of a synthetic cap is added to the mRNA during the mRNA vaccine development.

5′ Untranslated Regions

The structures, length, and regulatory elements are significant for mRNA, and these factors regulate the translational efficiency of mRNA.

Coding Sequence

The coding sequence of the mRNA vaccine is crucial. Therefore, the coding sequences are modified through codon optimization, which improves the expression of the CDS.

Codon usage is a significant factor that influences protein translation. However, codon optimization is performed during mRNA vaccine CDS design, which can replace rare codons with synonymous codons. In this case, abundantly available codons are used, which are frequent cognate tRNAs accessible in the cytosol. However, this method can augment the production of protein from mRNA (21), and the accuracy of this model has been questioned (22). Another method can be used during mRNA vaccine development that can enhance steady-state mRNA through the enrichment of G:C content (guanine and cytosine content) *in vitro* (23).

3′ Untranslated Regions

The 3' UTR is a significant part of the mRNA structure that also helps regulate the translational efficiency, similar to 5' UTRs.

3′ Poly(A) Tail

The 3′ poly(A) tail is important for translation as it protects mRNA molecules and it is also a significant part of the mRNA vaccine structure. The poly(A) tail plays a crucial role in the translation of mRNA by regulating the stability of mRNA (24). Therefore, the best possible length of the poly(A) tail must be included in mRNA through various options, such as using poly(A) polymerase or from the encoding DNA template (25).

Types of mRNA Vaccine Constructs

Non-Replicating mRNA Vaccine Constructs

The NRM is also called the conventional mRNA vaccine (26, 27). The NRM vaccine constructs contain conventional mRNA vaccine sequences, such as the universal 5′ Cap, 5′ untranslated regions (UTRs), an open reading frame (ORF), 3′ untranslated regions (UTRs) and a 3′ poly(A) tail (28, 29). The significant advantage of the NRM vaccine is simplicity. Another advantage of the vaccine is the comparatively small size of the mRNA molecule. Alternatively, limited activity and stability have been observed with the NRM vaccine construct *in vivo*, which is one of the disadvantages of the vaccine (26). However, the optimization of the structural elements of RNA molecules can augment antigen expression and the durability of antigen expression (Figure 4) (30).

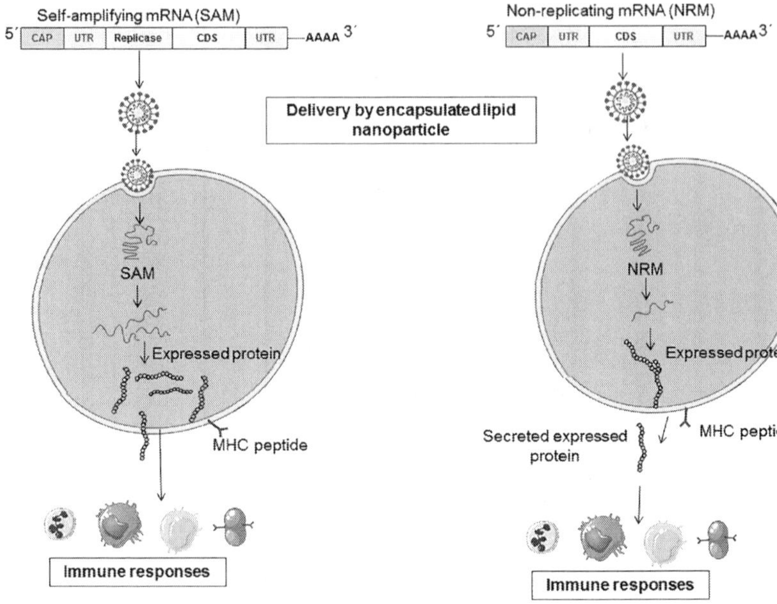

Figure 4: A schematic diagram which explains the pathways from vaccination to immune system activation for the two categories of mRNA vaccine [One for SAM vaccine constructs and other for NRM vaccine constructs].

Self-Amplifying mRNA Vaccine Constructs

SAM vaccines are normally constructed and obtained from the engineered RNA of positive-sense single-stranded RNA viruses, such as picornaviruses, flaviviruses, and alphaviruses (31, 32). Alternatively, negative-sense ssRNA viruses have been used to develop mRNA vaccines to protect against measles viruses and rhabdoviruses (33). However, the reverse genetics technique is required to construct the mRNA vaccine from negative-sense RNA genomes. In this case, cell culture-based systems are needed.

The SAM construct contains all of the components of the NRM construct. Additionally, there is an extra component in the SAM construct that encodes an extra replicase component. This component can direct intracellular mRNA amplification (6, 29).

The replication component of the SAM vaccine is generated by substituting viral structural genes that are inserted with antigen-specific genes. Therefore, after delivery, the SAM vaccine components are capable of high levels of amplification in the cytoplasm of target cells, causing high levels of antigen expression (**Figure 4**) (17). The RNA-dependent RNA polymerase (RdRP) complex is in the gene of interest for SAM vaccine constructs as a replicase component (34).

Trans-Amplifying mRNA Vaccine

Another type of mRNA vaccine that has recently been studied is called trans-amplifying RNA vaccines. In this type, a replicase can amplify the RNAs "in trans". The replicase component is offered "in trans", meaning that two genes act together but on dissimilar RNAs. Self-amplifying RNAs or nonreplicating mRNAs are involved (35). Recently, researchers developed a trans-amplifying RNA system (taRNA system). In this case, replicase-encoding RNA is produced first. Then, a trans-replicon of the antigen of interest is added. Using this influenza vaccine at low doses, animals were protected with the virus (36).

DIFFERENT DELIVERY SYSTEM FOR MRNA VACCINE

Efficient mRNA delivery is a significant factor in the therapeutic success of mRNA vaccines. A good delivery system helps mRNA vaccines achieve full therapeutic potential. Naked RNA is prone to nuclease degradation and has difficulty crossing cross the cell membrane because it is negatively charged and is a large molecule. However, several mRNA vaccine delivery strategies have used different delivery strategies: naked mRNA delivery, mRNA delivery through viral vectors, mRNA delivery through polymer-based vectors, mRNA delivery through lipid-based vectors, mRNA delivery through lipid-polymer hybrid nanoparticles, and mRNA delivery through peptide-based vectors (Table 1) (16).

Table 1: Different delivery system for the different type of mRNA vaccine

Sl. no	miRNA vaccine type		Route of entry	Disease	Refer-ence
1.	Vector based mRNA	Lipid nanoparticles	Intravenous	Anaemia	(37)
2.				Melanoma	(38)
3.			Subcutaneous	AIDS	(39)
4.			Nasal pumping	Cystic fibrosis	(40)
5.			Intramuscular	Respiratory syncy-tial virus infection	(41)
6.				ZIKV	(42)
7.				H10N8 and H7N9	(43, 44)
8.		Polymer-based	Subretinal injections	Retina diseases	(45)
9.			Intravenous	Pulmonary vascu-lar disease	(46)
10.				Anaemia, myelo-dysplasia	(47)
11.			Subcutaneous	AIDS	(48)
12.				Muscle atrophy	(49)
13.		Lipid and polymer hybrid	Intravenous	Ornithine trans-carbamylase deficiency	(50)
14.				Lymphoma	(51)
15.		Protamine-formu-lated	Intradermal	Melanoma	(52)
16.				Prostate cancer	(53)
17.				Non-small cell lung cancer	(54, 55)
18.			Intradermal, intramuscular	Rabies	(56)
19.				Infectious diseases, cancer	(57, 58)
20.			Intradermal	Ovarian cancer	(59)
21.		*Ex vivo* loading of dendritic cells	Subcutaneous	Different tumors	(60)
22.			Intradermal	Acute myeloid leukaemia	(61)

23.	Naked mRNA Intradermal Intranodal	Subcutaneous tumors, intranodal	Cervical cancer	(62)
24.	Intradermal (Electroporation)	Melanoma	(63)	
25.	Gene gun Intradermal (Microneedles)	Cancer	(64)	
26.		–	(65)	
27.		Melanoma	(66)	
28.		–	(67)	

Naked mRNA Delivery

The simplest strategy for the delivery of naked mRNA is administration through intramuscular (i.m.) injection (16), while other administration routes include intradermal (i.d.) injection (68) or subcutaneous (s.c.) injection (27). Additionally, researchers have administered mRNA through dissolvable microneedles (RNA patches) (69), which may be an important method for RNA vaccine delivery. However, there are some disadvantages to naked mRNA delivery. The plasma half-life is short, and naked mRNA is prone to ribonuclease degradation. Alternatively, negatively charged mRNA should be neutralized. Otherwise, the mRNA molecule will not be able to pass through cell membranes (70).

mRNA Delivery Through Viral Vectors

For gene/mRNA delivery, genetically modified viruses have been used for some time. For the delivery of viral RNA, there has been a significant amount of interest in genetic engineering (71). An adeno-associated virus is a viral vector-based delivery system that carries different therapeutic nucleic acid molecules (72). The advantage of RNA viruses is that the virus can replicate and be expressed in the cytoplasm locally and easily. Positive strand RNA viruses can be translated into proteins of interest with the host ribosomal machinery. Therefore, several virus vectors have been developed for mRNA delivery of genes other than adeno-associated viruses, such as flavivirus (73) (e.g., Kunjin virus), picornaviruses (74), and alphaviruses (e.g., Semliki Forest virus and Sindbis) (75). However, there are some disadvantages for mRNA delivery through viral vectors, such as some difficulties with host genome integration and the possibility of host rejection, cytotoxicity, and immunogenicity (76). Wadhwa et al. have also described the drawback of genome integration of mRNA-based vaccine and possible host rejection

(16). Ura et al. stated that sometimes genome integration could lead to cancer (77).

mRNA Delivery Through Lipid Vectors

Lipid-based vectors or lipid-like compounds (lipidoids) are regularly used in mRNA delivery. Naturally occurring lipids and synthetic lipid molecules, such as liposomes or lipid nanoparticles (LNPs), have been used to deliver mRNAs. Liposomes are membrane-bound structures that can be produced through the self-assembly process (78). Dhaliwal et al. used liposomes for mRNA delivery, and cationic liposomes are the preferred system (79). Due to rapid elimination, cationic lipid delivery systems are very challenging (80). However, there are some safety issues of cationic lipids due to their quaternary ammonium head group. Safety issues include immunogenicity and toxicity *in vivo* systems (81) and *in vitro* systems (82). These issues can be solved through the replacement of quaternary ammonium head groups using tripeptide-based lipids (DAO3). Researchers observed that this tripeptide-based lipid has no toxicity *in vivo* or *in vitro* (81).

mRNA Delivery Through Polymer-Based Vectors

The different polymer-based vectors are used from time to time for mRNA delivery. The first polymer used for mRNA delivery was diethylaminoethyl (DEAE) dextran (83, 84). Later, it was observed that lipid-mediated mRNA delivery is more efficient than the DEAE-dextran-mediated mRNA delivery system (85). However, Siewert et al. observed that charge ratio variation can improve mRNA delivery through the mRNA-DEAE-dextran polyplex system (86). Biodegradable polymers are also used for mRNA delivery, such as biopolymeric nanoparticles [e.g., poly(lactic-co-glycolic acid) (PLGA)]. PLGA is a compatible nanostructure (87, 88), and a PLGA-based mRNA delivery system was developed with tolerable toxicity and substantial transfection efficiency (89).

Alternatively, several multifunctional block copolymers have also been used for mRNA delivery, such as poly(ethylene glycol) methacrylate, dimethylaminoethyl methacrylate (DEAEMA), and DEAEMA-co-n-butyl methacrylate (90). However, due to the high molecular weight of some polymers, polymer-based delivery system design is challenging for mRNA delivery system development (70).

mRNA Delivery Through Lipid-Polymer Hybrid Nanoparticles

Lipid-polymer hybrid nanoparticles (LPNs) are effective molecules for mRNA delivery. Zhao et al. developed lipid-like nanoparticle TT3-LLN (N1,N3,N5-tris(2-aminoethyl) benzene-1,3,5-tricarboxamide (TT)-derived lipid-like nanomaterial), which was used for efficient mRNA delivery (91). In another study, successful mRNA delivery was performed in the lungs using LPNs, which consist of the degradable polymer poly(β-amino esters) (PBAEs). In this study, PBAEs with PEG and mRNA formulations were developed, which increased *in vitro* potency and serum stability (92). Conversely, using a lipidoid polymer hybrid, co-delivery of siRNA and mRNA was performed, augmenting the co-delivery (93).

mRNA Delivery Through the Peptide-Based Delivery System

There is increased interest in using peptide-based systems for mRNA delivery due to their versatility (94). Recently, using cell-penetrating peptides (PepFect14), Cerrato et al. performed mRNA delivery for mitochondrial disorders. Researchers have designed a peptide-based delivery technology called mitochondrial peptide-based oligonucleotide technologies, and the technology is promising for treating patients with mitochondrial disorders (95). Conversely, peptide-based delivery systems have some disadvantages, such as targeted cell delivery (59).

DENDRITIC CELLS: A POTENTIAL TARGET FOR DELIVERY OF MRNA VACCINES

Dendritic cells are significant antigen-presenting cells (APCs) in the immune system. Dendritic cells instigate the adaptive immune response through antigen processing. Dendritic cells process antigens offered by major histocompatibility complex (MHC) molecules (MHC class I or MHC class II molecules) to T cells (CD^{8+} T cells or CD^{4+} T cells). MHC class I molecules interact with CD^{8+} T cells, and MHC class II molecules interact with CD^{4+} T cells. Furthermore, dendritic cells may process antigens to B-cells to induce the antibody response (96). Wykes and MacPherson (2000) have described that dendritic cells play a significant role in dendritic cells B-cell interaction, thereby activating B-cell to its function and producing antibodies (97). Harvey et al. illustrated a mechanism by which B-cells interact with dendritic cells during the time of antigen processing. This interaction of B-cells with dendritic cells results in the transfer of B-cell receptors with antigen to the APCs. This antigen transfer may result in immunologic response in a more committed way (98). Heesters et al. also described that follicular dendritic cells might play a significant role in the antigen presentation to B-cells. At the same time, they also illustrated that the membrane-bound antigen presentation has a considerable influence on the B-cell activation and its subsequent stages of B-cell responses (99). Therefore, dendritic cells are a significant target for both *ex vivo* and *in vivo* delivery of mRNA vaccines through the transfection process (100). For cancer vaccination, the *ex vivo* dendritic cell loading is being studied to produce cell-mediated immunity (101).

mRNA VACCINES AND THE IMMUNE RESPONSE LANDSCAPE

mRNA vaccines can activate both the adaptive immune response and innate immune response (Figure 5).

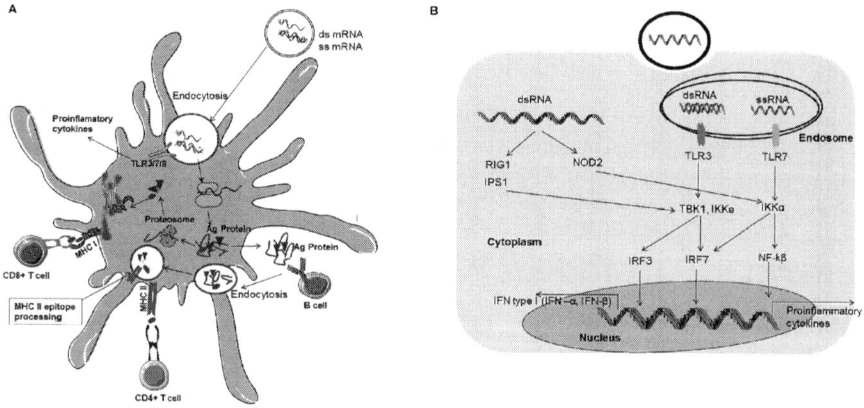

Figure 5: mRNA vaccines based immune response (A) mRNA vaccine based adaptive immune response (B) mRNA vaccine based innate immune response.

mRNA Vaccine and the Adaptive Immune Response

The immune response is activated through mRNA vaccines by two routes. First, after vaccination, mRNAs enter the cytoplasm through endocytosis. Dendritic cells are the most important cells for antigen protein presentation. Therefore, these cells are called antigen-presenting cells (APCs) in the immune system. Several mRNAs unite with host cell ribosomes, and translation occurs effectively. After antigen protein synthesis, the antigen is degraded in the cytoplasm into small antigenic peptides *via* the proteasome. Then, these small antigenic peptides are presented *via* the major histocompatibility complex (MHC) to cytotoxic T lymphocytes (CTLs). Alternatively, antigenic proteins can be released by the host cell. Then, these antigenic proteins can be taken up by dendritic cells, and they are degraded and presented to helper T cells and B-cells through MHC. Finally, MHC class I interacts with CD^{8+} T cells, and MHC class II interact with CD^{4+} T cells to activate them. B-cells can also recognize antigen proteins that are released by dendritic cells. Finally, B-cells release antibodies (70, 102).

mRNA Vaccine and the Innate Immune Response

A self-adjuvant effect has also been noted for mRNA vaccines. In this effect, APCs recognize mRNA, subsequently triggering PRRs (pattern recognition receptors). Pattern recognition receptor members include TLR family members, such as TLR3, TLR7 and TLR8 (58, 103), which are localized in the endolysosomal area of the cell. Receptors in the cytosol can detect nucleic acids in the cytoplasm. ssRNA molecules are recognized by two TLRs: TLR7 and TLR8 receptors. Auridine-rich tetramers are a minimum requirement for both receptors for activation. TLR7-mediated downstream pathway activation aids in type I IFN production (104). AU-rich sequences induce TLR8-mediated downstream pathway activation, leading to a tumour necrosis factor (TNF) response (105). Additionally, dsRNA triggers immune system activation through TLR3 recognition (57, 106). Binding with the TLR3 receptor requires a minimum length of 45 bp dsRNA (107).

Protein families, such as RIG-I, LGP2, and MDA5, function as pattern recognition receptors (108). RIG-I can also recognize dsRNA and ssRNA and activate the downstream pathway, which stimulates IFN production (109–113). mRNA can stimulate the immune response through the TLR pathway, which further stimulates cells to produce augmented amounts of proinflammatory cytokines, type I IFN, and other interferons. These interferons or proinflammatory cytokine molecules degrade RNA, inhibit the translation of mRNA, cause a reduction in CD^{8+} T cells, and ultimately terminate the immune response (5, 7, 10, 114, 115). However, this cascade may produce negative effects for some mRNA vaccines. Therefore, the self-adjuvant property of mRNA vaccines has both disadvantages and advantages.

RECENT CLINICAL TRIAL LANDSCAPE MRNA VACCINES

Several mRNA vaccines are currently registered for clinical trials. Different mRNA vaccines have been developed against cancer and different types of infectious diseases. mRNA vaccine can balance both adaptive as well as innate immune responses. Therefore, the mRNA vaccine can be used for cancer and infectious diseases (116). These mRNA vaccines have been well studied in animal models and human subjects.

mRNA Vaccines for Cancer and Their Pre-Clinical and Clinical Update

Different mRNA-based cancer vaccines were designed to target tumor-associated antigens. These antigens are more prevalent in cancerous cells. The majority of cancer vaccines are therapeutic rather than prophylactic (117). These vaccines may stimulate cell-mediated immune responses. Two decades ago, the proposal of RNA-based cancer vaccines was published (118). Recently, several mRNA-based cancer vaccines have been developed that are registered for different phases of clinical trials (**Table 2**). Nevertheless, few trials were terminated due to lack of efficiency, immunogenicity, toxicity, and other side effects. The terminated clinical trials are also listed in **Table 2**. Due to lack of efficiency, one clinical trial (clinical trial no. NCT01582672) was terminated. The trial was conducted using an mRNA vaccine against carcinoma. Similarly, another clinical trial, an mRNA-based prostate cancer vaccine (clinical trial no. NCT01817738) was terminated as the study's outcome was not as impactful as expected.

Table 2: Different mRNA based cancer vaccines which are registered for different phases of the clinical trial

Vaccine targets	Clinical trials no.	Status (Phase)	Vaccine type	Sponsor/Organization
Prostate cancer	NCT01817738	Terminated (I/II)	RNActivetumour-associated antigen mRNA	CureVac AG
	NCT00831467	Completed (I/II)		
	NCT02140138	Terminated (II)		
	NCT01446731	Completed (II)	Dendritic cell loaded with tumour-associated antigen mRNA	Herlev Hospital
	NCT01197625	Recruiting (I/II)		Oslo University Hospital
	NCT01278914	Completed (I/II)		
	NCT00906243	Terminated (I/II)	RNActivetumour-associated antigen mRNA	University of Florida
Glioblastoma	NCT02649582	Recruiting (I/II)	Dendritic cell electroporated with tumour-associated antigen mRNA	Antwerp University Hospital
	NCT02465268	Recruiting (II)	Dendritic cell loaded with cytomegalovirus antigen mRNA with granulocyte–macrophage colony-stimulating factor protein	University of Florida
	NCT02366728	Active (II)	Dendritic cell loaded with cytomegalovirus antigen mRNA	Duke University
	NCT00626483	Completed (I)		
	NCT02529072	Completed (I)		
	NCT00639639	Active (I)		
	NCT00890032	Completed (I)	Dendritic cell loaded with autologous tumour mRNA	
	NCT02709616	Active (I)	Dendritic cell loaded with tumour-associated antigen mRNA	Guangdong 999 Brain Hospital
	NCT02808364			
	NCT00846456	Completed(I/II)	Dendritic cell loaded with autologous tumour or tumour-associated antigen mRNA	Oslo University Hospital
Pancreatic cancer	NCT00664482	Completed (Not applicable)	Dendritic cell electroporated with autologous tumour mRNA with or without CD40L mRNA	Argos Therapeutics

Melanoma	NCT02035956	Completed (I)	Naked tumour-associated antigen or neo–Ag mRNA	BioNTech RNA Pharmaceuticals GmbH
	NCT01684241			
	NCT02410733	Active (I)	Liposome-complexed tumour-associated antigen mRNA	
	NCT01216436	Terminated	Dendritic cell, matured, loaded with tumour-associated antigen mRNA	Duke University
	NCT01456104	Active (I)	Dendritic cell (Langerhans) electroporated with tumour-associated antigen mRNA	Memorial Sloan Kettering Cancer Center
	NCT01278940	Completed (I/II)	Dendritic cell loaded with autologous tumour or tumour-associated antigen mRNA	Oslo University Hospital
	NCT00961844	Terminated (I/II)		
	NCT02285413	Completed (II)	Dendritic cellelectroporated with tumour or tumour-associated antigen mRNA	Radboud University
	NCT01530698	Completed (I/II)		
	NCT00940004	Completed (I/II)		
	NCT00243529	Completed (I/II)		
	NCT00929019	Terminated (I/II)		
Melanoma	NCT00204516	Completed (I/II)	Autologous tumour mRNA with granulocyte–macrophage colony-stimulating factor protein	University Hospital Tübingen
	NCT00204607	Completed (I/II)	Protamine-complexed tumour-associated antigen mRNA with macrophage colony-stimulating factor protein	
	NCT01983748	Recruiting (III)	Matured Dendritic cell, loaded with autologous tumour RNA	University Hospital Erlangen
	NCT01676779	Completed (II)	Dendritic cellelectroporated with tumour-associated antigen and TriMix mRNA	UniversitairZiekenhuis-Brussel
	NCT01066390	Completed (I)		
	NCT01302496	Completed (II)		
Colorectal cancer	NCT00228189	Completed (I/II)	Dendritic cellelectroporated with tumour-associated antigen mRNA	Radboud University

Ovarian cancer	NCT01334047	Terminated (I/II)	Dendritic cell loaded with autologous tumour or tumour-associated antigen mRNA	Oslo University Hospital
	NCT01456065	Unknown	Matured Dendritic cell, loaded with autologous tumour RNA	Life Research Technologies GmbH
Breast cancer z	NCT02316457	Active (I)	Liposome-formulated tumour-associated antigen and neo−antigen mRNA	BioNTech RNA Pharmaceuticals GmbH
	NCT00978913	Completed (I)	Dendritic cellloaded with tumour-associated antigen mRNA	Herlev Hospital
Acute Myeloid Leukemia	NCT00514189	Terminated (I)	Dendritic cell loaded with Acute Myeloid Leukemia lysate and mRNA	MD Anderson Cancer Center
	NCT01734304	Completed (I/II)	Dendritic cell loaded with tumour-associated antigen and cytomegalovirus antigenmRNA	Ludwig-Maximilian-University of Munich
	NCT00510133	Completed (II)	Dendritic cell loaded with tumour-associated antigen mRNA	AsteriasBiotherapeutics
	NCT00965224	Unknown (II)	Dendritic cellelectroporated with tumour-associated antigen mRNA	Antwerp University Hospital
	NCT01686334	Recruiting (II)		
	NCT00834002	Completed (I)		
	NCT03083054	Active (I/II)	Dendritic cell loaded with tumour-associated antigen mRNA	University of Campinas, Brazil
Brain metastases	NCT02808416	Active (I)	Dendritic cell loaded with tumour-associated antigen mRNA	Guangdong 999 Brain Hospital
Non-small-cell lung cancer	NCT01915524	Terminated (I)	RNActivetumour-associated antigen mRNA	CureVac AG
	NCT00923312	Completed (I/II)		
Renal cell carcinoma	NCT00087984	Completed (I/II)	Dendritic cell electroporated with autologous tumour mRNA with or without CD40L mRNA	Argos Therapeutics
	NCT01482949	Terminated (II)		
	NCT01582672	Terminated (III)		
	NCT00678119	Completed (II)		
	NCT00272649	Completed (I/II)		
Mesothelioma	NCT02649829	Recruiting (I/II)	Dendritic cell electroporated with tumour-associated antigen mRNA	Antwerp University Hospital

Slam et al. developed an mRNA vaccine NP (nanoparticle) with a C16-R848 adjuvant for cancer immunotherapy (119). *Ex vivo* dendritic cell loading is currently a method of interest for mRNA-based cancer vaccine development. This vaccine produces cell-mediated immunity efficiently against cancer. Dendritic cell-based mRNA vaccines for cancers have shown promising results in different phases of clinical trials (120). Recently, Ary and colleagues developed mRNA-lipid nanocomplexes that provide strong immune responses to inhibit B16-OVA tumour progression. The mRNA vaccine was tested in tissue culture and mice by direct local injection and increased mouse survival (121). Presently, researchers are trying to develop different mRNA vaccines for different types of cancer.

mRNA Vaccines for Different Infectious Diseases and Their Pre-Clinical and Clinical Update

Similar to cancer mRNA vaccines, mRNA-based vaccines for different infectious diseases have been studied extensively during the previous two decades. The mRNA-based vaccines developed for different infectious diseases are currently registered for different phases of clinical trials (Table 3). In the case of infectious diseases, mRNA vaccines were designed both for therapeutic use and prophylactic use (27). In a Phase-I clinical trial, an mRNA-based vaccine against rabies virus showed that the mRNA could be complexed with protamine. The *in vivo* study showed that it was well-tolerated and safe. Vaccine efficacy depends on the route of administration and the dose. This study demonstrated that the effectiveness of the vaccine was better when administered with a needle-free intramuscular or intradermal device compared to a direct intramuscular or intradermal needle injection (56). Similarly, Bahl and colleagues formulated lipid nanoparticle (LNP)-based mRNA vaccines using the haemagglutinin proteins H10N8 or H7N9, which were tested in ferrets, mice, and nonhuman primates. In a Phase-I clinical trial, the results showed that mRNA vaccines for H10N8 induced robust prophylactic immunity in human volunteers. In this clinical trial, mild or moderate adverse events were noted, while no serious events were documented (43). Presently, several other mRNA vaccines against different infectious diseases are being developed, which are in phases of pre-clinical and clinical trials.

Table 3: Different mRNA vaccines for different diseases which are registered for clinical trial

Vaccine targets	Clinical trials no.	Status (Phase)	Vaccine type	Sponsor/Organization
HIV−1	NCT02888756	Terminated (II)	Dendritic cell; loaded with viral antigenic mRNA with TriMix	Erasmus Medical Center
HIV−1	NCT00833781	Completed (II)	Dendritic cell loaded with viral antigenic mRNA	Massachusetts General Hospital
HIV−1	NCT00381212	Completed (I/II)	Dendritic cellelectroporated with autologous viral antigen and CD40L mRNAs	McGill University Health Centre
Zika Virus	NCT03014089	Completed (I)	Nucleoside-modified viral antigenic mRNA	Moderna Therapeutics
Influenza	NCT03076385	Completed (I)		
HIV−1	NCT02042248	Completed (I)	Dendritic cellelectroporated with autologous viral antigen and CD40L mRNAs	ArgosTherapeutics
	NCT00672191	Completed (II)		
	NCT01069809	Completed (II)		
HIV−1	NCT02413645	Completed (I)	Viral Antigenic mRNA with TriMix	FundacióClínic per la Recerca Biomèdica
Rabies virus	NCT02241135	Completed (I)	Viral antigenic mRNA (RNActive®)	CureVac AG

Toxicity Issues Related to mRNA Vaccines and Some Approved Vaccines Scenario

Toxicity is the other side of mRNA vaccines. Sometimes therapeutic nucleosides show toxicity. For example, some nucleoside-based anti-cancer drugs and antivirals drugs containing unnatural nucleoside analogs show toxicity (122–124). During pre-clinical studies, liver toxicity was observed in mRNA therapeutic while delivering them through the lipid nanoparticles. The mRNA therapeutic was developed using a lipid nanoparticle-based delivery system for de Crigler-Najjar syndrome. However, during the delivery of the mRNA vaccine, some delivery system or some formulation might have generated toxicity (125). On the other hand, some systemic adverse events (AEs) were observed for a rabies mRNA vaccine during a human clinical trial. It highlights another side of mRNA, which is the inflammatory nature of the mRNA (56).

In general, toxicity is observed during any vaccine or therapeutics development, especially during pre-clinical studies. It may generally occur due to some issues like developing a delivery system or the formulation of the mRNA vaccine. These are the part of the process of mRNA vaccine development which provides stability/more efficacy to the mRNA molecule. However, if toxicity occurs during development processes, it can be rectified by altering the delivery system or changing the vaccine formulation. Therefore, the therapeutic or vaccine development process is a trial and error method where issues like toxicity may happen during the development process.

The toxicity of an mRNA vaccine is verified through the human clinical trials (Phase-I, Phase-II, and Phase-III), a general rule for the drug development process. If the toxicity is not statistically significant, then only the mRNA vaccine receives approval. Occasionally, in very few cases, toxicity is observed during the clinical trial, which is statistically significant. If the safety profile of an mRNA vaccine is not proper or considerable toxicity of the vaccine is found during the clinical trial, the mRNA vaccine is withdrawn from the clinical trial.

Conversely, there are several success stories of mRNA vaccines. In these cases, three RNA vaccines have shown excellent safety profiles in clinical trials. Recent success stories of two COVID-19 mRNA vaccines (Moderna (mRNA-1273) and Pfizer/BioNTech (BNT162b2)) have demonstrated excellent safety and efficacy profile. Another COVID-19 mRNA vaccine, CureVac, has been registered for Phase-III clinical trial, which CureVac AG has developed. It has shown an ideal safety profile during the Phase-I and Phase-II clinical trials. All the safety profiles of mRNA vaccines are illustrated in Table 4. These promising case studies of safety and efficacy profiles will help future manufacturers develop mRNA vaccines with more focus.

Table 4: mRNA vaccines (approved/last part of Phase-III trial) which showing very good safety profile

Sl no.	mRNA vaccine	Disease	Developer	Status	Remark
1.	BNT162b2	COVID-19	BioNTech and Pfizer	The vaccine is approved, several country have started the vaccination program with this vaccine	This vaccine shows excellent safety profile
2.	mRNA-1273	COVID-19	ModernaTX, Inc.	The vaccine is approved, several country have started the vaccination program with this vaccine	This vaccine shows excellent safety profile
3.	CureVac AG	COVID-19	CureVac N.V.	Registered under Phase-III clinical trial (NCT04860258)	The vaccine showed very good safety profile in Phase-I and Phase-II

COVID-19 mRNA VACCINES

Recently, two significant COVID-19 mRNA vaccines have been approved, and others are in development (Figure 6). The two approved mRNA vaccine candidates were developed quickly and changed the developmental history of vaccines.

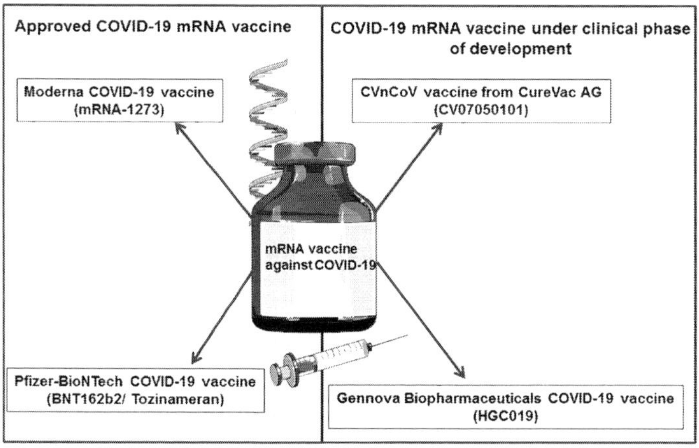

Figure 6: Different significant COVID-19 mRNA vaccines.

BNT162b2 mRNA Vaccine from Pfizer and BioNTech

One mRNA-based vaccine was developed jointly by BioNTech and Pfizer against COVID-19, marked as the brand name tozinameran, and the vaccine is also called BNT162b2 (126). It is a modified mRNA vaccine that was developed from the optimized version of the entire spike protein of SARS-CoV-2. This vaccine was formulated with lipid nanoparticles and can be administered through intramuscular injection in two doses, and the two doses can be administered three weeks apart. This vaccine was evaluated through clinical trials conducted in Germany (clinical trial no. NCT04380701) and the USA (clinical trial no. NCT04368728).

In a Phase-I clinical trial, the immunogenicity and safety of the vaccine candidate was evaluated. The results showed neutralizing antibody titres against SARS-CoV-2 and strong antigen-specific T-cell responses (127, 128). The Phase-III clinical trial result was published by Polack et al. (127). In this study, 43,448 human subjects received injections. Among them, 21,728 human subjects received placebo injections, and 21,720 human subjects received BNT162b2 injections. The results show 95% protection for people 16 years of age or older. The vaccine was analysed through a safety profile, which showed headache, fatigue, and mild-to-moderate pain at the injection site (127). This mRNA vaccine was approved by the USA for emergency use. Later, the vaccine was approved by the UK and Canada (126, 129–131).

mRNA Vaccine (mRNA-1273) From ModernaTX

mRNA-1273 is a mRNA vaccine against COVID-19 that was developed by Moderna and two other research institutes (NIAID and BARDA) of the USA. The vaccine is encapsulated with a lipid nanoparticle carrier that encodes the full-length spike protein of the virus. The vaccine is administered by intramuscular injection. This vaccine is given in two doses that can be administered four weeks apart.

The vaccine was studied through pre-clinical research in nonhuman primate animal models, and the animals received 10 or 100 μg of the vaccine. The results showed that the vaccine augments Th1 (type 1 helper T-cell)-biased CD4 T-cell responses. Additionally, undetectable or very low Th2 or CD8 T-cell responses were noted (132).

A Phase-I clinical trial (NCT04283461) was conducted for this vaccine, which was an open-label and dose-ranging clinical trial performed by the NIH, USA (133).

A Phase-I clinical trial (ClinicalTrials.gov number, NCT04470427) was a randomized, placebo-controlled, observer-blinded trial in which persons 18 years of age or older were vaccinated. The participants received a 1:1 ratio of two intramuscular injections of this vaccine (100 μg). This mRNA vaccine showed 94.1% efficacy (134). This mRNA vaccine was first given emergency use authorization (EUA) in the USA (126, 135) and was later approved by Canada and the UK (135, 136).

CVnCoV From CureVac AG

CVnCoV is a mRNA vaccine against COVID-19 that was developed by CureVac AG. This vaccine was enrolled in a Phase-IIb/III placebo-controlled, observer-blinded, randomized, multicentre clinical trial (ClinicalTrials.gov number NCT04652102). A total of 36,500 participants were enrolled in this clinical trial. The vaccine was administered through intramuscular (i.m.) injection on day 1 and day 29 (12 μg of vaccine).

HGC019 From Gennova Biopharmaceuticals

This mRNA vaccine was developed by Gennova Biopharmaceuticals in collaboration with HDT Biotech. This vaccine uses lipid inorganic nanoparticles as a delivery system. This vaccine is currently in a pre-clinical trial stage and will be starting a clinical trial.

Different Leading mRNA Vaccine Developers

Several leading mRNA vaccine developers have been noted, such as Moderna Therapeutics, Argos Therapeutics, Fundació Clínic per la Recerca Biomèdica, CureVac AG, Gennova Biopharmaceuticals, etc.

Moderna Therapeutics

This biopharmaceutical company is well known for its mRNA vaccine development. They have developed several vaccines, such as mRNA-1647 against cytomegalovirus, mRNA-1273 against COVID-19, and mRNA-1893 against Zika. mRNA-1647 was registered for a Phase-II clinical trial, and mRNA-1893 was registered for a Phase-I clinical trial. mRNA-1273 received approval for COVID-19.

Argos Therapeutics

This biopharmaceutical company was founded in 1997 and developed mRNA vaccines against metastatic renal cell carcinoma and HIV-1. Their first vaccine, AGS-003, was developed against metastatic renal cell carcinoma. Their second vaccine, AGS-004, was developed to treat HIV-1 and is registered for a Phase-IIb clinical trial.

CureVac AG

CureVac AG was founded in 2000, and it is a German biopharmaceutical company that has developed several mRNA vaccines. CV7202 is a mRNA vaccine that uses the rabies virus glycoprotein RABV-G. The vaccine was developed for immunization against the rabies virus. They have also developed CVnCoV against COVID-19, which is registered for Phase-III clinical trials (ClinicalTrials.gov number, NCT04860258). CureVac generated mRNA-based prostate cancer vaccines (CV9103 and CV9104) using RNActive®-derived technology (58, 137). CV9103 completed a Phase-I/IIa clinical study for prostate cancer. The vaccine was administered at a dose of 1280 μg (ClinicalTrials.gov number, NCT00906243) (53).

CURRENT UPDATES ON MRNA VACCINE IN LIGHT OF COVID-19 VACCINE AND VACCINATION

The first approved COVID-19 vaccine is an mRNA vaccine, approved by the USA. Two mRNA vaccines were approved in this direction: Pfizer-BioNTech mRNA-based vaccine (BNT162b2) and Moderna mRNA-based vaccine (mRNA-1273). BioNTech and Pfizer, Inc. developed the first one, and the second one was developed by ModernaTX, Inc. Both of the vaccines were developed at breakneck speed. Presently, both of these vaccines are being used by different countries (Canada, United Kingdom, Israel, and Singapore), including the USA, for mass vaccination to their people (138–140).

ANTIGEN ACTIVATION AFTER MRNA VACCINATION: A VIEW IN COVID-19 MRNA VACCINATION SCENARIO

Both of the two COVID-19 mRNA vaccines [Moderna (mRNA-1273) and Pfizer/BioNTech (BNT162b2)] have shown excellent safety and efficacy profile and also achieved approximately 90 to 95% efficacy. These vaccines generated considerable neutralizing antibody(nAb) titres during the clinical trial, a study of over 100,000 participants (141). mRNA may act as an immunogen, and adjuvant has immunostimulatory characteristics and can stimulate innate immunity without severe side effects. RNA strands (ssRNA and dsRNA) are recognized by a variety of cytosolic sensors and endosomal sensors. For example, endosomal TLR7 or TLR3 can attach to endosomal ssRNA. While, inflammasome such as PKR, RIG-I, MDA5, and NOD2 may bind to dsRNA/ssRNA present in the cytosol. It can cause further lead to cellular activation. Consequently, various inflammatory mediators and type I IFN are produced (5). Moreover, these mRNA vaccines can activate MDA5 and TLR7, activating the dendritic Cells. Activation of dendritic cells activates the naive T-cells, which helps to generate antibody-secreting plasma cells (5, 141).

The CDC data demonstrated that very few people had side effects after the first doses of BNT162b2 COVID-19 vaccines. It was noted that among 1,893,360 individuals who received first doses of BNT162b2, only 0.2% reported side effects or unpleasant reactions (4393 individuals) (142, 143).

The side effects are recorded as viral enhanced disease (VED) after COVID vaccination with COVID-19 mRNA vaccines. VED is associated with vaccine-associated enhanced respiratory disease (VAERD) and antibody-dependent enhancement (ADE) (142, 143). VAERD is a distinctive medical syndrome, and it is associated with a considerable amount of non-neutralizing antibody production. It usually activates T_H2 associated immune responses, leading to activation of the complement system. Meanwhile, it also contributes to the generation of the immune complex deposition (144, 145).

On the other hand, during the trial of BNT162b2 (NCT04368728 trial) using 733 patients, malignancies were activated in few patients, which was about 3.9%. In this clinical trial, solid metastatic tumor (4 patients), lymphoma (22 patients), and leukaemia (12 patients) were recorded (128). Hence, several questions are still remains unanswered. For example, does the vaccine is produced after intramuscular or subcutaneous injection leaks to other organs such as the liver or endothelial cells in other sites? Do the antigens spill over to the bloodstream, and which types of cells produce the antigens? These questions need to be resolved immediately to understand more about the COVID mRNA vaccine.

CHALLENGES AND FUTURE PROSPECTS

After the first publications on the delivery of mRNA into animal models, this field has progressed quickly and has shown promise as a next-generation biopharmaceutical for the vaccine development landscape. mRNA vaccines are considered one of the most significant and promising next-generation vaccines due to their rapid development capacity, high potency, safety profile, and low cost of manufacturing. This biopharmaceutical has gained important momentum in the recent few years. Several significant achievements have been observed in the field of mRNA vaccines in the past decade. After the recent launch of two COVID-19 vaccines (mRNA-1273 and BNT162b2), this next-generation mRNA vaccine development landscape has been widely recognized. Human trials with data from both mRNA cancer vaccines and mRNA infectious disease vaccines have provided encouraging results. Therefore, the next few years will be extremely crucial for the development of new mRNA-based therapeutics as the technology is rapidly refined. For the biopharmaceutical industry, investors/sponsor organizations will be more aligned for the generation of mRNA vaccines for new diseases. More public and private partnerships will create more favourable conditions for mRNA-based vaccine development.

However, several points in the area of mRNA vaccine technologies need to be improved. First, there is a need for further advancement of delivery platforms or delivery materials for mRNA vaccines. Second, more extended clinical trials are needed for mRNA vaccines to better understand immunogenicity and safety. Third, the long-term consequences of these vaccine candidates need to be evaluated. Fourth, further optimization is required in the manufacturing processes of mRNA vaccines. Fifth, the stability of formulated mRNA vaccines at normal temperatures is still challenging for vaccine distribution.

The opportunity for mRNA vaccines is considerable, and the future of this vaccine technology is bright. Finally, mRNA vaccines will overcome these obstacles, and more mRNA vaccines will enter the clinic as next-generation vaccines.

AUTHOR CONTRIBUTIONS

Conception and design by CC. Writing, reviewing, and/or revising the manuscript by CC, and AS. Review, table, and figure preparations by MB. Overall supervision of the study by CC and S-SL. All authors contributed to the article and approved the submitted version.

REFERENCES

1. Le TT, Cramer JP, Chen R, Mayhew S. Evolution of the COVID-19 Vaccine Development Landscape. *Nat Rev Drug Discovery* (2020) 19:667–8. 10.1038/d41573-020-00151-8
2. SARS-CoV-2 Vaccine Development: Current Status. *Mayo Clinic Proc* (2020) 95:2172–88. 10.1016/j.mayocp.2020.07.021
3. Wang J, Peng Y, Xu H, Cui Z, Williams RO. The COVID-19 Vaccine Race: Challenges and Opportunities in Vaccine Formulation. *AAPS Pharm Sci Tech* (2020) 21:1–12. 10.1208/s12249-020-01744-7
4. Plotkin SA. Vaccines: The Fourth Century. *Clin Vaccine Immunol* (2009) 16:1709–19. 10.1128/CVI.00290-09
5. Pardi N, Hogan MJ, Porter FW, Weissman D. mRNA Vaccines—A New Era in Vaccinology. *Nat Rev Drug Discov* (2018) 17:261–79. 10.1038/nrd.2017.243
6. Jackson NA, Kester KE, Casimiro D, Gurunathan S, DeRosa F. The Promise of mRNA Vaccines: A Biotech and Industrial Perspective. *NPJ Vaccines* (2020) 5:1–6. 10.1038/s41541-020-0159-8
7. Xu S, Yang K, Li R, Zhang L. mRNA Vaccine Era—Mechanisms, Drug Platform and Clinical Prospection. *Int J Mol Sci* (2020) 21:6582. 10.3390/ijms21186582
8. Wolff JA, Malone RW, Williams P, Chong W, Acsadi G, Jani A, et al.. Direct Gene Transfer Into Mouse Muscle *In Vivo* . *Science* (1990) 247:1465–8. 10.1126/science.1690918
9. Jirikowski GF, Sanna PP, Maciejewski-Lenoir D, Bloom FE. Reversal of Diabetes Insipidus in Brattleboro Rats: Intrahypothalamic Injection of Vasopressin mRNA. *Science* (1992) 255:996–8. 10.1126/science.1546298
10. Karikó K, Muramatsu H, Welsh FA, Ludwig J, Kato H, Akira S, et al.. Incorporation of Pseudouridine Into mRNA Yields Superior Nonimmunogenic Vector With Increased Translational Capacity and Biological Stability. *Mol Ther* (2008) 16:1833–40. 10.1038/mt.2008.200
11. Thess A, Grund S, Mui BL, Hope MJ, Baumhof P, Fotin-Mleczek M, et al.. Sequence-Engineered mRNA Without Chemical Nucleoside Modifications Enables an Effective Protein Therapy in Large Animals. *Mol Ther* (2015) 23:1456–64. 10.1038/mt.2015.103

12. Kariko K, Muramatsu H, Ludwig J, Weissman D. Generating the Optimal mRNA for Therapy: HPLC Purification Eliminates Immune Activation and Improves Translation of Nucleoside-Modified, Protein-Encoding mRNA. *Nucleic Acids Res* (2011) 39:e142–e. 10.1093/nar/gkr695
13. Pardi N, Muramatsu H, Weissman D, Karikó K. *In Vitro* Transcription of Long RNA Containing Modified Nucleosides. *Methods Mol Biol* (2013) 969:29–42. 10.1007/978-1-62703-260-5_2
14. Ramanathan A, Robb GB, Chan S-H. mRNA Capping: Biological Functions and Applications. *Nucleic Acids Res* (2016) 44:7511–26. 10.1093/nar/gkw551
15. Cowling VH, Cole MD. Myc Regulation of mRNA Cap Methylation. *Genes Cancer* (2010) 1:576–9. 10.1177/1947601910378025
16. Wadhwa A, Aljabbari A, Lokras A, Foged C, Thakur A. Opportunities and Challenges in the Delivery of mRNA-Based Vaccines. *Pharmaceutics* (2020) 12:102. 10.3390/pharmaceutics12020102
17. Bloom K, van den Berg F, Arbuthnot P. Self-Amplifying RNA Vaccines for Infectious Diseases. *Gene Ther* (2021) 28:117–29. 10.1038/s41434-020-00204-y
18. Devarkar SC, Wang C, Miller MT, Ramanathan A, Jiang F, Khan AG, et al.. Structural Basis for M7g Recognition and 2′-O-Methyl Discrimination in Capped RNAs by the Innate Immune Receptor RIG-I. *Proc Natl Acad Sci USA* (2016) 113:596–601. 10.1073/pnas.1515152113
19. Fechter P, Brownlee GG. Recognition of mRNA Cap Structures by Viral and Cellular Proteins. *J Gen Virol* (2005) 86:1239–49. 10.1099/vir.0.80755-0
20. Martin S, Paoletti E, Moss B. Purification of mRNA Guanylyltransferase and mRNA (Guanine-7-) Methyltransferase From Vaccinia Virions. *J Biol Chem* (1975) 250:9322–9. 10.1016/S0021-9258(19)40646-7
21. Gustafsson C, Govindarajan S, Minshull J. Codon Bias and Heterologous Protein Expression. *Trends Biotechnol* (2004) 22:346–53. 10.1016/j.tibtech.2004.04.006

22. Mauro VP, Chappell SA. A Critical Analysis of Codon Optimization in Human Therapeutics. *Trends Mol Med* (2014) 20:604–13. 10.1016/j.molmed.2014.09.003
23. Kudla G, Lipinski L, Caffin F, Helwak A, Zylicz M. High Guanine and Cytosine Content Increases mRNA Levels in Mammalian Cells. *PloS Biol* (2006) 4:e180. 10.1371/journal.pbio.0040180
24. Gallie DR. The Cap and Poly (A) Tail Function Synergistically to Regulate mRNA Translational Efficiency. *Genes Dev* (1991) 5:2108–16. 10.1101/gad.5.11.2108
25. Proudfoot NJ. Ending the Message: Poly (A) Signals Then and Now. *Genes Dev* (2011) 25:1770–82. 10.1101/gad.17268411
26. Maruggi G, Zhang C, Li J, Ulmer JB, Yu D. mRNA as a Transformative Technology for Vaccine Development to Control Infectious Diseases. *Mol Ther* (2019) 27(4):757–72. 10.1016/j.ymthe.2019.01.020
27. Zhang C, Maruggi G, Shan H, Li J. Advances in mRNA Vaccines for Infectious Diseases. *Front Immunol* (2019) 10:594. 10.3389/fimmu.2019.00594
28. Kester KE, Danilo C, Sanjay G, DeRosa F. The Promise of mRNA Vaccines: A Biotech and Industrial Perspective. *NPJ Vaccines* (2020) 5(1)1–6. 10.1038/s41541-020-0159-8
29. Rinaldi A. RNA to the Rescue: RNA Is One of the Most Promising Targets for Drug Development Given its Wide Variety of Uses. *EMBO Rep* (2020) 21(7):e51013. 10.15252/embr.202051013
30. Ross J. mRNA Stability in Mammalian Cells. *Microbiol Rev* (1995) 59:423–50. 10.1128/mr.59.3.423-450.1995
31. Lundstrom K. Replicon RNA Viral Vectors as Vaccines. *Vaccines* (2016) 4:39. 10.3390/vaccines4040039
32. Tews BA, Meyers G. Self-Replicating RNA. In: *RNA Vaccines*. Humana Press, New York, NY: Springer; (2017). p. 15–35.
33. Humphreys IR, Sebastian S. Novel Viral Vectors in Infectious Diseases. *Immunology* (2018) 153:1–9. 10.1111/imm.12829
34. Vogel AB, Lambert L, Kinnear E, Busse D, Erbar S, Reuter KC, et al.. Self-Amplifying RNA Vaccines Give Equivalent Protection Against Influenza to mRNA Vaccines But at Much Lower Doses. *Mol Ther* (2018) 26:446–55. 10.1016/j.ymthe.2017.11.017

35. Fuller DH, Berglund P. Amplifying RNA Vaccine Development. *N Engl J Med* (2020) 382:2469–71. 10.1056/NEJMcibr2009737
36. Beissert T, Perkovic M, Vogel A, Erbar S, Walzer KC, Hempel T, et al.. A Trans-Amplifying RNA Vaccine Strategy for Induction of Potent Protective Immunity. *Mol Ther* (2020) 28:119–28. 10.1016/j.ymthe.2019.09.009
37. Arteta MY, Kjellman T, Bartesaghi S, Wallin S, Wu X, Kvist AJ, et al.. Successful Reprogramming of Cellular Protein Production Through mRNA Delivered by Functionalized Lipid Nanoparticles. *Proc Natl Acad Sci USA* (2018) 115:E3351–E60. 10.1073/pnas.1720542115
38. Miao L, Li L, Huang Y, Delcassian D, Chahal J, Han J, et al.. Delivery of mRNA Vaccines With Heterocyclic Lipids Increases Anti-Tumor Efficacy by STING-Mediated Immune Cell Activation. *Nat Biotechnol* (2019) 37:1174–85. 10.1038/s41587-019-0247-3
39. Pollard C, Rejman J, De Haes W, Verrier B, Van Gulck E, Naessens T, et al.. Type I IFN Counteracts the Induction of Antigen-Specific Immune Responses by Lipid-Based Delivery of mRNA Vaccines. *Mol Ther* (2013) 21:251–9. 10.1038/mt.2012.202
40. Robinson E, MacDonald KD, Slaughter K, McKinney M, Patel S, Sun C, et al.. Lipid Nanoparticle-Delivered Chemically Modified mRNA Restores Chloride Secretion in Cystic Fibrosis. *Mol Ther* (2018) 26:2034–46. 10.1016/j.ymthe.2018.05.014
41. Geall AJ, Verma A, Otten GR, Shaw CA, Hekele A, Banerjee K, et al.. Nonviral Delivery of Self-Amplifying RNA Vaccines. *Proc Natl Acad Sci USA* (2012) 109:14604–9. 10.1073/pnas.1209367109
42. Richner JM, Himansu S, Dowd KA, Butler SL, Salazar V, Fox JM, et al.. Modified mRNA Vaccines Protect Against Zika Virus Infection. *Cell* (2017) 168:1114–25. 10.1016/j.cell.2017.02.017
43. Bahl K, Senn JJ, Yuzhakov O, Bulychev A, Brito LA, Hassett KJ, et al.. Preclinical and Clinical Demonstration of Immunogenicity by mRNA Vaccines Against H10N8 and H7N9 Influenza Viruses. *Mol Ther* (2017) 25:1316–27. 10.1016/j.ymthe.2017.03.035
44. Feldman RA, Fuhr R, Smolenov I, Ribeiro AM, Panther L, Watson M, et al.. mRNA Vaccines Against H10N8 and H7N9 Influenza Viruses of Pandemic Potential Are Immunogenic and Well Tolerated in Healthy Adults in Phase 1 Randomized Clinical Trials. *Vaccine* (2019) 37:3326–34. 10.1016/j.vaccine.2019.04.074

45. Sunshine JC, Sunshine SB, Bhutto I, Handa JT, Green JJ. Poly (β-Amino Ester)-Nanoparticle Mediated Transfection of Retinal Pigment Epithelial Cells *In Vitro* and *In Vivo* . *PloS One* (2012) 7:e37543. 10.1371/journal.pone.0037543
46. Dunn AW, Kalinichenko VV, Shi D. Highly Efficient *In Vivo* Targeting of the Pulmonary Endothelium Using Novel Modifications of Polyethylenimine: An Importance of Charge. *Adv Healthc Mater* (2018) 7:1800876. 10.1002/adhm.201800876
47. Dong Y, Dorkin JR, Wang W, Chang PH, Webber MJ, Tang BC, et al.. Poly (Glycoamidoamine) Brushes Formulated Nanomaterials for Systemic siRNA and mRNA Delivery. *Vivo Nano Lett* (2016) 16:842–8. 10.1021/acs.nanolett.5b02428
48. Zhao M, Li M, Zhang Z, Gong T, Sun X. Induction of HIV-1 Gag Specific Immune Responses by Cationic Micelles Mediated Delivery of Gag mRNA. *Drug Deliv* (2016) 23:2596–607. 10.3109/10717544.2015.1038856
49. Schumann C, Nguyen DX, Norgard M, Bortnyak Y, Korzun T, Chan S, et al.. Increasing Lean Muscle Mass in Mice *via* Nanoparticle-Mediated Hepatic Delivery of Follistatin mRNA. *Theranostics* (2018) 81:5276. 10.7150/thno.27847
50. Prieve MG, Harvie P, Monahan SD, Roy D, Li AG, Blevins TL, et al.. Targeted mRNA Therapy for Ornithine Transcarbamylase Deficiency. *Mol Ther* (2018) 26:801–13. 10.1016/j.ymthe.2017.12.024
51. Fan Y-N, Li M, Luo Y-L, Chen Q, Wang L, Zhang H-B, et al.. Cationic Lipid-Assisted Nanoparticles for Delivery of mRNA Cancer Vaccine. *Biomater Sci* (2018) 6:3009–18. 10.1039/C8BM00908B
52. Weide B, Pascolo S, Scheel B, Derhovanessian E, Pflugfelder A, Eigentler TK, et al.. Direct Injection of Protamine-Protected mRNA: Results of a Phase 1/2 Vaccination Trial in Metastatic Melanoma Patients. *J Immunother* (2009) 32:498–507. 10.1097/CJI.0b013e3181a00068
53. Kübler H, Scheel B, Gnad-Vogt U, Miller K, Schultze-Seemann W, Vom Dorp F, et al.. Self-Adjuvanted mRNA Vaccination in Advanced Prostate Cancer Patients: A First-in-Man Phase I/IIa Study. *J Immunother Cancer* (2015) 3:26. 10.1186/s40425-015-0068-y
54. Papachristofilou A, Hipp MM, Klinkhardt U, Früh M, Sebastian M, Weiss C, et al.. Phase Ib Evaluation of a Self-Adjuvanted

Protamine Formulated mRNA-Based Active Cancer Immunotherapy, BI1361849 (CV9202), Combined With Local Radiation Treatment in Patients With Stage IV Non-Small Cell Lung Cancer. *J Immunother Cancer* (2019) 7:38. 10.1186/s40425-019-0520-5

55. Sebastian M, Schröder A, Scheel B, Hong HS, Muth A, von Boehmer L, et al.. A Phase I/IIa Study of the mRNA-Based Cancer Immunotherapy CV9201 in Patients With Stage IIIB/IV non-Small Cell Lung Cancer. *Cancer Immunol Immunother* (2019) 68:799–812. 10.1007/s00262-019-02315-x
56. Alberer M, Gnad-Vogt U, Hong HS, Mehr KT, Backert L, Finak G, et al.. Safety and Immunogenicity of a mRNA Rabies Vaccine in Healthy Adults: An Open-Label, non-Randomised, Prospective, First-In-Human Phase 1 Clinical Trial. *Lancet* (2017) 390:1511–20. 10.1016/S0140-6736(17)31665-3
57. Schlake T, Thess A, Fotin-Mleczek M, Kallen K-J. Developing mRNA-Vaccine Technologies. *RNA Biol* (2012) 9:1319–30. 10.4161/rna.22269
58. Kallen K-J, Heidenreich R, Schnee M, Petsch B, Schlake T, Thess A, et al.. A Novel, Disruptive Vaccination Technology: Self-Adjuvanted RNActive® Vaccines. *Hum Vaccin Immunother* (2013) 9:2263–76. 10.4161/hv.25181
59. van den Brand D, Gorris MA, van Asbeck AH, Palmen E, Ebisch I, Dolstra H, et al.. Peptide-Mediated Delivery of Therapeutic mRNA in Ovarian Cancer. *Eur J Pharm Biopharm* (2019) 141:180–90. 10.1016/j.ejpb.2019.05.014
60. Dewitte H, Van Lint S, Heirman C, Thielemans K, De Smedt SC, Breckpot K, et al.. The Potential of Antigen and TriMix Sonoporation Using mRNA-Loaded Microbubbles for Ultrasound-Triggered Cancer Immunotherapy. *J Control Release* (2014) 194:28–36. 10.1016/j.jconrel.2014.08.011
61. Khoury HJ, Collins RH, Jr, Blum W, Stiff PS, Elias L, Lebkowski JS, et al.. Immune Responses and Long-Term Disease Recurrence Status After Telomerase-Based Dendritic Cell Immunotherapy in Patients With Acute Myeloid Leukemia. *Cancer* (2017) 123:3061–72. 10.1002/cncr.30696
62. Bialkowski L, Van Weijnen A, van der Jeught K, Renmans D, Daszkiewicz L, Heirman C, et al.. Intralymphatic mRNA Vaccine

Induces CD8 T-Cell Responses That Inhibit the Growth of Mucosally Located Tumours. *Sci Rep* (2016) 6:22509. 10.1038/srep22509

63. Weide B, Carralot J-P, Reese A, Scheel B, Eigentler TK, Hoerr I, et al.. Results of the First Phase I/II Clinical Vaccination Trial With Direct Injection of mRNA. *J Immunother* (2008) 31:180–8. 10.1097/CJI.0b013e31815ce501
64. Kreiter S, Selmi A, Diken M, Koslowski M, Britten CM, Huber C, et al.. Intranodal Vaccination With Naked Antigen-Encoding RNA Elicits Potent Prophylactic and Therapeutic Antitumoral Immunity. *Cancer Res* (2010) 70:9031–40. 10.1158/0008-5472.CAN-10-0699
65. Johansson DX, Ljungberg K, Kakoulidou M, Liljeström P. Intradermal Electroporation of Naked Replicon RNA Elicits Strong Immune Responses. *PloS One* (2012) 7:e29732. 10.1371/journal.pone.0029732
66. Sahin U, Derhovanessian E, Miller M, Kloke B-P, Simon P, Löwer M, et al.. Personalized RNA Mutanome Vaccines Mobilize Poly-Specific Therapeutic Immunity Against Cancer. *Nature* (2017) 547:222–6. 10.1038/nature23003
67. Golombek S, Pilz M, Steinle H, Kochba E, Levin Y, Lunter D, et al.. Intradermal Delivery of Synthetic mRNA Using Hollow Microneedles for Efficient and Rapid Production of Exogenous Proteins in Skin. *Mol Ther Nucleic Acids* (2018) 11:382–92. 10.1016/j.omtn.2018.03.005
68. Pardi N, Hogan MJ, Pelc RS, Muramatsu H, Andersen H, DeMaso CR, et al.. Zika Virus Protection by a Single Low-Dose Nucleoside-Modified mRNA Vaccination. *Nature* (2017) 543:248–51. 10.1038/nature214283
69. Koh KJ, Liu Y, Lim SH, Loh XJ, Kang L, Lim CY, et al.. Formulation, Characterization and Evaluation of mRNA-Loaded Dissolvable Polymeric Microneedles (RNApatch). *Sci Rep* (2018) 8:1–11. 10.1038/s41598-018-30290-3
70. Kowalski PS, Rudra A, Miao L, Anderson DG. Delivering the Messenger: Advances in Technologies for Therapeutic mRNA Delivery. *Mol Ther* (2019) 27:710–28. 10.1016/j.ymthe.2019.02.012
71. Sung Y, Kim S. Recent Advances in the Development of Gene Delivery Systems. *Biomater Res* (2019) 23:8. 10.1186/s40824-019-0156-z

72. Chou JY, Mansfield BC. Recombinant AAV-Directed Gene Therapy for Type I Glycogen Storage Diseases. *Expert Opin Biol Ther* (2011) 11:1011–24. 10.1517/14712598.2011.578067
73. Schott JW, Morgan M, Galla M, Schambach A. Viral and Synthetic RNA Vector Technologies and Applications. *Mol Ther* (2016) 24:1513–27. 10.1038/mt.2016.143
74. Rozovics JM, Chase AJ, Cathcart AL, Chou W, Gershon PD, Palusa S, et al.. Picornavirus Modification of a Host mRNA Decay Protein. *MBio* (2012) 3(6):1–11. 10.1128/mBio.00431-12
75. Ehrengruber MU, Schlesinger S, Lundstrom K. Alphaviruses: Semliki Forest Virus and Sindbis Virus Vectors for Gene Transfer Into Neurons. *Curr Protoc Neurosci* (2011) 57(4.22):1–4. 10.1002/0471142301.ns0422s57
76. Tezel A, Dokka S, Kelly S, Hardee GE, Mitragotri S. Topical Delivery of Anti-Sense Oligonucleotides Using Low-Frequency Sonophoresis. *Pharm Res* (2004) 21:2219–25. 10.1007/s11095-004-7674-6
77. Ura T, Okuda K, Shimada M. Developments in Viral Vector-Based Vaccines. *Vaccines* (2014) 2:624–41. 10.3390/vaccines2030624
78. Akbarzadeh A, Rezaei-Sadabady R, Davaran S, Joo SW, Zarghami N, Hanifehpour Y, et al.. Liposome: Classification, Preparation, and Applications. *Nanoscale Res Lett* (2013) 8:102. 10.1186/1556-276X-8-102
79. Dhaliwal HK, Fan Y, Kim J, Amiji MM. Intranasal Delivery and Transfection of mRNA Therapeutics in the Brain Using Cationic Liposomes. *Mol Pharm* (2020) 17:1996–2005. 10.1021/acs.molpharmaceut.0c00170
80. Viger-Gravel J, Schantz A, Pinon AC, Rossini AJ, Schantz S, Emsley L. Structure of Lipid Nanoparticles Containing siRNA or mRNA by Dynamic Nuclear Polarization-Enhanced NMR Spectroscopy. *J Phys Chem B* (2018) 122:2073–81. 10.1021/acs.jpcb.7b10795
81. Zhao Y, Piao Y, Zhang C, Jiang Y, Liu A, Cui S, et al.. Replacement of Quaternary Ammonium Headgroups by Tri-Ornithine in Cationic Lipids for the Improvement of Gene Delivery *In Vitro* and *In Vivo* . *J Mater Chem B* (2017) 5:7963–73. 10.1039/C7TB01915G
82. Filion MC, Phillips NC. Toxicity and Immunomodulatory Activity of Liposomal Vectors Formulated with Cationic Lipids Toward Immune

Effector Cells. *Biochim Biophys Acta* (1997) 1329:(2)345–56. 10.1007/128_2010_68

83. Sizovs A, McLendon PM, Srinivasachari S, Reineke TM. Carbohydrate Polymers for Nonviral Nucleic Acid Delivery. *Top Curr Chem* (2010) 296:131–90. 10.1007/128_2010_68
84. Zhang P, Wagner E. History of Polymeric Gene Delivery Systems. *Top Curr Chem* (2017) 375(26)1–39. 10.1007/s41061-017-0112-0
85. Malone RW, Felgner PL, Verma IM. Cationic Liposome-Mediated RNA Transfection. *Proc Natl Acad Sci USA* (1989) 86:6077–81. 10.1073/pnas.86.16.6077
86. Siewert C, Haas H, Nawroth T, Ziller A, Nogueira S, Schroer M, et al.. Investigation of Charge Ratio Variation in mRNA–DEAE-Dextran Polyplex Delivery Systems. *Biomaterials* (2019) 192:612–20. 10.1016/j.biomaterials.2018.10.020
87. Makadia HK, Siegel SJ. Poly Lactic-Co-Glycolic Acid (PLGA) as Biodegradable Controlled Drug Delivery Carrier. *Polymers* (2011) 3:1377–97. 10.3390/polym3031377
88. Kim K-T, Lee J-Y, Kim D-D, Yoon I-S, Cho H-J. Recent Progress in the Development of Poly (Lactic-Co-Glycolic Acid)-Based Nanostructures for Cancer Imaging and Therapy. *Pharmaceutics* (2019) 11:280. 10.3390/pharmaceutics11060280
89. Lungwitz U, Breunig M, Blunk T, Göpferich A. Polyethylenimine-Based non-Viral Gene Delivery Systems. *Eur J Pharm Biopharm* (2005) 60:247–66. 10.1016/j.ejpb.2004.11.011
90. Cheng C, Convertine AJ, Stayton PS, Bryers JD. Multifunctional Triblock Copolymers for Intracellular Messenger RNA Delivery. *Biomaterials* (2012) 33:6868–76. 10.1016/j.biomaterials.2012.06.020
91. Zhao W, Zhang C, Li B, Zhang X, Luo X, Zeng C, et al.. Lipid Polymer Hybrid Nanomaterials for mRNA Delivery. *Cell Mol Bioeng* (2018) 11(5):397–406. 10.1007/s12195-018-0536-9
92. Kaczmarek JC, Patel AK, Kauffman KJ, Fenton OS, Webber MJ, Heartlein MW, et al.. Polymer–lipid Nanoparticles for Systemic Delivery of mRNA to the Lungs. *Angewandte Chem* (2016) 128:14012–6. 10.1002/ange.201608450

93. Ball RL, Hajj KA, Vizelman J, Bajaj P, Whitehead KA. Lipid Nanoparticle Formulations for Enhanced Co-Delivery of siRNA and mRNA. *Nano Lett* (2018) 18:3814–22. 10.1021/acs.nanolett.8b01101
94. Kang Z, Meng Q, Liu K. Peptide-Based Gene Delivery Vectors. *J Mater Chem B* (2019) 7:1824–41. 10.1039/C8TB03124J
95. Corbett KS, Flynn B, Foulds KE, Francica JR, Boyoglu-Barnum S, Werner AP, et al.. Evaluation of the mRNA-1273 Vaccine Against SARS-CoV-2 in Nonhuman Primates. *N Engl J Med* (2020) 383:1544–55. 10.1056/NEJMoa2024671
96. Wykes M, Pombo A, Jenkins C, MacPherson GG. Dendritic Cells Interact Directly With Naive B Lymphocytes to Transfer Antigen and Initiate Class Switching in a Primary T-Dependent Response. *J Immunol* (1998) 161:1313–9. [PubMed]
97. Wykes M, MacPherson G. Dendritic Cell–B-Cell Interaction: Dendritic Cells Provide B Cells With CD40-Independent Proliferation Signals and CD40-Dependent Survival Signals. *Immunology* (2000) 100:1–3. 10.1046/j.1365-2567.2000.00044.x
98. Harvey BP, Raycroft MT, Quan TE, Rudenga BJ, Roman RM, Craft J, et al.. Transfer of Antigen From Human B Cells to Dendritic Cells. *Mol Immunol* (2014) 58(1):56–65. 10.1016/j.molimm.2013.10.013
99. Heesters BA, van der Poel CE, Das A, Carroll MC. Antigen Presentation to B Cells. *Trends Immunol* (2016) 37:844–54. 10.1016/j.it.2016.10.003
100. Zeng C, Zhang C, Walker PG, Dong Y. Formulation and Delivery Technologies for mRNA Vaccines. *Curr Top Microbiol Immunol* (2020) 1–40. 10.1007/82_2020_217
101. Akita H, Kogure K, Moriguchi R, Nakamura Y, Higashi T, Nakamura T, et al.. Nanoparticles for Ex Vivo siRNA Delivery to Dendritic Cells for Cancer Vaccines: Programmed Endosomal Escape and Dissociation. *J Control Release* (2010) 143:311–7. 10.1016/j.jconrel.2010.01.012
102. Pardi N, Hogan MJ, Weissman D. Recent Advances in mRNA Vaccine Technology. *Curr Opin Immunol* (2020) 65:14–20. 10.1016/j.coi.2020.01.008
103. Weissman D. mRNA Transcript Therapy. *Expert Rev Vaccines* (2015) 14:265–81. 10.1586/14760584.2015.973859
104. Akira S, Saitoh T, Kawai T. Nucleic Acids Recognition by Innate Immunity. *Uirusu* (2012) 62:39–45. 10.2222/jsv.62.39

105. Forsbach A, Nemorin J-G, Montino C, Müller C, Samulowitz U, Vicari AP, et al.. Identification of RNA Sequence Motifs Stimulating Sequence-Specific TLR8-Dependent Immune Responses. *J Immunol* (2008) 180:3729–38. 10.4049/jimmunol.180.6.3729

106. Verbeke R, Lentacker I, De Smedt SC, Dewitte H. Three Decades of Messenger RNA Vaccine Development. *Nano Today* (2019) 28:100766. 10.1016/j.nantod.2019.100766

107. Botos I, Liu L, Wang Y, Segal DM, Davies DR. The Toll-Like Receptor 3: dsRNA Signaling Complex. *Biochim Biophys Acta* (2009) 1789:667–74. 10.1016/j.bbagrm.2009.06.005

108. Kato H, Oh S-W, Fujita T. RIG-I-Like Receptors and Type I Interferonopathies. *J Interferon Cytokine Res* (2017) 37:207–13. 10.1089/jir.2016.0095

109. Gregorio J, Meller S, Conrad C, Di Nardo A, Homey B, Lauerma A, et al.. Plasmacytoid Dendritic Cells Sense Skin Injury and Promote Wound Healing Through Type I Interferons. *J Exp Med* (2010) 207:2921–30. 10.1084/jem.20101102

110. Goubau D, Deddouche S. E Sousa CR. Cytosolic Sensing of Viruses. *Immunity* (2013) 38:855–69. 10.1016/j.immuni.2013.05.007

111. Goubau D, Schlee M, Deddouche S, Pruijssers AJ, Zillinger T, Goldeck M, et al.. Antiviral Immunity *via* RIG-I-Mediated Recognition of RNA Bearing 5′-Diphosphates. *Nature* (2014) 514:372–5. 10.1038/nature13590

112. Schlee M. Master Sensors of Pathogenic RNA–RIG-I Like Receptors. *Immunobiology* (2013) 218:1322–35. 10.1016/j.imbio.2013.06.007

113. Rehwinkel J, Tan CP, Goubau D, Schulz O, Pichlmair A, Bier K, et al.. RIG-I Detects Viral Genomic RNA During Negative-Strand RNA Virus Infection. *Cell* (2010) 140:397–3408. 10.1016/j.cell.2010.01.020

114. Pollard C, De Koker S, Saelens X, Vanham G, Grooten J. Challenges and Advances Towards the Rational Design of mRNA Vaccines. *Trends Mol Med* (2013) 19:705–13. 10.1016/j.molmed.2013.09.002

115. De Beuckelaer A, Pollard C, Van Lint S, Roose K, Van Hoecke L, Naessens T, et al.. Type I Interferons Interfere With the Capacity of mRNA Lipoplex Vaccines to Elicit Cytolytic T Cell Responses. *Mol Ther* (2016) 24:2012–20. 10.1038/mt.2016.161

116. Linares-Fernández S, Lacroix C, Exposito J-Y, Verrier B. Tailoring mRNA Vaccine to Balance Innate/Adaptive Immune Response. *Trends Mol Med* (2020) 26:311–23. 10.1016/j.molmed.2019.10.002
117. Guo C, Manjili MH, Subjeck JR, Sarkar D, Fisher PB, Wang X-Y. Therapeutic Cancer Vaccines: Past, Present, and Future. *Adv Cancer Res* (2013) 119:421–75. 10.1016/B978-0-12-407190-2.00007-1
118. Conry RM, LoBuglio AF, Wright M, Sumerel L, Pike MJ, Johanning F, et al.. Characterization of a Messenger RNA Polynucleotide Vaccine Vector. *Cancer Res* (1995) 55:1397–400.
119. Islam MA, Rice J, Reesor E, Zope H, Tao W, Lim M, et al.. Adjuvant-Pulsed mRNA Vaccine Nanoparticle for Immunoprophylactic and Therapeutic Tumor Suppression in Mice. *Biomaterials* (2020) 266:120431. 10.1016/j.biomaterials.2020.120431
120. Benteyn D, Heirman C, Bonehill A, Thielemans K, Breckpot K. mRNA-Based Dendritic Cell Vaccines. *Expert Rev Vaccines* (2015) 14:161–76. 10.1586/14760584.2014.957684
121. Arya S, Lin Q, Zhou N, Gao X, Huang J-D. Strong Immune Responses Induced by Direct Local Injections of Modified mRNA-Lipid Nanocomplexes. *Mol Ther Nucleic Acids* (2020) 19:1098–109. 10.1016/j.omtn.2019.12.044
122. Moyle G. Toxicity of Antiretroviral Nucleoside and Nucleotide Analogues. *Drug Safety* (2000) 23:467–81. 10.2165/00002018-200023060-00001
123. Feng JY, Johnson AA, Johnson KA, Anderson KS. Insights Into the Molecular Mechanism of Mitochondrial Toxicity by AIDS Drugs. *J Biol Chem* (2001) 276:23832–7. 10.1074/jbc.M101156200
124. Johnson AA, Ray AS, Hanes J, Suo Z, Colacino JM, Anderson KS, et al.. Toxicity of Antiviral Nucleoside Analogs and the Human Mitochondrial DNA Polymerase. *J Biol Chem* (2001) 276:40847–57. 10.1074/jbc.M106743200
125. Liu MA. A Comparison of Plasmid DNA and mRNA as Vaccine Technologies. *Vaccines* (2019) 7:37. 10.3390/vaccines7020037
126. Oliver SE, Gargano JW, Marin M, Wallace M, Curran KG, Chamberland M, et al.. The Advisory Committee on Immunization Practices' Interim Recommendation for Use of Pfizer-BioNTech COVID-19 Vaccine—United States, December 2020. *MMWR Morb Mortal Wkly Rep* (2020) 69:1922–4. 10.15585/mmwr.mm6950e2

127. Polack FP, Thomas SJ, Kitchin N, Absalon J, Gurtman A, Lockhart S, et al.. Safety and Efficacy of the BNT162b2 mRNA Covid-19 Vaccine. *N Engl J Med* (2020) 383:2603–15. 10.1056/NEJMoa2034577
128. Walsh EE, Frenck RW, Jr., Falsey AR, Kitchin N, Absalon J, Gurtman A, et al.. Safety and Immunogenicity of Two RNA-Based Covid-19 Vaccine Candidates. *N Engl J Med* (2020) 383:2439–50. 10.1056/NEJMoa2027906
129. Mahase E. Covid-19: Pfizer and BioNTech Submit Vaccine for US Authorisation. *BMJ* (2020) 371:m4552. 10.1136/bmj.m4552
130. Mahase E. Covid-19: UK Approves Pfizer and BioNTech Vaccine With Rollout Due to Start Next Week. *BMJ* (2020) 371:m4714. 10.1136/bmj.m4714
131. Tanne JH. Covid-19: FDA Panel Votes to Approve Pfizer BioNTech Vaccine. *BMJ* (2020) 371:m4799. 10.1136/bmj.m4799
132. Cerrato CP, Kivijärvi T, Tozzi R, Lehto T, Gestin M, Langel Ü. Intracellular Delivery of Therapeutic Antisense Oligonucleotides Targeting mRNA Coding Mitochondrial Proteins by Cell-Penetrating Peptides. *J Mater Chem B* (2020) 8:10825–36. 10.1039/D0TB01106A
133. Nichol AA. Potential Implications of Testing an Experimental mRNA-Based Vaccine During an Emerging Infectious Disease Pandemic. *Am J Bioeth* (2020) 20:W2–W3. 10.1080/15265161.2020.1763696
134. Baden LR, El Sahly HM, Essink B, Kotloff K, Frey S, Novak R, et al.. Efficacy and Safety of the mRNA-1273 SARS-CoV-2 Vaccine. *N Engl J Med* (2020) 384:403–16. 10.1056/NEJMoa2035389
135. Tumban E. Lead SARS-CoV-2 Candidate Vaccines: Expectations From Phase III Trials and Recommendations Post-Vaccine Approval. *Viruses* (2021) 13:54. 10.3390/v13010054
136. Li Q, Lu H. Latest Updates on COVID-19 Vaccines. *Biosci Trends* (2020) 14:463–6. 10.5582/bst.2020.03445
137. Rausch S, Schwentner C, Stenzl A, Bedke J. mRNA Vaccine CV9103 and CV9104 for the Treatment of Prostate Cancer. *Hum Vaccin Immunother* (2014) 10:3146–52. 10.4161/hv.29553
138. Shimabukuro TT, Cole M, Su JR. Reports of Anaphylaxis After Receipt of mRNA COVID-19 Vaccines in the US—December 14, 2020-January 18, 2021. *JAMA* (2021) 325:1101–2. 10.1001/jama.2021.1967

139. Dagan N, Barda N, Kepten E, Miron O, Perchik S, Katz MA, et al.. BNT162b2 mRNA Covid-19 Vaccine in a Nationwide Mass Vaccination Setting. *N Engl J Med* (2021) 384:1412–23. 10.1056/NEJMoa2101765
140. Bradley T, Grundberg E, Selvarangan R, LeMaster C, Fraley E, Banerjee D, et al.. Antibody Responses After a Single Dose of SARS-CoV-2 mRNA Vaccine. *N Engl J Med* (2021) 384:1959–61. 10.1056/NEJMc2102051
141. Teijaro JR, Farber DL. COVID-19 Vaccines: Modes of Immune Activation and Future Challenges. *Nat Rev Immunol* (2021) 21:195–7. 10.1038/s41577-021-00526-x
142. COVID C, Team R Allergic Reactions Including Anaphylaxis After Receipt of the First Dose of Pfizer-BioNTech COVID-19 Vaccine—United States, December 14–23, 2020. *MMWR Morb Mortal Wkly Rep* (2021) 70:46–51. 10.15585/mmwr.mm7002e1
143. Hendaus MA, Jomha FA. mRNA Vaccines for COVID-19: A Simple Explanation. *Qatar Med J* (2021) 7:1–5. 10.5339/qmj.2021.07
144. Graham BS. Rapid COVID-19 Vaccine Development. *Science* (2020) 368:945–6. 10.1126/science.abb8923
145. Lombardi A, Bozzi G, Ungaro R, Villa S, Castelli V, Mangioni D, et al.. Mini Review Immunological Consequences of Immunization With COVID-19 mRNA Vaccines: Preliminary Results. *Front Immunol* (2021) 12:657711. 10.3389/fimmu.2021.657711

CHAPTER 10

Vaccine Design and Vaccination Strategies against Rickettsiae

Anke Osterloh

Research Center Borstel, Parkallee 22, 23845 Borstel, Germany

ABSTRACT

Rickettsioses are febrile, potentially lethal infectious diseases that are a serious health threat, especially in poor income countries. The causative agents are small obligate intracellular bacteria, rickettsiae. Rickettsial infections are emerging worldwide with increasing incidence and geographic distribution. Nonetheless, these infections are clearly underdiagnosed because methods of diagnosis are still limited and often not available. Another problem is that the bacteria respond to only a few antibiotics, so delayed or wrong antibiotic treatment often leads to a more severe outcome of the disease. In addition to that, the development of antibiotic resistance is a serious threat because alternative antibiotics are missing. For these reasons, prophylactic vaccines against rickettsiae are urgently needed. In the past years, knowledge about protective immunity against rickettsiae and immunogenic determinants has

Citation: Osterloh, A. Vaccine Design and Vaccination Strategies against Rickettsiae. Vaccines 2021, 9, 896. https://doi.org/10.3390/ vaccines9080896

been increasing and provides a basis for vaccine development against these bacterial pathogens. This review provides an overview of experimental vaccination approaches against rickettsial infections and perspectives on vaccination strategies.

Keywords: rickettsiae, orientia, immunity, vaccination

INTRODUCTION

Rickettsiae are small obligate intracellular bacteria of the family of Rickettsiacea that cause febrile potentially fatal diseases in humans. The family of Rickettsiacea consists of two genera, Rickettsia and Orientia. *Orientia* (*O.*) *tsutsugamushi* has long been considered the only member of the latter. Meanwhile, two additional species have been recently identified (candidatus *O. chuto* [1] and candidatus *O. chiloensis* [2]). The genus Rickettsia is further divided into four major groups of species according to phylogenetic relationship and way of transmission. The majority of rickettsial species belong to the spotted fever group (SFG). More than 20 species of this group have been identified so far. Prominent representatives of this group are *R. rickettsii*, the causative agent of Rocky Mountain spotted fever (RMSF), and *R. conorii* that causes Mediterranean spotted fever (MSF). The second group is the typhus group (TG) of rickettsiae that has two members, *R. prowazekii*, the causative agent of epidemic typhus, which was and is a serious threat in times of war, and *R. typhi* that causes endemic or murine typhus. The third group of pathogenic rickettsiae is the transitional group (*R. felis*, *R. akari*, *R. australis*), while the fourth group, the ancestral group (*R. bellii* and *R. canadensis*), is non-pathogenic.

Transmission of rickettsiae to vertebrates generally occurs through arthropods that carry the bacteria as endosymbionts in the gut epithelium. Once infected, the arthropod stays infected for life and transmits the bacteria transovarially and transstadially to the next generation. *R. prowazekii* is transmitted from human to human via the body louse (*Pediculus humanus*), while all other so-far-known rickettsial species are transmitted to humans via ectoparasites from rodents, predominantly from rats and mice. Rodents serve as a natural reservoir and play a key role in the distribution of rickettsial infections. It is, therefore, not surprising that rickettsial infections generally occur worldwide, which is especially true for TG rickettsiae while other rickettsial species appear endemic in certain areas of the world (Table 1). All SFG rickettsiae are transmitted by ticks (genera Dermacentor, Rhipicephalus,

Amblyomma, Hyalomma, or Ixodes), while mites are the vectors for the transmission of orientia species (*Leptotrombidium deliense*) and *R. akari* (house mouse mite Liponyssoides sanguineus). The only so-far-known rickettsial species that are transmitted by fleas are *R. typhi* (predominantly the rat flea *Xenopsylla cheopis*) and *R. felis*, the causative agent of cat-flea typhus (the cat flea *Ctenocephalides felis*) [3].

Table 1: Family of *Rickettsiaceae*, rickettsial diseases and distribution. The table gives an overview of so-far-identified members of the family of Rickettsiaceae. SFG: spotted fever group, TG: typhus group, SF: spotted fever

Genus	Group	Species	Disease	Distribution
Rickettsia	**SFG**	*R. rickettsii*	Rocky Mountain SF	North America (Midwest and Southeastern U.S.), Central and South America (Mexico, Panama, Costa Rica, Brazil, Argentina, Colombia)
		R. conorii ssp. *conorii*	Mediterranean SF	Europe (Mediterranean Basin), North Africa (Tunisia, Algeria, Morocco), multiple sub-Saharan countries
		R. conorii ssp. *indica*	Indian tick typhus	Middle East, India
		R. conorii ssp. *israelensis*	Israeli SF	Israel, North Africa (Tunisia)
		R. conorii ssp. *israelensis*	Astrakhan fever	Astrakhan region, France
		R. conorii ssp. *caspia*	Astrakhan fever	Africa (Chad)
		364D (R. phillipi)	Unnamed rickettsiosis	U.S. (southern California)
		R. honei	Flinder's Island SF/ Thailand SF	Australia, Tasmania, Thailand

		R. helvetica	Tick-bite fever	Europe (Denmark, Austria, France, Italy), Asia (Laos)
		R. japonica	Japanese SF	Japan, detected in ticks in South Korea and Northern Thailand
		R. heilongjiangensis	Far-Eastern SF	Northern China, Russia far east, Japan, Eastern Asia
		R. parkeri	Maculatum infection/ American boutonneuse fever/Tidewater SF	North and South America
		R. africae	African tick-bite fever	Sub-Saharan Africa, Caribbean, West Indies
		R. sibirica	Siberian tick typhus/North Asian tick typhus	Russia, China, Mongolia
		R. sibiria ssp. *mongolotimonae*	Tickborne lymphadenopathy (TIBOLA)	Southern Europe (France, Greece, Portugal, Spain), Asia, South Africa
		R. massiliae	Mediterranean SF-like disease	Southern Europe, South America (Argentina)
		R. monacensis	Tick-bite fever	Europe

		R. slovaca	Tickborne lymphadenopathy (TIBOLA)/Dermacentor-borne necrosis and lymphadenopathy (DEBONEL)/ scalp eschar and neck lymphadenopathy after tick bite (SENLAT)	Europe (France, Slovakia, Italy, Germany, Hungary, Spain, Poland), Georgia, Russia
		R. raoultii		Europe (France, Slovakia, Poland)
		R. aeschlimannii	Tick-bite fever	North Africa (Tunisia, Morocco), South Africa
		364D (*R. phillipi*)	Unnamed rickettsiosis	Southern California
	TG	*R. prowazekii*	Epidemic typhus	Worldwide, sporadic in Africa, Asia, Central and South America, Russia
		R. typhi	Endemic typhus/Murine typhus	Worldwide
	transitional	*R. felis*	Cat-flea typhus	Probably worldwide
		R. akari	Rickettsialpox	Probably worldwide
		R. australis	Queensland tick typhus	Australia, Tasmania
	ancestral	*R. bellii*		
		R. canadensis		

Orientia		*O. tsutsugamushi*	Scrub typhus/ tsutsugamushi fever	Asia, Northern Australia, serological evidence in sub-Saharan countries (Cameroon, Congo, Kenya)
		candidatus O. chuto	Scrub typhus/ tsutsugamushi fever	Arabian Peninsula (Dubai), other areas in Middle East?
		candidatus O. chiloensis	Scrub typhus/ tsutsugamushi fever	Chile

In the case of TG rickettsiae, transmission occurs via deposit of feces from an arthropod that has been infected by ingesting blood from an infected rodent. The bacteria are then scratched into the wound. SFG rickettsiae can also be transmitted via the bite of an infected tick. After entering the vertebrate skin, rickettsiae first infect phagocytic cells and then spread into endothelial cells (ECs) that form the inner wall of the blood vessels and represent the dominant target cells [4,5]. Rickettsiae replicate free in the cytosol of infected ECs and are released by different mechanisms. SFG rickettsiae can induce focal lysis of the cellular membrane to be released and spread to adjacent cells [6,7]. TG rickettsiae multiply until lysis or burst of the cell [7], and orientia performs a kind of budding similar to viruses [8]. Free bacteria then infect adjacent ECs leading to local blood leakages and inflammatory reactions that can become visible as an eschar at the site of entry in the infection with some SFG rickettsiae (e.g., *R. conorii* but not *R. rickettsii*), orientia, and *R. prowazekii* but not *R. typhi*. The bacteria further systemically spread throughout the body via the bloodstream and can enter nearly all organs, where they also infect other cells, predominantly monocytes/macrophages (MΦ) [4,5,9]. These cells are considered to function as a niche for bacterial replication and to play a central role as a vehicle for dissemination [10,11,12]. Disseminated rickettsial infections cause a characteristic skin rash in approximately 60% of the patients, which is due to local blood leakages and inflammatory reactions.

The disease that is caused by the infection with different rickettsial species appears quite similar but with different severity. Patients usually develop high fever and suffer from headaches and abdominal pain. Because of the systemic distribution of the bacteria in the body, the infections can lead to multiple organ pathology, including pneumonia, meningoencephalitis, nephritis, myocarditis, hepatic damage, and other complications that can be fatal. The highest mortality is observed for the infection with *R. rickettsii* (>20% without antibiotic treatment and nowadays 1–7% [13]) followed by epidemic typhus caused by *R. prowazekii* that usually appears under poor hygienic conditions and lack of medical care (15–30%) [14,15]. Due to the relatively high mortality, *R. rickettsii* and *R. prowazekii* are classified as potential biological weapons.

Disease surveillance data are rare because diagnosis is still problematic. In a recent study from the Chinese Center for Disease Control, it was described that the incidence of orientia infections increased from 0.09/100,000 in 2006 to 1.6/100,000 in 2016 and that the disease that was originally endemic to southern China expanded to all provinces [16]. In the U.S., an increase in spotted fever rickettsioses is observed, and also endemic typhus caused by *R. typhi* appears with steadily increasing incidence predominantly in Southern California, Texas, and Hawaii [17,18,19]. The Centers for Disease Control (CDC) recorded 738 cases of endemic typhus in Texas in 2018 compared to 222 in 2013 and 157 in 2008 (https://www.dshs.texas.gov/IDCU/disease/murine_typhus/Typhus-2008-2018.pdf, accessed 6 August 2021).

Generally, infections with *R. typhi* are highly endemic in coastal tropical and subtropical regions in low-income countries in Asia [20,21,22,23], Africa [24], and South America, e.g., Mexico [25]. In Laos, it was found that the infection with *O. tsutsugamushi* and *R. typhi* was responsible for a high proportion of central nervous system infections (12% and 11%, respectively) with a high mortality rate [26]. In Europe, rickettsial infections are predominantly detected in travelers who acquired the infection abroad. Nonetheless, e.g., *R. typhi* also appears in Europe (Greece, Cypres, Spain, Portugal [27,28,29,30,31,32,33,34,35]). Especially homeless people are at enhanced risk. In France, the seropositivity of the homeless in Marseille dramatically increased in the past 20 years from 0.054% in the years 2000–2003 to 22% in the years 2010–2013 [36]. In addition to that, sporadic outbreaks of rickettsial infections occur that are usually associated with poor hygienic conditions. After the earthquake in Nepal in 2015, increasing cases of febrile illnesses were observed in refugee camps. These were revealed to be caused by the infection with *O. tsutsugamushi* and had relatively high mortality of 5.7% [37]. Since then, the transmission of *O. tsutsugamushi* is ongoing in Nepal. Other examples are the large outbreak of epidemic typhus during civil war in Burundi in 1995 in a jail and in refugee camps [38,39,40] and a smaller outbreak of epidemic typhus in Russia in 1997 [41].

A major problem in the recognition and treatment of rickettsial diseases is that diagnostic methods are still limited, not standardized, expensive, and often not accessible. Misdiagnosis, delayed antibiotic treatment, and treatment with inappropriate antibiotics often result in more severe or fatal disease. This was also the reason for the high mortality of *O. tsutsugamushi* infections after the earthquake in Nepal, where clinicians treated the patients with cefexime or ceftraiaxone and imipenem [37]. Rickettsiae and orientiae respond to only a few antibiotics (doxycycline, rifampin, chloramphenicol), with doxycycline being the treatment of choice. The development of antibiotic resistance is a great threat, and also, in cases of doxycycline intolerance, effective alternative antibiotics are missing. Finally, some rickettsial species can persist and reoccur despite antibiotic treatment, which is well known for *R. prowazekii*. These bacteria can reappear several years after primary infection and cause Brill-Zinsser disease [42]. The persistence of other rickettsial species in humans (*R. rickettsii* [43,44], *R. typhi* [12], as well as of *O. tsutsugamushi* [45]) is anticipated but has not yet been proven. At least for *O. tsutsugamushi*, relapse of patients that had been initially treated with antibiotics and recovered from the disease was observed months to years after the infection [45].

For the reasons mentioned above, a prophylactic vaccine against rickettsial infections is urgently needed. Vaccine design requires the understanding of protective immunity and also immunopathological reactions. This review provides a brief overview of current knowledge on immunity and immunopathology in rickettsial infections and focuses on experimental vaccination approaches and perspectives on vaccine design against these bacteria.

ADAPTIVE IMMUNITY IS ESSENTIAL FOR DEFENSE AGAINST RICKETTSIAL INFECTIONS

Immunodeficient mice that lack T and B cells are highly susceptible to the infection with various rickettsiae [12,46,47], while common laboratory wild-type mouse strains such as C57BL/6 and BALB/c mice are resistant. This clearly indicates that adaptive immunity is essential for defense against rickettsiae.

Cytotoxic $CD8^+$ T cells generally play a major role in defense against most intracellular pathogens and are also activated in the infection of mice with various rickettsiae as well as *O. tsutsugamushi*. The cells express IFNγ and show enhanced cytotoxic activity [48,49,50,51,52]. In case of the infection with SFG (*R. rickettsii, R. conorii*), transitional rickettsiae (*R. australis*), and *O. tsutsugamushi*, $CD8^+$ T cells seem to be indispensable for defense as reflected by the observations that $CD8^+$ T cell-deficient or depleted mice show reduced survival, enhanced bacterial burden and pathology in the infection with *R. conorii, R. australis* and *O. tsutsugamushi* [48,51,52,53]. Furthermore, mice are protected against the infection with *R. conorii* as well as with *O. tsutsugamushi* after the adoptive transfer of immune $CD8^+$ T cells [51,53]. In the case of *O. tsutsugamushi*, a long-lasting $CD8^+$ T cell response seems to be important for the control of persisting bacteria as the depletion of $CD8^+$ T cells months after the infection leads to reactivation of the bacteria in mice [51]. Similar seems to be true for the infection with *R. typhi* that has also been demonstrated to persist in C57BL/6 mice as well as in BALB/c mice [12]. In concordance with the persistence of *R. typhi*, both mouse strains show a long-lasting $CD8^+$ as well as a $CD4^+$ T cell response that is even sporadically reactivated in BALB/c mice over time [49,50], indicating that both cell populations are needed for protection against recurrence.

The protective activity of $CD8^+$ T cells in defense against the infection with SFG and transitional rickettsiae as well as *O. tsutsugamushi* mainly relies on the cytotoxic activity of $CD8^+$ T cells rather than the production of IFNγ. Perforin-knockout mice that lack the cytotoxic activity of $CD8^+$ T cells show higher susceptibility to *R. australis* than mice that lack the expression of IFNγ [48]. These mice also succumb to the infection with *O. tsutsugamushi* with enhanced bacterial burden in several organs [51].

The importance of $CD8^+$ T cells and cytotoxic activity of these cells, however, may vary in defense against different rickettsial species. $CD8^+$ T cells from *R. typhi*-infected mice also express IFNγ and granzyme B [49,50], indicating enhanced cytotoxic activity. Although it has been described that the depletion of $CD8^+$ T cells leads to enhanced bacterial burden and pathology in *R. typhi*-infected C3H/HeN mice [54], other studies show that $CD4^+$ T cells are sufficient for protection against this pathogen. In contrast to the infection with *R. australis*, $CD8^+$ T cell-deficient C57BL/6 mice are not susceptible to the infection with *R. typhi* [50]. Moreover, the adoptive transfer of either $CD8^+$ or $CD4^+$ T cells into immunodeficient mice of different genetic backgrounds (C57BL/6, BALB/c) protects the animals from *R. typhi*-mediated disease and leads to bacterial elimination [49,50]. In contrast to the infection with SFG rickettsiae or *R. australis*, however, the cytotoxic activity seems to be dispensable for the protective effect of $CD8^+$ T cells against *R. typhi*. This is demonstrated by the observation that BALB/c perforin-knockout mice are not susceptible to infection with this pathogen. Moreover, the transfer of Perforin-knockout $CD8^+$ T cells still protects immunodeficient mice BALB/c CB17 SCID mice against *R. typhi*. Interestingly, $CD8^+$ $IFN\gamma^{-/-}$ T cells were revealed to be less efficient than $CD8^+$ $Perforin^{-/-}$ T cells to keep persisting *R. typhi* below the qPCR detection limit in this infection model [49,50], suggesting that the production of IFNγ by $CD8^+$ T cells is more important than the cytotoxic activity for long-term control of the bacteria.

The protective capacity of $CD4^+$ T cells has been demonstrated in murine infection models of SFG as well as of TG rickettsiae. Although the depletion of $CD4^+$ T cells does not alter the course of the disease in the infection of C3H/HeN mice with a sublethal dose of *R. conorii*, adoptive transfer of immune $CD4^+$ T cells protects these animals against challenge with a normally lethal dose of this pathogen [53]. Similarly, adoptive transfer of $CD4^+$ T cells protects *R. typhi*-infected immunodeficient BALB/c CB17 SCID mice as well C57BL/6 $RAG1^{-/-}$ mice [49,50]. In these experiments, adoptive transfer of either $CD8^+$ or $CD4^+$ T cells was comparably protective against *R. typhi,* although $CD8^+$ T cells were clearly quicker in bacterial elimination. These findings indicate that $CD4^+$ T cells are sufficient for protection against TG rickettsiae, at least *R. typhi*.

$CD4^+$ T cells differentiate into T_H1 cells that produce IFNγ and TNFα in the infection with rickettsiae and orientiae. Both cytokines activate the expression of inducible nitric oxide synthase (iNOS) in infected target cells such as ECs and MΦ, which leads to the production of bactericidal nitric oxide (NO) and bacterial killing [49,55,56]. Both cytokines are important in defense against SFG as well as TG rickettsiae and orientiae [49,57], whereby IFNγ may play a more critical role in the infection with SFG and transitional rickettsiae. For example, IFNγ-deficient C57BL/6 mice show enhanced susceptibility to *R. conorii* as well as *R. australis* [48,57]. In contrast, IFNγ-knockout BALB/c mice were shown to be resistant to *R. typhi* [49]. In this case, it can be assumed that $CD8^+$ T cells and the cytotoxic activity of these cells may compensate for the absence of IFNγ. However, adoptive transfer of $CD4^+$ T cells from BALB/c IFNγ-knockout mice still protects congenic immunodeficient mice against *R. typhi* [49]. These cells show a T_H17 phenotype and produce IL-17, TNFα, and IL-22, indicating that also T_H17 cells can confer protection against *R. typhi*. However, T_H17 cells were also shown to have pathological effects via the coproduction of IL-17 and TNFα. Neutralization of one or the other cytokine led to enhanced survival of $CD4^+$ T_H17 recipients [49].

The role of the humoral response is considered to be less important than the cellular arm of adaptive immunity in primary defense because B cells start to produce high-affinity antibodies relatively late in the infection with rickettsiae (>day 15 in the infection with *R. typhi*, >day 16 and >day 25 in the infection with *R. conorii* and *R. africae*) [58,59]. Nonetheless, antibodies can contribute to the protection, as demonstrated by passive immunization experiments. Administration of polyclonal immune serum from *R. conorii*-infected C3H/HeN mice into C3H SCID mice protects the animals against a lethal challenge with *R. conorii* [47]. Even in already infected C3H SCID mice, the application of immune serum leads to prolonged survival and reduced bacterial load [47]. Targets of the humoral response are likely surface proteins of the bacteria that are easily accessible for antibodies. Bound to surface proteins, antibodies can opsonize the bacteria for the uptake by phagocytes, inhibit the receptor-mediated uptake of the bacteria into target cells, or induce complement activation and bacterial destruction.

IMMUNOPATHOLOGY IN RICKETTSIAL INFECTIONS

Only a few descriptions of immunopathological mechanisms in rickettsial infections are found in the literature. These relate to the infection with *O. tsutsugamushi* and *R. typhi*.

O. tsutsugamushi enters MΦ and replicates in these cells. Unlike many other intracellular bacteria, it induces an M1 phenotype. *O. tsutsugamushi*-infected human as well as murine MΦ produce NO as well as enhanced levels of inflammatory cytokines including IL-1β and TNFα [60,61,62]. It has recently further been shown that *O. tsutsugamushi* not only survives and replicates in murine MΦ despite the presence of NO, but that NO even enhances bacterial replication in MΦ [63]. This M1 polarization likely depends on TLR2 as *O. tsutsugamushi* has been demonstrated to use this receptor to induce the secretion of TNFα and IL-6 in DCs [64]. Although IL-1β, TNFα, and other pro-inflammatory cytokines that are produced by *O. tsutsugamushi*-infected MΦ and DCs contribute to a protective T_H1-polarized immune response, they also induce inflammatory reactions in the tissue environment. Therefore, M1 MΦ are considered to be largely responsible for tissue pathology that is observed in scrub typhus patients [62]. In support of this, it was shown that *O. tsutsugamushi*-infected Toll-like receptor 2 $(TLR2)^{-/-}$ C57BL/6 mice were even better protected from lethal infection compared to wild-type mice and showed lower bacterial burden and milder symptoms of disease [64]. These observations indicate that the inflammatory effects of MΦ and maybe also DCs are responsible for more severe disease and that TLR2 is dispensable for the induction of protective immunity.

In addition to that, also $CD8^+$ T cells have been shown to be involved in MΦ-mediated tissue pathology in experimentally infected mice. *O. tsutsugamushi*-infected C57BL/6 mice show lung inflammation and hepatic injury. The latter was shown to be dependent on the infiltration of $CD8^+$ T cells, followed by MΦ infiltration. Furthermore, inflammation of the lung could be attributed to $CD8^+$ T cells [51]. These observations indicate a positive feedback mechanism between activated $CD8^+$ T cells and MΦ, most likely via $CD8^+$ T cell-derived IFNγ as an activator of MΦ, that accelerates the inflammatory response and leads to enhanced pathology in scrub typhus disease. On the other hand, *O. tsutsugamushi*-infected $CD8^+$ T cell-deficient mice show enhanced lethality and uncontrolled bacterial growth, although these mice produce enhanced IFNγ levels and show stronger MΦ responses in the organs, which is correlated to enhanced tissue damage [51]. In this case, it is speculated that the absence of $CD8^+$ T cells results in enhanced activation of $CD4^+$ T cells as a compensatory mechanism and that IFNγ that is produced by these cells drives MΦ activation and pathology. Together these observations indicate that especially IFNγ, produced by either $CD8^+$ or $CD4^+$ T cells, can enhance tissue pathology by activating MΦ.

Other, yet unclear mechanisms, that may contribute to pathology are the development of anti-nuclear antibodies (ANAs) that are observed in around 40% of scrub typhus patients [65] and the release of IL-17. Levels of IL-17 are generally enhanced in scrub typhus patients and higher in patients who suffer from headaches [66], suggesting a causal relationship.

Inflammatory MΦ can clearly enhance tissue damage. Other cells of the innate immune system that can be involved in pathology also include neutrophils. *R. typhi*-infected immunodeficient BALB/c CB17 SCID mice develop severe liver necrosis. In the absence of neutrophils upon depletion of this cell population, the mice succumb to the infection with the same kinetics and develop comparable bacterial loads in all organs as control groups, but the depletion of neutrophils completely prevents liver damage [46]. In addition to neutrophils, also MΦ play clearly a role in pathology in the infection with *R. typhi*. In contrast to the infection with *O. tsutsugamushi*, however, MΦ hardly respond to *R. typhi* in vitro and do not show an M1 phenotype per se. They do not produce inflammatory cytokines or NO upon infection with *R. typhi* but exclusively upregulate major histocompatibility class I (MHCI) [46]. This indicates that the bacteria are either not recognized in a classical manner, e.g., via TLR, or that the bacteria actively suppress MΦ activation. In BALB/c CB17 SCID mice, *R. typhi* predominantly resides in MΦ [46]. MΦ also expressed iNOS and produced, together with NK cells, high amounts of IFNγ. The expression of iNOS, however, was restricted to those MΦ that did not harbor *R. typhi* [46], indicating that activation of these cells appears through indirect mechanisms, probably endogenous danger signals that are released from damaged tissue or IFNγ produced by NK cells. In another model of *R. typhi* infection (immunodeficient C57BL/6 RAG1$^{-/-}$ mice), the bacteria are also found predominantly in MΦ. These mice develop severe central nervous system (CNS) inflammation, which is due to massive accumulation and activation of microglia as well as to the presence of infiltrating MΦ. In contrast to BALB/c CB17 SCID mice, these infiltrating MΦ carry *R. typhi* and express iNOS [12]. Here, the expression of iNOS and CNS inflammation is largely enhanced by the adoptive transfer of immune CD4^{+} T cells but not CD8^{+} T cells relatively late in the infection, although the bacteria were efficiently eliminated by both cell populations [50]. This observation suggests that brain inflammation in this model is mainly due to immunopathology rather than cellular destruction by the bacteria themselves. This MΦ activation can be put down to the release of IFNγ by CD4^{+} T cells [50] and again, as in the infection with *O. tsutsugamushi*, demonstrates that the T cell-derived IFNγ-MΦ axis, although essential for protection, has pathological side effects.

Whether immunopathology plays a role in the infection with SFG rickettsiae is unclear. *R. conorii* has been demonstrated to induce an MΦ M2 phenotype with reduced production of reactive oxygen species (ROS), among other effects that inhibit pro-inflammatory signaling and M1 polarization [10]. This argues against a major contribution of MΦ to pathology in the infection with SFG rickettsiae agents.

VACCINATION AGAINST RICKETTSIAE WITH WHOLE-CELL ANTIGEN (WCA)

The first attempts of immunization against rickettsiae were made with inactivated intact bacteria that were either produced in arthropods, embryonated chicken eggs, embryonal chicken fibroblasts, or infected animals. The first whole-cell antigen (WCA) vaccines against *R. prowazekii* and *R. rickettsii* were produced already in the 1920s. R. L. Weigl produced *R. prowazekii* by intrarectal injection of the bacteria into lice and fed the arthropods on humans. The bacteria were prepared from the gut of the lice and inactivated in phenol. This vaccine not only protected guinea pigs from disease [67] but was also used for the vaccination of German soldiers during World War II [68]. A similar vaccine was produced by the U.S. military at the same time where *R. prowazekii* was grown in chicken egg yolk sacs and inactivated in formalin. Vaccination of U.S. soldiers during World War II ameliorated disease [69]. Another vaccine against epidemic typhus was produced by isolation of *R. prowazekii* from the lungs of infected rabbits (Castaneda vaccine) [69] or the tunica vaginalis and the peritoneum from infected rats (*Zinsser-Castaneda* vaccine) and inactivation of the bacteria in formalin [70]. The application of three doses of the Zinsser-Castaneda vaccine was sufficient to protect guinea pigs from the disease [71].

Similarly, in 1924 the first WCA vaccine against RMSF was developed by growing *R. rickettsii* in ticks that were fed on guinea pigs. The bacteria were isolated by triturating the arthropods and inactivated in phenol and formalin [72]. In another attempt, *R. rickettsii* was grown and isolated from embryonated chicken eggs. In this way, the *Cox* vaccine, also inactivated *R. rickettsii*, was produced [73]. Administration of these inactivated bacteria leads to milder disease in humans [73] and the production of antibodies but does not completely prevent infection and disease [74]. Similarly, formalin-inactivated *R. rickettsii* that were produced by the U.S. military in the 1970s in embryonal chicken fibroblasts [75,76] protected cynomolgus and rhesus monkeys after two times immunization [77,78]. This vaccine ameliorated disease in humans but did not prevent the infection [79].

Formalin-inactivated bacteria were also used for first attempts for vaccination against scrub typhus. Here, *O. tsutsugamushi* was isolated from homogenized formalin-fixed lungs from infected cotton rats [80,81] or purified, followed by formalin-inactivation [82]. The vaccination with such inactivated *O. tsutsugamushi* only partially protected mice against challenge with the homologous bacterial strain in early studies from the 1940s [83,84], while a more recent study described protection of C3H/HeN mice against challenge with the homologous strain and long-term immunity (>8 months) [85]. Vaccination of humans, however, did not prevent infection and disease [82].

Phenol or formalin treatment of the bacteria can result in the modification of antigenic determinants, which could explain the ineffectiveness of such vaccines. Other possibilities of inactivation that can preserve antigenic structures include heat-inactivation at 56 °C or irradiation. Irradiated *O. tsutsugamushi* was found to protect mice against challenge with homologous bacteria [86,87,88], and heat-inactivated *R. rickettsii* protected dogs from severe RMSF [89], indicating a higher protective capacity compared to formalin-fixed bacteria.

Another possibility is the use of avirulent or attenuated bacteria. Examples are the vaccination with a low-virulence strain of *O. tsutsugamushi* that efficiently induces immunity in humans [90] and vaccination with *R. prowazekii* strain Madrid E. This strain was isolated during World War II and lost virulence during several passages through embryonated chicken eggs. *R. prowazekii* Madrid E has been successfully used for the vaccination of humans and induces long-term immunity up to approximately five years [91,92]. The use of such strains, however, bears the risk of reversion to a pathogenic form. Avirulence of *R. prowazekii* Madrid E is due to a mutation in the methyltransferase that is responsible for methylation of surface proteins, including OmpB. In *R. prowazekii* Madrid E, this protein, as well as other surface proteins, is hypomethylated [93]. After passage through mice, *R. prowazekii* Madrid E shows a reversion of this mutation, and reisolates of these bacteria (*R. prowazekii* Evir) are pathogenic again [94].

Stably attenuated rickettsial strains that can be produced by the introduction of mutations or the deletion of virulence genes may provide a safer way of vaccination. Genetic manipulation of rickettsiae is possible, and an attenuated strain of *R. prowazekii* was produced by site-directed knockout of the gene encoding for phospholipase D [95] that is involved in phagosomal escape [96]. Guinea pigs that were immunized with these bacteria were protected against lethal challenge with virulent *R. prowazekii* [95].

Although attenuated mutant or knockout strains are promising vaccine candidates, virulence factors that are essential for infectivity and pathogenicity are largely unknown and still need to be identified. These may include other proteins that are involved in bacterial adherence and invasion, e.g., OmpA and OmpB. Knockout of OmpA, however, did not influence the infectivity of *R. rickettsii* in guinea pigs [97]. Another problem with this kind of vaccine is that large-scale production is time-consuming and expensive, and hardly applicable for the immunization of a larger portion of people in affected areas.

Therefore, other strategies and vaccines that can be produced much more easily at larger amounts are needed.

IMMUNOGENIC DETERMINANTS AND VACCINE CANDIDATES

The development of such vaccines requires the knowledge of immunodominant rickettsial antigens that can induce protective adaptive immune responses as well as the elucidation of the optimal way of antigen delivery.

So far, only a few rickettsial antigens have been described. Most of these have been identified because they are recognized by antibodies. The most prominent ones belong to the surface cell antigen (Sca) autotransporter family (Sca 0–5) that are involved in bacterial adherence and uptake into target cells. Among this family, especially the outer membrane protein A (OmpA/Sca0), which is not expressed by TG rickettsiae, and OmpB/Sca5 represent immunodominant surface antigens that are recognized by antibodies and also by T cells in infected mice and patients [98]. Passive immunization with antibodies against OmpA and OmpB protects C3H/HeN mice and guinea pigs from normally lethal challenges with *R. rickettsii* and *R. conorii* [99,100,101,102] and even C3H SCID mice against infection with *R. conorii* [47]. Antibodies against OmpA and OmpB have been shown to enhance the uptake of *R. conorii* by phagocytic cells [103], to inhibit adherence of *R. rickettsii* to L929 cells [104] as well as to mediate complement-mediated killing of the bacteria [102] so that all three mechanisms may contribute to protection.

The majority of other antigens described in the literature are also surface-expressed proteins that are predominantly recognized by antibodies. Exceptions are Sca4 and the molecular chaperone GroEL, both of which are cytosolic proteins. GroEL, however, has also been demonstrated to be surface-exposed in SFG as well as TG rickettsiae and to be recognized by antibodies [105,106] that can enhance bacterial uptake into phagocytes [106]. Table 2 provides an overview of all so-far-identified rickettsial antigens.

Table 2: Overview on experimentally identified rickettsial immunogens, their localization, and recognition by B and/or T cells. The table provides an overview of so-far-known antigens from rickettsiae and orientia, their localization and function, and whether they are recognized by B and/or T cells. OM: outer membrane, IM: inner membrane, C: cytoplasm, P: periplasm, EC: extracellular, √: experimentally proven recognition. Empty fields: not described

Rickettsial	**Localiza-tion**	**Function**	**Recognition by**		
Immunogens			**B**	**CD4+**	**CD8+**
Sca0 (OmpA)	OM	adhesion and invasion	√	√	
Sca1	OM	adhesion and invasion	√		
Sca2	OM	adhesion and invasion	√		
Sca3	OM	adhesion and invasion	√		
Sca4	C	binds and activates vinculin [107]	√		
Sca5 (OmpB)	OM	adhesion and invasion	√	√	√
Adr1	OM	adhesion and invasion, binds vitronectin, confers resistance to complement-mediated killing [108,109]	√		
Adr2	OM	adhesion and invasion, binds vitronectin, confers resistance to complement-mediated killing [110]	√	√	√
TolC	OM	adhesion and invasion of vascular endothelial cells [111]	√		
OmpW	OM	adhesion and invasion of vascular endothelial cells [111]			
Porin-4	IM/OM/EC	export of glycostructures (eg. LPS O-antigen)	√		
YbgF	OM/C	tol-pal system protein	√	√	√
GroEL	C/OM }	60 kDa heat shock protein, molecular chaperone; surface-exposed [105,106,112]	√		
PrsA	OM/C	Parvulin-like peptidyl-prolyl cis-trans isomerase (Parvulin-like PPIase), protein export protein	√		
RplY	C/OM	50S ribosomal protein L25/ general stress protein Ctc	√		

RpsB	C/OM	30S ribosomal protein S2	√		
SurA	C/OM	chaperone SurA, parvulin-like peptidyl-prolyl isomerase	√		
RP403	C/OM	RecB family exonuclease			√
RP598	C/OM	transcription repair coupling factor			√
RP739	IM	ADP/ATP carrier protein (tlc5)			√
RP778	C/OM	DNA polymerase III a chain (dnaE)			√
RP884	C	ferrochelatase (hemE)			√
Orientia immunogens	**Localization**	**Function**	**B**	**$CD4^+$**	**$CD8^+$**
Sta22	OM	TSA47, transposase/DegP-like serin protease	√	√	
Sta47	C/P	TSA56, multi-pass membrane protein	√		
Sta56	OM	autotransporter protein	√	√	
ScaA	OM	autotransporter protein	√	√	√
ScaC	OM	autotransporter protein	√		
ScaD	OM	autotransporter protein	√		
ScaE	C/OM	TSA47, transposase/DegP-like serin protease	√		

Only a few of these proteins (OmpA, OmpB, Adr2, YbgF, and ScaA from orientia) have been shown to be also detected by T cells. Experimental evidence for the recognition by B and T cells of these antigens is reviewed elsewhere in more detail [113]. Generally, data on antigen-specific T cell responses, however, are rare, and immunodominant antigens that are recognized by $CD4^+$ and/or $CD8^+$ T cells still need to be identified. Experimentally, this can be achieved by immunoprecipitation of MHCII from professional APCs such as DCs and MΦ treated with live or inactivated bacteria, or of MHCI from cells infected with rickettsiae. Bound peptides can then be identified by mass spectrometry. Such studies, however, are still missing.

Other possibilities include the use of bioinformatic tools. Meanwhile, several bioinformatic tools are available that can assist in the determination of antigenic proteins and vaccine design by predicting the general immunogenicity of a protein (Vaxign and Vaxitope [114], VaxiJen [115]), potential B cell epitopes (ANTIGENpro, APBpred, Epitome [116,117,118]), potential $CD4^+$ and $CD8^+$ T cell epitopes and the probability of MHCI or MHCII presentation (PREDBALB/c, PRED(TAP), MHCPred, NetMHCpan, NetMHCIIpan, IEBD Analysis Resource, RANKPEP and SYFPEITHI [119,120,121,122,123,124,125,126]). In addition, knowledge of the predicted localization of a protein (SOSUIGramN, pSORTb, SignalP, SecretomeP [127,128,129,130]) and its function can be helpful to estimate whether it might be accessible for protective antibodies or the MHCI and MHCII presentation pathways to be recognized by $CD4^+$ or $CD8^+$ T cells. Bioinformatic approaches have been successfully used for the identification of five antigens from *R. prowazekii* that are recognized by $CD8^+$ T cells (RP403, RP598, RP739, RP778, RP884) [131,132]. These antigens were expressed in SVEC 4–10 cells, and immunization of mice with these cells induced antigen-specific $CD8^+$ T cells that produced IFNγ and granzyme B and protected the mice from lethal challenge with *R. typhi* [131,132].

Except for RP884, which is a cytosolic protein, the other four proteins are surface-exposed. Generally, it can be assumed that surface-exposed proteins or proteins that are released by the bacteria are accessible for the proteasome in the cytosol for degradation and transport into the MHCI presentation pathway for recognition by $CD8^+$ T cells.

EXPERIMENTAL APPROACHES OF VACCINATION AGAINST RICKETTSIAE

Because of the important role of $CD8^+$ T and $CD4^+$ T cells in protection against rickettsiae, it stands to reason that a vaccine should address cellular immune responses, ideally in addition to the production of antibodies. While the induction of $CD4^+$ T cell responses can be easily achieved by the application of recombinant protein, the difficulty in addressing $CD8^+$ T cells with a vaccine lies in the delivery of the antigen into the cytosol of host cells to gain access to the MHCI presentation pathway. Antigen delivery into the cytoplasm of host cells can be achieved by different methods such as immunization with nucleotides, vector-based vaccines, or the use of APCs that express rickettsial antigens. Experimental approaches to vaccination against rickettsiae are described in the following. Figure 1 provides an overview of all experimental vaccination approaches described so far in the literature.

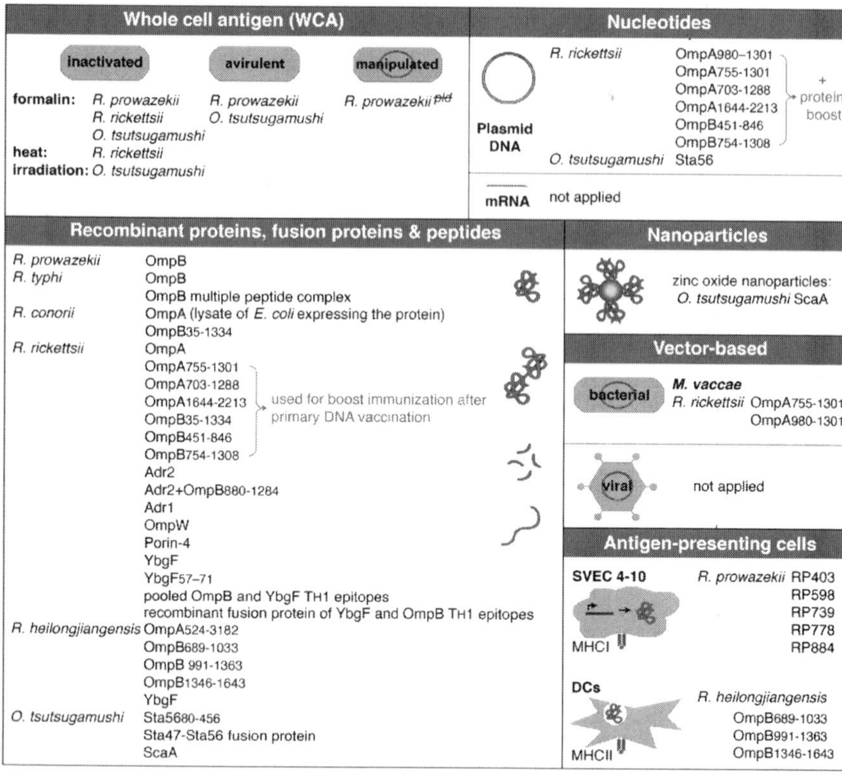

Figure 1: Experimental approaches of vaccination against rickettsiae. The figure shows all so far described approaches of vaccination against rickettsiae and names strains and antigens. WCA immunization was performed with either inactivated bacteria or avirulent strains. In addition, an attenuated knock-out strain of *R. prowazekii* that lacks phospholipase D was generated and used for immunization in experimental infection of mice. The vast majority of vaccinations were performed with recombinant proteins, fusion proteins, or peptides. Other methods include recombinant protein coupled to nanoparticles, bacteria (*M. vaccae*), or transfected cells that express rickettsial antigens and DCs that were pulsed with recombinant protein. mRNA vaccination and vaccination with adenoviral vectors as they are used now for the immunization against SARS-Cov2 with great success have not been applied yet, but represent great new tools that should be taken under consideration.

Immunization with Recombinant Proteins and Peptides

A conventional way of immunization is the application of recombinant proteins, and most approaches of vaccination against rickettsial infections in experimental animal models used either proteins, protein fragments, fusion proteins, or peptides.

OmpA and OmpB are clearly immunodominant antigens that have been extensively used for the experimental vaccination of mice. Both proteins are recognized by B as well as by T cells in the infection of animals and humans. T cells from *R. rickettsii*, *R. typhi* and *R. felis* react to MΦ that express fragments of *R. rickettsii* OmpB with the release of IL-2 and IFNγ, indicating the recognition of peptides presented by MHCI by $CD8^{+}$ T cells and cross-reaction of T cells to conserved OmpB epitopes between SFG and TG rickettsiae [133]. This is an important point with regard to vaccination because conserved proteins such as OmpB have the potential to mediate immunity against various rickettsial species. In the case of OmpA and OmpB, recombinant proteins or protein fragments have been used for experimental immunization of animals, mainly mice or guinea pigs. Vaccination of guinea pigs with recombinant OmpA from *R. rickettsii* or truncated OmpA from *R. heilongjiangensis* protects the animals against challenges with the homologous bacteria [134,135]. In the case of the immunization with *R. heilongjiangensis* OmpA, also cross-protection against *R. rickettsii* was achieved [135]. Here, antibody production may well play a role in protection. The transfer of monoclonal antibodies against OmpA has been shown to protect immunodeficient mice from fatal infection with *R. conorii* [47]. The same was true for the application of monoclonal antibodies against OmpB [47]. In another study, it was shown that guinea pigs were protected against challenge with *R. conorii* and partially protected against the infection *R. rickettsii* upon immunization with a lysate from *E. coli* that expressed OmpA [136].

Similarly, vaccination of guinea pigs with OmpB from *R. typhi* protected the animals against challenge with this pathogen [137]. The immunization of rabbits with OmpB from *R. prowazekii* induces antibody production, and B cell epitopes were identified by the analysis of antibody binding to overlapping synthetic peptides (Table 2) [138]. An additional B cell epitope was determined from *R. typhi* OmpB and from *R. typhi* Sca1, Sca2, Sca3, and Sca4 [139]. All of these peptides are recognized by antibodies upon immunization of rabbits with a multiple peptide antigen conjugate [139].

Further, also epitopes that are recognized by $CD4^+$ T cells and $CD8^+$ T cells have been identified from *R. rickettsii* and *R. conorii* OmpB and from another protein, YbgF, from *R. rickettsii* (Table 2). Immunization of C3H/HeN mice with pooled $CD4^+$ T cell epitopes from OmpB and YbgF or a fusion protein of these epitopes resulted in the induction of $CD4^+$ T_H1 cells that produced TNFα and IFNγ as well as in enhanced IgG1 and IgG2a production and reduced bacterial load upon infection with *R. rickettsii* [140]. The immunization of C3H/HeN mice with recombinant YbgF protein leads to enhanced proliferation and IFNγ release by both $CD4^+$ and $CD8^+$ T cells, prolonged IgG2a and IgG1 production, and reduced bacterial burden in the infection with *R. rickettsii* [141]. Vaccination with another recombinant immunogen from *R. rickettsii*, TolC, was less efficient than immunization with YbgF [141]. Similarly, the immunization of C3H/HeN mice with recombinant YbgF from *R. heilongjiangensis* results in reduced bacterial load upon infection with homologous bacteria [142]. The authors further demonstrate that YbgF is recognized by $CD4^+$ T cells as well as by B cells in the infection with *R. heilongjiangensis* [142]. Other immunogenic proteins that have been used for experimental vaccination are Adr1, Adr2, OmpW, and Porin-4 from *R. rickettsii*. The immunization of C3H/HeN mice with recombinant Adr1, TolC, OmpW, or Porin-4 results in reduced bacterial load upon challenge with *R. rickettsii* [111]. Similarly, the immunization with recombinant Adr2 protected the animals from *R. rickettsii* infection and led to enhanced production of IFNγ by $CD4^+$ T cells and TNFα by $CD8^+$ T cells and increased IgG2a and IgG1 production [143]. Adr2 and OmpB have also been used in combination for experimental vaccination with the same effect [144].

In the case of orientia that phylogenetically differs from rickettsiae, three proteins have been used for experimental vaccination: Sta47, Sta56, and ScaA. *O. tsutsugamushi*-infected mice, as well as humans, develop Sta56-specific antibodies and $CD4^+$ T cells [145,146,147] and antibodies against Sta47 [148]. vSta56-immunized mice produced Sta56-specific antibodies and showed enhanced proliferation of lymphocytes, which was associated with increased IFNγ and IL-2 production. Moreover, the mice were protected against challenge with the homologous *O. tsutsugamushi* strain, which produced enhanced antibody levels and lymphocytes showed increased proliferation and IFNγ and IL-2 release [149,150,151]. In a more recent study, conserved blocks of the Sta56 protein were used for the immunization of mice. This vaccination not only conferred protection against the infection with homologous bacteria but also heterologous orientia genotypes [152]. The authors further synthesized overlapping peptides from the Sta56 protein and could identify 39 peptides that are recognized by CD8⁺ T cells. Immunization with a mixture of these peptides also provided protection against lethal challenge with *O. tsutsugamushi* [152], underlining the protective activity of cytotoxic CD8⁺ T cells. In addition, a fusion protein of Sta56 and Sta47 has been used for experimental vaccination, which was, however, only partially protective against the infection with homologous orientia [153], and the vaccination of primates (*Macaca fascicularis*) with a recombinant Sta56 fragment ($Sta56_{80-456}$) was only weakly protective and did not prevent disease and rickettsemia [154].

Other immunogenic proteins from orientia are the surface proteins Sta22, ScaA, ScaC, ScaD, and ScaE. All of these antigens have been shown to be recognized by antibodies in the infection with *O. tsutsugamushi* with a stronger response to ScaA and ScaC compared to ScaE [155,156]. In the case of Sta22, it has been shown that *O. tsutsugamushi*-infected mice develop Sta22-specific CD4⁺ T cells [155]. Of these proteins, only recombinant ScaA and ScaC have been used for experimental vaccination. Of these, only the immunization with ScaA protected mice from challenge with homologous as well as heterologous orientia strains [157]. The authors further showed that ScaA-specific antibodies inhibit the uptake of orientia by non-phagocytic HeLa cells.

Immunization with Antigen-Coupled Nanoparticles

There is one description in the literature where antigen-coupled nanoparticles were used for the immunization of mice against *O. tsutsugamushi*. Recombinant ScaA protein from *O. tsutsugamushi* was coupled to zinc oxide nanoparticles. These particles were taken up by DCs in vitro. Immunization of C57BL/6 mice with these particles induced CD4⁺ as well as CD8⁺ T cells that produced IFNγ as well as the generation of antibodies. Moreover, ScaA/nanoparticle immunization protected the animals against lethal challenge with *O. tsutsugamushi* [158]. The use of nanoparticles for vaccine development has gained interest in the past years because nanoparticles can stabilize antigens and enhance the uptake of antigen by APCs, and in this way, overcome otherwise probably low immunogenicity. Furthermore, the use of nanoparticles allows targeted antigen delivery and slow release [159]. However, there are no further descriptions of the use of nanoparticles for experimental immunization against rickettsia.

Immunization with Nucleotides

One possibility of cytotoxic $CD8^+$ and $CD4^+$ T_H cell-oriented vaccination is the use of DNA. DNA immunization has been proven in various animal models of infections to efficiently induce cellular immunity. Upon intramuscular or intradermal application, the DNA is taken up by muscle cells and monocytes that then start to express the encoded protein. Intracellular cytosolic processing of the protein results in the generation of antigenic peptides that are presented by MHCI molecules for recognition by $CD8^+$ T cells. In addition, $CD4^+$ T cells can be induced by APCs that engulf the protein when released from the cells and present antigenic peptides via MHCII molecules [160,161].

DNA vaccination has been successfully applied for the induction of protective immunity against SFG rickettsiae and *O. tsutsugamushi* in experimental murine infection models. Heterologous prime-boost immunization was used for vaccination against SFG rickettsiae. DNA encoding for fragments of the OmpA protein ($OmpA_{703-1288}$, $OmpA_{755-1301}$ or $OmpA_{980-1301}$ or $OmpA_{1644-2213}$) from *R. conorii* in addition to a plasmid encoding for IL-12 was used for primary immunization followed by boost immunization with the corresponding recombinant protein fragments. In this way, lymphocytes were induced that produced IFNγ upon in vitro restimulation with *R. conorii* whole-cell antigen, and the mice were protected against normally lethal challenge with *R. conorii* [162,163]. Similarly, primary immunization with DNA encoding for fragments of the OmpB protein from *R. conorii* ($OmpB_{451-846}$ or $OmpB_{754-1308}$) followed by boost immunization with the corresponding recombinant protein fragments led to the same result, and combined DNA immunization with plasmids encoding for four protein fragments of OmpA and OmpB ($OmpA_{703-1288}$ and $OmpA_{1644-2213}$ or $OmpB_{451-846}$ and $OmpB_{754-1308}$) was protective against lethal challenge with the pathogen [163]. In the case of *O. tsutsugamushi*, DNA encoding for the Sta56 protein was used for the immunization of mice. Here, single immunization with the plasmid vector was not sufficient for protection, while partial protection (60% of the mice) was achieved after four immunizations with plasmid DNA [164].

The application of DNA, however, bears the risk of integration and persistence of the introduced DNA in the cellular genome. In addition, DNA vaccination can induce the generation of anti-DNA antibodies that can have serious side effects. A more elegant, safer, and modern way is the immunization with mRNA, which is shortly described in the following and reviewed in more detail elsewhere [165]. The mRNA encodes for an antigen or a part of an antigen and instructs the cells to transiently produce the encoded protein without the need to pass the nucleus membrane and without integration into the cellular genome. Conventional mRNAs carry the coding sequence of an antigen flanked by regulatory regions. Another form of mRNA vaccine is based on the modification of the genome of positive-stranded RNA viruses to obtain self-amplifying mRNA encoding for the antigen of choice, which ensures prolonged and robust expression of the antigen and subsequently better induction of adaptive immune responses. Conventional as well as self-amplifying mRNA vaccines are usually delivered packed with lipid nanoparticles (LNPs) as a vehicle that enhances the uptake of the material into the cell. Another method of delivery is the complexation of the mRNA with nucleotide-binding peptides such as protamine that stabilizes the mRNA and enhances its uptake into the cell. Apart from that, protamine acts as adjuvants by activating innate cells via the pattern recognition receptors (PRR) TLR7 and 8 [166,167,168]. In addition, bacterial and viral RNAs have been shown to be recognized by TLRs 3 and 7 [169], and in this way, possess an intrinsic adjuvant effect themselves [170,171]. This adjuvant effect, especially on professional APCs, is needed for efficient induction of adaptive immune response, and the complexation of mRNA with protamine has been demonstrated to enhance cytotoxic $CD8^+$ T cell and $CD4^+$ T_H1 responses [167].

As these responses are also desired in protection against rickettsiae, mRNA vaccination is a promising strategy but has not been applied yet. Generally, the design of DNA as well as mRNA vaccines is quite flexible and offers the opportunity to combine several antigenic determinants from different proteins to obtain a broader spectrum of antigen-specific adaptive immune responses.

Regarding DNA and also mRNA immunization, one could also think of constructs that encode for a combination of antigenic determinants from different proteins, probably as a fusion protein. For example, $CD4^+$ T cell epitopes have been identified from the *R. rickettsii* OmpB protein ($OmpB_{152–166}$ (QNVVVQFNNGAAIDN), $OmpB_{399–413}$ (NTDFGNLAAQIKVPN), $OmpB_{563–577}$(TIDLQANGGTIKLTS), $OmpB_{698–712}$(TNPLAEINFGSKGVN) and $OmpB_{1411–1425}$ (NLMIGAAIGITKTDI)) and from the *R. rickettsii* YbgF protein ($YbgF_{57–71}$ (LQHKIDLLTQNSNIS) [140]. These peptides, used alone or pooled or expressed as a recombinant fusion protein, induced protective immunity in C3H/HeN mice with the induction of $CD4^+$ T_H1 cells and antibody response [140]. A comparable vaccine that primarily addresses

$CD4^+$ T cells may also be sufficient for protection against *R. typhi*, and DNA or mRNA constructs could be designed for the expression of such fusion proteins.

Furthermore, five $CD8^+$ T cell epitopes from the *R. conorii* OmpB protein have been described ($OmpB_{708-716}$ (SKGVNVDTV), $OmpB_{789-797}$ (ANSTLQIGG), $OmpB_{812-820}$ (IVEFVNTGP), $OmpB_{735-743}$ (ANVGSFVFN), and $OmpB_{749-757}$ (IVSGTVGGQ) [172] that could also be integsrated into a fusion construct to obtain a $CD8^+$ T cell response in addition to $CD4^+$ T cells and antibody production.

The development of efficient DNA or mRNA vaccines may require codon-optimization to enable robust expression of rickettsial antigens in eukaryotic cells because the rickettsial genome possesses a very high A/T content. In addition, the efficacy of DNA and mRNA immunization can be generally significantly improved by several methods [173], e.g., the use of liposomes that facilitate the uptake into the cells after injection, the use of adjuvants or bicistronic constructs encoding for the antigen of choice in addition to costimulatory molecules or cytokines such as IL-12 that contribute to more efficient immune induction.

Vector-Based Immunization: Adenoviral Vectors

Genetically engineered replication-incompetent adenoviral vectors allow the efficient introduction of the transported genetic material into eukaryotic cells and have the potential to induce potent humoral as well as cellular responses. Different adenoviral vectors are in use for vaccination against SARS-Cov2, and adenoviral vectors have been studied as carriers for vaccinating antigens from several pathogens such as human immunodeficiency virus type I (HIV-1), *Plasmodium falciparum,* and *Mycobacterium* (*M.*) *tuberculosis* [174,175,176]. Different types of human adenoviral vectors have been studied. Human adenoviral vectors may be recognized by preexisting antibodies that are found in a very large proportion of the population. These antibodies can reduce the vector uptake and expression of the transgene, leading to reduced specific immune responses [177,178,179]. A chimpanzee adenoviral vector is an alternative. The immunizing effect of adenoviral vectors is generally very promising. Use for vaccination against rickettsiae has not been described yet, but should be taken into consideration in the future. This method offers similar opportunities as the design of DNA or mRNA vaccines with regard to flexibility in the combination of different antigens.

Vaccination with Genetically Modified Bacterial Vectors

Mycobacterium (*M.*) *vaccae* is an environmental member of the mycobacterial family and non-pathogenic for humans. It belongs to the same genus as *M. tuberculosis*, contains many homologous antigens, and is a promising vaccine in humans to prevent tuberculosis (e.g., Vaccae™ vaccine) used in an irradiation-killed or heat-killed form [180]. Immunization of mice with heat-killed *M. vaccae* itself induces cytotoxic $CD8^+$ T cells that react to *M. tuberculosis*-infected MΦ and produce IFNγ and [181] and triggers a $CD4^+$ T_H1 response [182]. *M. vaccae* was also used for the expression of *M. tuberculosis* antigens. Applied to mice, such vaccine induces a T_H1-biased *M. tuberculosis* antigen-specific response [183].

Similarly, genetically modified *M. vaccae* can potentially be used to induce immunity against other pathogens, including rickettsiae, as described in one study. Here, a plasmid encoding for fragments of the OmpA protein from *R. rickettsii* ($OmpA_{755-1301}$ or $OmpA_{980-1301}$) was introduced into *M. vaccae*. The engineered bacteria were then used for the immunization of C3H/HeN mice, followed by a boost immunization with recombinant $OmpA_{755-1301}$ or $OmpA_{980-1301}$ protein. In this way, IFNγ-producing rickettsia-specific T cells were induced, and the immunization mediated partial protection against challenge with *R. conorii* at a normally lethal dose [162].

Immunization with Antigen-Expressing Cells or Antigen-Pulsed APCs

Recently, $CD8^+$ T cell antigens from R. prowazekii have been identified by bioinformatic approaches in reverse vaccinology (RP403, RP598, RP739, RP778, RP884) [131,132]. These antigens were recombinantly expressed in SVEC 4–10 cells to be presented by MHCI. Transfected SVEC 4–10 cells were then used for the immunization of C3H/HeN mice. The antigens were recognized by $CD8^+$ T cells, and the immunization induced protective immunity to lethal challenge with *R. typhi* [131,132]. These are the only descriptions of the use of antigen-expressing cells for immunization, however.

Another possibility to induce $CD4^+$ T cell responses is the use of professional APCs that are pulsed with recombinant antigenic proteins. This approach has been applied for immunization against the infection with *R. heilongjiangensis*. Bone marrow-derived DCs (bmDCs) from C3H/HeN mice were pulsed with recombinant fragments of the OmpB protein from the bacteria ($OmpB_{689-1033}$, $OmpB_{991-1363}$, $OmpB_{1346-1643}$) and transferred into naïve C3H/HeN mice followed by challenge with *R. heilongjiangensis* 14 days afterward. The immunization resulted in reduced bacterial load and led to the activation of $CD4^+$ as well as $CD8^+$ T cells that produced IFNγ and TNFα, indicating a $CD4^+$ T_H1-biased and cytotoxic $CD8^+$ T cell response [184].

These methods are highly interesting for the determination of immunogenic parts of a protein but would only be applicable for the immunization of individuals because the MHC haplotype must match for the recognition by T cells.

CONCLUSIONS

Several animal models for the infection with various rickettsiae have been used for experimental vaccination against these pathogens with different methods and some success. A limiting factor still is missing knowledge about immunogenic rickettsial determinants in general and especially about those that are recognized by T cells. As the cellular arm of the adaptive immune response clearly plays a dominant role in defense against rickettsial infections, such antigens need to be identified. Another focus should be the way of antigen delivery. So far, recombinant proteins and plasmid DNA immunization have been predominantly used for experimental vaccination of animals. Other promising ways of antigen delivery include the use of mRNA and adenoviral vectors, both of which are now successfully in use against the SARS-Cov2 pandemic. Other aspects that should be addressed and taken into consideration include the use of appropriate adjuvants and heterologous or homologous prime/boost regimens.

ACKNOWLEDGMENTS

I thank Bernhard Fleischer for carefully reading the manuscript and discussions.

REFERENCES

1. Izzard L., Fuller A., Blacksell S.D., Paris D.H., Richards A.L., Aukkanit N., Nguyen C., Jiang J., Fenwick S., Day N.P., et al. Isolation of a novel Orientia species (*O. chuto* sp. nov.) from a patient infected in Dubai. *J. Clin. Microbiol.* 2010;48:4404–4409. doi: 10.1128/JCM.01526-10.
2. Abarca K., Martinez-Valdebenito C., Angulo J., Jiang J., Farris C.M., Richards A.L., Acosta-Jamett G., Weitzel T. Molecular Description of a Novel Orientia Species Causing Scrub Typhus in Chile. *Emerg. Infect. Dis.* 2020;26:2148–2156. doi: 10.3201/eid2609.200918.
3. Abdad M.Y., Abou Abdallah R., Fournier P.E., Stenos J., Vasoo S. A Concise Review of the Epidemiology and Diagnostics of Rickettsioses: Rickettsia and Orientia spp. *J. Clin. Microbiol.* 2018;56:e01728-17. doi: 10.1128/JCM.01728-17.
4. Mansueto P., Vitale G., Cascio A., Seidita A., Pepe I., Carroccio A., di Rosa S., Rini G.B., Cillari E., Walker D.H. New insight into immunity and immunopathology of Rickettsial diseases. *Clin. Dev. Immunol.* 2012;2012:967852. doi: 10.1155/2012/967852.
5. Sahni S.K., Rydkina E. Host-cell interactions with pathogenic Rickettsia species. *Future Microbiol.* 2009;4:323–339. doi: 10.2217/fmb.09.6.
6. Silverman D.J., Bond S.B. Infection of human vascular endothelial cells by Rickettsia rickettsii. *J. Infect. Dis.* 1984;149:201–206. doi: 10.1093/infdis/149.2.201.
7. Hackstadt T. The biology of rickettsiae. *Infect. Agents Dis.* 1996;5:127–143.
8. Kim M.J., Kim M.K., Kang J.S. Involvement of lipid rafts in the budding-like exit of Orientia tsutsugamushi. *Microb. Pathog.* 2013;63:37–43. doi: 10.1016/j.micpath.2013.06.002.
9. Radulovic S., Price P.W., Beier M.S., Gaywee J., Macaluso J.A., Azad A. Rickettsia-macrophage interactions: Host cell responses to Rickettsia akari and Rickettsia typhi. *Infect. Immun.* 2002;70:2576–2582. doi: 10.1128/IAI.70.5.2576-2582.2002.
10. Curto P., Riley S.P., Simoes I., Martinez J.J. Macrophages Infected by a Pathogen and a Non-pathogen Spotted Fever Group Rickettsia Reveal Differential Reprogramming Signatures Early in Infection. *Front. Cell. Infect. Microbiol.* 2019;9:97. doi: 10.3389/fcimb.2019.00097.

11. Drevets D.A., Leenen P.J., Greenfield R.A. Invasion of the central nervous system by intracellular bacteria. *Clin. Microbiol. Rev.* 2004;17:323–347. doi: 10.1128/CMR.17.2.323-347.2004.
12. Osterloh A., Papp S., Moderzynski K., Kuehl S., Richardt U., Fleischer B. Persisting Rickettsia typhi Causes Fatal Central Nervous System Inflammation. *Infect. Immun.* 2016;84:1615–1632. doi: 10.1128/IAI.00034-16.
13. Regan J.J., Traeger M.S., Humpherys D., Mahoney D.L., Martinez M., Emerson G.L., Tack D.M., Geissler A., Yasmin S., Lawson R., et al. Risk factors for fatal outcome from rocky mountain spotted Fever in a highly endemic area-Arizona, 2002–2011. *Clin. Infect. Dis.* 2015;60:1659–1666. doi: 10.1093/cid/civ116.
14. Kuloglu F. Rickettsial infections. *Dis. Mol. Med.* 2013;1:39–45. doi: 10.5455/dmm.20130704110837.
15. Raoult D., Woodward T., Dumler J.S. The history of epidemic typhus. *Infect. Dis. Clin. N. Am.* 2004;18:127–140. doi: 10.1016/S0891-5520(03)00093-X.
16. Li Z., Xin H., Sun J., Lai S., Zeng L., Zheng C., Ray S.E., Weaver N.D., Wang L., Yu J., et al. Epidemiologic Changes of Scrub Typhus in China, 1952–2016. *Emerg. Infect. Dis.* 2020;26:1091–1101. doi: 10.3201/eid2606.191168.
17. Centers for Disease Control and Prevention Murine typhus—Hawaii, 2002. *MMWR Morb. Mortal. Wkly. Rep.* 2003;52:1224–1226.
18. Centers for Disease Control and Prevention Outbreak of Rickettsia typhi infection—Austin, Texas, 2008. *MMWR Morb. Mortal. Wkly. Rep.* 2009;58:1267–1270.
19. Adjemian J., Parks S., McElroy K., Campbell J., Eremeeva M.E., Nicholson W.L., McQuiston J., Taylor J. Murine typhus in Austin, Texas, USA, 2008. *Emerg. Infect. Dis.* 2010;16:412–417. doi: 10.3201/eid1603.091028.
20. Parola P., Miller R.S., McDaniel P., Telford S.R., 3rd, Rolain J.M., Wongsrichanalai C., Raoult D. Emerging rickettsioses of the Thai-Myanmar border. *Emerg. Infect. Dis.* 2003;9:592–595. doi: 10.3201/eid0905.020511.
21. Suputtamongkol Y., Suttinont C., Niwatayakul K., Hoontrakul S., Limpaiboon R., Chierakul W., Losuwanaluk K., Saisongkork W. Epidemiology and clinical aspects of rickettsioses in Thailand.

Ann. N. Y. Acad. Sci. 2009;1166:172–179. doi: 10.1111/j.1749-6632.2009.04514.x.

22. Phongmany S., Rolain J.M., Phetsouvanh R., Blacksell S.D., Soukkhaseum V., Rasachack B., Phiasakha K., Soukkhaseum S., Frichithavong K., Chu V., et al. Rickettsial infections and fever, Vientiane, Laos. *Emerg. Infect. Dis.* 2006;12:256–262. doi: 10.3201/eid1202.050900.
23. Maude R.R., Maude R.J., Ghose A., Amin M.R., Islam M.B., Ali M., Bari M.S., Majumder M.I., Tanganuchitcharnchai A., Dondorp A.M., et al. Serosurveillance of Orientia tsutsugamushi and Rickettsia typhi in Bangladesh. *Am. J. Trop. Med. Hyg.* 2014;91:580–583. doi: 10.4269/ajtmh.13-0570.
24. Dupont H.T., Brouqui P., Faugere B., Raoult D. Prevalence of antibodies to Coxiella burnetti, Rickettsia conorii, and Rickettsia typhi in seven African countries. *Clin. Infect. Dis.* 1995;21:1126–1133. doi: 10.1093/clinids/21.5.1126.
25. Acuna-Soto R., Calderon-Romero L., Romero-Lopez D., Bravo-Lindoro A. Murine typhus in Mexico City. *Trans. R. Soc. Trop. Med. Hyg.* 2000;94:45. doi: 10.1016/S0035-9203(00)90432-2.
26. Dittrich S., Rattanavong S., Lee S.J., Panyanivong P., Craig S.B., Tulsiani S.M., Blacksell S.D., Dance D.A., Dubot-Peres A., Sengduangphachanh A., et al. Orientia, rickettsia, and leptospira pathogens as causes of CNS infections in Laos: A prospective study. *Lancet. Glob. Health.* 2015;3:e104–e112. doi: 10.1016/S2214-109X(14)70289-X.
27. Koliou M., Psaroulaki A., Georgiou C., Ioannou I., Tselentis Y., Gikas A. Murine typhus in Cyprus: 21 paediatric cases. *Eur. J. Clin. Microbiol. Infect. Dis.* 2007;26:491–493. doi: 10.1007/s10096-007-0310-8.
28. Psaroulaki A., Christou C., Chochlakis D., Tsiligianni I., Sandalakis V., Georgalis L., Ioannou I., Giorgalas G., Tselentis Y. Murine typhus in Cyprus: A 9-year survey. *Trans. R. Soc. Trop. Med. Hyg.* 2012;106:489–495. doi: 10.1016/j.trstmh.2012.02.010.
29. Chaliotis G., Kritsotakis E.I., Psaroulaki A., Tselentis Y., Gikas A. Murine typhus in central Greece: Epidemiological, clinical, laboratory, and therapeutic-response features of 90 cases. *Int. J. Infect. Dis.* 2012;16:e591–e596. doi: 10.1016/j.ijid.2012.03.010.

30. Lledo L., Gegundez M.I., Medina J., Gonzalez J.V., Alamo R., Saz J.V. Epidemiological study of Rickettsia typhi infection in two provinces of the north of Spain: Analysis of sera from the general population and sheep. *Vector Borne Zoonotic Dis.* 2005;5:157–161. doi: 10.1089/vbz.2005.5.157.
31. Bernabeu-Wittel M., del Toro M.D., Nogueras M.M., Muniain M.A., Cardenosa N., Marquez F.J., Segura F., Pachon J. Seroepidemiological study of Rickettsia felis, Rickettsia typhi, and Rickettsia conorii infection among the population of southern Spain. *Eur. J. Clin. Microbiol. Infect. Dis.* 2006;25:375–381. doi: 10.1007/s10096-006-0147-6.
32. Nogueras M.M., Cardenosa N., Sanfeliu I., Munoz T., Font B., Segura F. Serological evidence of infection with Rickettsia typhi and Rickettsia felis among the human population of Catalonia, in the northeast of Spain. *Am. J. Trop. Med. Hyg.* 2006;74:123–126. doi: 10.4269/ajtmh.2006.74.123.
33. Espinosa N., Canas E., Bernabeu-Wittel M., Martin A., Viciana P., Pachon J. The changing etiology of fever of intermediate duration. *Enferm. Infecc. Y Microbiol. Clin.* 2010;28:416–420. doi: 10.1016/j.eimc.2009.07.014.
34. Andre E., Correia R., Castro P., Neto M., Rola J., Bacelar F., Oliveira I., Velosa I., Feio A., Filipe A. Murine typhus in Portugal. *Acta Med. Port.* 1998;11:81–85.
35. Bacellar F., Lencastre I., Filipe A.R. Is murine typhus re-emerging in Portugal? *Eurosurveillance.* 1998;3:18–20. doi: 10.2807/esm.03.02.00124-en.
36. Badiaga S., Benkouiten S., Hajji H., Raoult D., Brouqui P. Murine typhus in the homeless. *Comp. Immunol. Microbiol. Infect. Dis.* 2012;35:39–43. doi: 10.1016/j.cimid.2011.09.008.
37. Dhimal M., Dumre S.P., Sharma G.N., Khanal P., Ranabhat K., Shah L.P., Lal B.K., Jha R., Upadhyaya B.P., Acharya B., et al. An outbreak investigation of scrub typhus in Nepal: Confirmation of local transmission. *BMC Infect. Dis.* 2021;21:193. doi: 10.1186/s12879-021-05866-6.
38. Bise G., Coninx R. Epidemic typhus in a prison in Burundi. *Trans. R. Soc. Trop. Med. Hyg.* 1997;91:133–134. doi: 10.1016/S0035-9203(97)90198-X.

39. Raoult D., Roux V., Ndihokubwayo J.B., Bise G., Baudon D., Marte G., Birtles R. Jail fever (epidemic typhus) outbreak in Burundi. *Emerg. Infect. Dis.* 1997;3:357–360. doi: 10.3201/eid0303.970313.

40. Raoult D., Ndihokubwayo J.B., Tissot-Dupont H., Roux V., Faugere B., Abegbinni R., Birtles R.J. Outbreak of epidemic typhus associated with trench fever in Burundi. *Lancet.* 1998;352:353–358. doi: 10.1016/S0140-6736(97)12433-3.

41. Tarasevich I., Rydkina E., Raoult D. Outbreak of epidemic typhus in Russia. *Lancet.* 1998;352:1151. doi: 10.1016/S0140-6736(05)79799-3.

42. Lutwick L.I. Brill-Zinsser disease. *Lancet.* 2001;357:1198–1200. doi: 10.1016/S0140-6736(00)04339-7.

43. Parker R.T., Menon P.G., Merideth A.M., Snyder M.J., Woodward T.E. Persistence of Rickettsia rickettsii in a patient recovered from Rocky Mountain spotted fever. *J. Immunol.* 1954;73:383–386.

44. Hove M.G., Walker D.H. Persistence of rickettsiae in the partially viable gangrenous margins of amputated extremities 5 to 7 weeks after onset of Rocky Mountain spotted fever. *Arch. Pathol. Lab. Med.* 1995;119:429–431.

45. Chung M.H., Lee J.S., Baek J.H., Kim M., Kang J.S. Persistence of Orientia tsutsugamushi in humans. *J. Korean Med. Sci.* 2012;27:231–235. doi: 10.3346/jkms.2012.27.3.231.

46. Papp S., Moderzynski K., Rauch J., Heine L., Kuehl S., Richardt U., Mueller H., Fleischer B., Osterloh A. Liver Necrosis and Lethal Systemic Inflammation in a Murine Model of Rickettsia typhi Infection: Role of Neutrophils, Macrophages and NK Cells. *PLoS Negl. Trop. Dis.* 2016;10:e0004935. doi: 10.1371/journal.pntd.0004935.

47. Feng H.M., Whitworth T., Olano J.P., Popov V.L., Walker D.H. Fc-dependent polyclonal antibodies and antibodies to outer membrane proteins A and B, but not to lipopolysaccharide, protect SCID mice against fatal Rickettsia conorii infection. *Infect. Immun.* 2004;72:2222–2228. doi: 10.1128/IAI.72.4.2222-2228.2004.

48. Walker D.H., Olano J.P., Feng H.M. Critical role of cytotoxic T lymphocytes in immune clearance of rickettsial infection. *Infect. Immun.* 2001;69:1841–1846. doi: 10.1128/IAI.69.3.1841-1846.2001.

49. Moderzynski K., Heine L., Rauch J., Papp S., Kuehl S., Richardt U., Fleischer B., Osterloh A. Cytotoxic effector functions of T cells are not required for protective immunity against fatal Rickettsia typhi infection in a murine model of infection: Role of TH1 and TH17 cytokines in protection and pathology. *PLoS Negl. Trop. Dis.* 2017;11:e0005404. doi: 10.1371/journal.pntd.0005404.
50. Moderzynski K., Papp S., Rauch J., Heine L., Kuehl S., Richardt U., Fleischer B., Osterloh A. CD4+ T Cells Are as Protective as CD8+ T Cells against Rickettsia typhi Infection by Activating Macrophage Bactericidal Activity. *PLoS Negl. Trop. Dis.* 2016;10:e0005089. doi: 10.1371/journal.pntd.0005089.
51. Hauptmann M., Kolbaum J., Lilla S., Wozniak D., Gharaibeh M., Fleischer B., Keller C.A. Protective and Pathogenic Roles of CD8+ T Lymphocytes in Murine Orientia tsutsugamushi Infection. *PLoS Negl. Trop. Dis.* 2016;10:e0004991. doi: 10.1371/journal.pntd.0004991.
52. Xu G., Mendell N.L., Liang Y., Shelite T.R., Goez-Rivillas Y., Soong L., Bouyer D.H., Walker D.H. CD8+ T cells provide immune protection against murine disseminated endotheliotropic Orientia tsutsugamushi infection. *PLoS Negl. Trop. Dis.* 2017;11:e0005763. doi: 10.1371/journal.pntd.0005763.
53. Feng H., Popov V.L., Yuoh G., Walker D.H. Role of T lymphocyte subsets in immunity to spotted fever group Rickettsiae. *J. Immunol.* 1997;158:5314–5320.
54. Walker D.H., Popov V.L., Feng H.M. Establishment of a novel endothelial target mouse model of a typhus group rickettsiosis: Evidence for critical roles for gamma interferon and CD8 T lymphocytes. *Lab. Investig. A J. Tech. Methods Pathol.* 2000;80:1361–1372. doi: 10.1038/labinvest.3780144.
55. Feng H.M., Walker D.H. Mechanisms of intracellular killing of Rickettsia conorii in infected human endothelial cells, hepatocytes, and macrophages. *Infect. Immun.* 2000;68:6729–6736. doi: 10.1128/IAI.68.12.6729-6736.2000.
56. Turco J., Winkler H.H. Gamma-interferon-induced inhibition of the growth of Rickettsia prowazekii in fibroblasts cannot be explained by the degradation of tryptophan or other amino acids. *Infect. Immun.* 1986;53:38–46. doi: 10.1128/iai.53.1.38-46.1986.
57. Feng H.M., Popov V.L., Walker D.H. Depletion of gamma interferon

and tumor necrosis factor alpha in mice with Rickettsia conorii-infected endothelium: Impairment of rickettsicidal nitric oxide production resulting in fatal, overwhelming rickettsial disease. *Infect. Immun.* 1994;62:1952–1960. doi: 10.1128/iai.62.5.1952-1960.1994.

58. Dumler J.S., Taylor J.P., Walker D.H. Clinical and laboratory features of murine typhus in south Texas, 1980 through 1987. *JAMA.* 1991;266:1365–1370. doi: 10.1001/jama.1991.03470100057033.
59. Fournier P.E., Jensenius M., Laferl H., Vene S., Raoult D. Kinetics of antibody responses in Rickettsia africae and Rickettsia conorii infections. *Clin. Diagn. Lab. Immunol.* 2002;9:324–328. doi: 10.1128/CDLI.9.2.324-328.2002.
60. Tantibhedhyangkul W., Prachason T., Waywa D., El Filali A., Ghigo E., Thongnoppakhun W., Raoult D., Suputtamongkol Y., Capo C., Limwongse C., et al. Orientia tsutsugamushi stimulates an original gene expression program in monocytes: Relationship with gene expression in patients with scrub typhus. *PLoS Negl. Trop. Dis.* 2011;5:e1028. doi: 10.1371/journal.pntd.0001028.
61. Koo J.E., Hong H.J., Dearth A., Kobayashi K.S., Koh Y.S. Intracellular invasion of Orientia tsutsugamushi activates inflammasome in asc-dependent manner. *PLoS ONE.* 2012;7:e39042. doi: 10.1371/journal.pone.0039042.
62. Tantibhedhyangkul W., Ben Amara A., Textoris J., Gorvel L., Ghigo E., Capo C., Mege J.L. Orientia tsutsugamushi, the causative agent of scrub typhus, induces an inflammatory program in human macrophages. *Microb. Pathog.* 2013;55:55–63. doi: 10.1016/j.micpath.2012.10.001.
63. Ogawa M., Satoh M., Kataoka M., Ando S., Saijo M. Nitric oxide enhanced the growth of an obligated intracellular bacterium Orientia tsutsugamushi in murine macrophages. *Microb. Pathog.* 2017;107:335–340. doi: 10.1016/j.micpath.2017.04.012.
64. Gharaibeh M., Hagedorn M., Lilla S., Hauptmann M., Heine H., Fleischer B., Keller C. Toll-Like Receptor 2 Recognizes Orientia tsutsugamushi and Increases Susceptibility to Murine Experimental Scrub Typhus. *Infect. Immun.* 2016;84:3379–3387. doi: 10.1128/IAI.00185-16.
65. Peter J.V., Griffith M.F., Prakash J.A., Chrispal A., Pichamuthu K., Varghese G.M. Anti-nuclear antibody expression in severe scrub typhus

infection: Preliminary observations. *J. Glob. Infect. Dis.* 2014;6:195–196. doi: 10.4103/0974-777X.145260.

66. Yoon H.J., Lee M.S., Ki M., Ihm C., Kim D., Kim Y., Yoo S.M. Does IL-17 play a role in hepatic dysfunction of scrub typhus patients? *Vector Borne Zoonotic Dis.* 2010;10:231–235. doi: 10.1089/vbz.2008.0149.
67. Weigl R.L. Die Methoden der aktiven Fleckfieberimmunisierung. *Bull. Int. Acad. Pol. Sci. Et Lett.* 1930;7:25–62.
68. Weigl R. Immunization against typhus fever in Poland during World War II. *Tex. Rep. Biol. Med.* 1947;5:177–179.
69. Walker D.H. The realities of biodefense vaccines against Rickettsia. *Vaccine.* 2009;27(Suppl. 4):D52–D55. doi: 10.1016/j.vaccine.2009.07.045.
70. Zinsser H., Castaneda M.R. Studies on Typhus Fever: Vii. Active Immunization against Mexican Typhus Fever with Dead Virus. *J. Exp. Med.* 1931;53:493–497. doi: 10.1084/jem.53.4.493.
71. Veintemillas F. Vaccination against typhus fever with the Zinsser-Castaneda Vaccine. *J. Immunol.* 1939;36:339–348.
72. Spencer R.R., Parker R.R. Rocky Mountain spotted fever: Vaccination of monkeys and man. *Public Health Rep.* 1925;40:2159–2167. doi: 10.2307/4577679.
73. Ecke R.S., Gilliam A.G., Snyder J.C., Yeomans A., Zarafonetis C.J., Murray E.S. The effect of Cox-type vaccine on louse-borne typhus fever; an account of 61 cases of naturally occurring typhus fever in patients who had previously received one or more injections of Cox-type vaccine. *Am. J. Trop. Med. Hyg.* 1945;25:447–462. doi: 10.4269/ajtmh.1945.s1-25.447.
74. DuPont H.L., Hornick R.B., Dawkins A.T., Heiner G.G., Fabrikant I.B., Wisseman C.L., Jr., Woodward T.E. Rocky Mountain spotted fever: A comparative study of the active immunity induced by inactivated and viable pathogenic Rickettsia rickettsii. *J. Infect. Dis.* 1973;128:340–344. doi: 10.1093/infdis/128.3.340.
75. Kenyon R.H., Pedersen C.E., Jr. Preparation of Rocky Mountain spotted fever vaccine suitable for human immunization. *J. Clin. Microbiol.* 1975;1:500–503. doi: 10.1128/jcm.1.6.500-503.1975.
76. Kenyon R.H., Sammons L.S., Pedersen C.E., Jr. Comparison of three rocky mountain spotted fever vaccines. *J. Clin. Microbiol.* 1975;2:300–304. doi: 10.1128/jcm.2.4.300-304.1975.

77. Gonder J.C., Kenyon R.H., Pedersen C.E., Jr. Evaluation of a killed Rocky Mountain spotted fever vaccine in cynomolgus monkeys. *J. Clin. Microbiol.* 1979;10:719–723. doi: 10.1128/jcm.10.5.719-723.1979.
78. Maugh T.H., 2nd Rickettsiae: A new vaccine for Rocky Mountain spotted fever. *Science.* 1978;201:604. doi: 10.1126/science.97783.
79. Clements M.L., Wisseman C.L., Jr., Woodward T.E., Fiset P., Dumler J.S., McNamee W., Black R.E., Rooney J., Hughes T.P., Levine M.M. Reactogenicity, immunogenicity, and efficacy of a chick embryo cell-derived vaccine for Rocky Mountain spotted fever. *J. Infect. Dis.* 1983;148:922–930. doi: 10.1093/infdis/148.5.922.
80. Buckland F.E., Dudgeon A. Scrub typhus vaccine; large-scale production. *Lancet.* 1945;2:734–737. doi: 10.1016/S0140-6736(45)91070-1.
81. Card W.I., Walker J.M. Scrub-typhus vaccine; field trial in South-east Asia. *Lancet.* 1947;1:481–483. doi: 10.1016/S0140-6736(47)91989 2.
82. Berge T.O., Gauld R.L., Kitaoka M. A field trial of a vaccine prepared from the Volner strain of Rickettsia tsutsugamushi. *Am. J. Hyg.* 1949;50:337–342. doi: 10.1093/oxfordjournals.aje.a119366.
83. Bailey C.A., Diercks F.H., Proffitt J.E. Preparation of a serological antigen and a vaccine for experimental tsutsugamushi disease. *J. Immunol.* 1948;60:431–441.
84. Rights F.L., Smadel J.E. Studies on scrib typhus; tsutsugamushi disease; heterogenicity of strains of R. tsutsugamushi as demonstrated by cross-vaccination studies. *J. Exp. Med.* 1948;87:339–351. doi: 10.1084/jem.87.4.339.
85. Choi Y., Kim K.S., Kim T.Y., Cheong H.S., Ahn B.Y. Long-term egg-yolk adaptation of the Orientia tsutsugamushi for preparation of a formalinized immunogen. *Vaccine.* 2006;24:1438–1445. doi: 10.1016/j.vaccine.2005.07.113.
86. Eisenberg G.H., Jr., Osterman J.V. Gamma-irradiated scrub typhus immunogens: Broad-spectrum immunity with combinations of rickettsial strains. *Infect. Immun.* 1979;26:131–136. doi: 10.1128/iai.26.1.131-136.1979.
87. Eisenberg G.H., Jr., Osterman J.V. Gamma-irradiated scrub typhus immunogens: Development and duration of immunity. *Infect. Immun.* 1978;22:80–86. doi: 10.1128/iai.22.1.80-86.1978.

88. Eisenberg G.H., Jr., Osterman J.V. Experimental scrub typhus immunogens: Gamma-irradiated and formalinized rickettsiae. *Infect. Immun.* 1977;15:124–131. doi: 10.1128/iai.15.1.124-131.1977.
89. Alhassan A., Liu H., McGill J., Cerezo A., Jakkula L., Nair A.D.S., Winkley E., Olson S., Marlow D., Sahni A., et al. Rickettsia rickettsii Whole-Cell Antigens Offer Protection against Rocky Mountain Spotted Fever in the Canine Host. *Infect. Immun.* 2019;87:e00628-18. doi: 10.1128/IAI.00628-18.
90. Kawamura R., Kasahar S., Toyama T., Nishinarita F., Tsubaki S. On the prevention of tsutsugamushi. Results of preventive inoculations for people in the endemic region, and laboratory tests with the Pescadores strain. *Trop. Dis. Bull.* 1940;37:269–270.
91. Fox J.P., Jordan M.E., Gelfand H.M. Immunization of man against epidemic typhus by infection with avirulent Rickettsia prowazeki strain E. IV. Persistence of immunity and a note as to differing complement-fixation antigen requirements in post-infection and post-vaccination sera. *J. Immunol.* 1957;79:348–354.
92. Wisseman C.L., Jr. Concepts of louse-borne typhus control in developing countries: The use of the living attenuated E strain typhus vaccine in epidemic and endemic situations. *Adv. Exp. Med. Biol.* 1972;31:97–130.
93. Ching W.M., Wang H., Davis J., Dasch G.A. Amino acid analysis and multiple methylation of lysine residues in the surface protein antigen of Rickettsia prowazekii. In: Angeletti R.H., editor. *Techniques in Protein Chemistry IV.* Academic Press, Inc.; San Diego, CA, USA: 1993. pp. 307–314.
94. Zhang J.Z., Hao J.F., Walker D.H., Yu X.J. A mutation inactivating the methyltransferase gene in avirulent Madrid E strain of Rickettsia prowazekii reverted to wild type in the virulent revertant strain Evir. *Vaccine.* 2006;24:2317–2323. doi: 10.1016/j.vaccine.2005.11.044.
95. Driskell L.O., Yu X.J., Zhang L., Liu Y., Popov V.L., Walker D.H., Tucker A.M., Wood D.O. Directed mutagenesis of the Rickettsia prowazekii pld gene encoding phospholipase D. *Infect. Immun.* 2009;77:3244–3248. doi: 10.1128/IAI.00395-09.
96. Whitworth T., Popov V.L., Yu X.J., Walker D.H., Bouyer D.H. Expression of the Rickettsia prowazekii pld or tlyC gene in Salmonella enterica serovar Typhimurium mediates phagosomal escape. *Infect. Immun.* 2005;73:6668–6673. doi: 10.1128/IAI.73.10.6668-6673.2005.

97. Noriea N.F., Clark T.R., Hackstadt T. Targeted knockout of the Rickettsia rickettsii OmpA surface antigen does not diminish virulence in a mammalian model system. *MBio.* 2015;6:e00323-15. doi: 10.1128/mBio.00323-15.

98. Teysseire N., Raoult D. Comparison of Western immunoblotting and microimmunofluorescence for diagnosis of Mediterranean spotted fever. *J. Clin. Microbiol.* 1992;30:455–460. doi: 10.1128/jcm.30.2.455-460.1992.

99. Anacker R.L., List R.H., Mann R.E., Hayes S.F., Thomas L.A. Characterization of monoclonal antibodies protecting mice against Rickettsia rickettsii. *J. Infect. Dis.* 1985;151:1052–1060. doi: 10.1093/infdis/151.6.1052.

100. Anacker R.L., McDonald G.A., List R.H., Mann R.E. Neutralizing activity of monoclonal antibodies to heat-sensitive and heat-resistant epitopes of Rickettsia rickettsii surface proteins. *Infect. Immun.* 1987;55:825–827. doi: 10.1128/iai.55.3.825-827.1987.

101. Lange J.V., Walker D.H. Production and characterization of monoclonal antibodies to Rickettsia rickettsii. *Infect. Immun.* 1984;46:289–294. doi: 10.1128/iai.46.2.289-294.1984.

102. Chan Y.G., Riley S.P., Chen E., Martinez J.J. Molecular basis of immunity to rickettsial infection conferred through outer membrane protein B. *Infect. Immun.* 2011;79:2303–2313. doi: 10.1128/IAI.01324-10.

103. Feng H.M., Whitworth T., Popov V., Walker D.H. Effect of antibody on the rickettsia-host cell interaction. *Infect. Immun.* 2004;72:3524–3530. doi: 10.1128/IAI.72.6.3524-3530.2004.

104. Li H., Walker D.H. rOmpA is a critical protein for the adhesion of Rickettsia rickettsii to host cells. *Microb. Pathog.* 1998;24:289–298. doi: 10.1006/mpat.1997.0197.

105. Qi Y., Xiong X., Wang X., Duan C., Jia Y., Jiao J., Gong W., Wen B. Proteome analysis and serological characterization of surface-exposed proteins of Rickettsia heilongjiangensis. *PLoS ONE.* 2013;8:e70440. doi: 10.1371/journal.pone.0070440.

106. Rauch J., Barton J., Kwiatkowski M., Wunderlich M., Steffen P., Moderzynski K., Papp S., Höhn K., Schwanke H., Witt S., et al. GroEL is an immunodominant surface-exposed antigen of Rickettsia typhi. *PLoS ONE.* 2021;16:e0253084. doi: 10.1371/journal.pone.0253084.

107. Park H., Lee J.H., Gouin E., Cossart P., Izard T. The rickettsia surface cell antigen 4 applies mimicry to bind to and activate vinculin. *J. Biol. Chem.* 2011;286:35096–35103. doi: 10.1074/jbc.M111.263855.

108. Fish A.I., Riley S.P., Singh B., Riesbeck K., Martinez J.J. The Rickettsia conorii Adr1 Interacts with the C-Terminus of Human Vitronectin in a Salt-Sensitive Manner. *Front. Cell. Infect. Microbiol.* 2017;7:61. doi: 10.3389/fcimb.2017.00061.

109. Riley S.P., Patterson J.L., Nava S., Martinez J.J. Pathogenic Rickettsia species acquire vitronectin from human serum to promote resistance to complement-mediated killing. *Cell. Microbiol.* 2014;16:849–861. doi: 10.1111/cmi.12243.

110. Garza D.A., Riley S.P., Martinez J.J. Expression of Rickettsia Adr2 protein in E. coli is sufficient to promote resistance to complement-mediated killing, but not adherence to mammalian cells. *PLoS ONE.* 2017;12:e0179544. doi: 10.1371/journal.pone.0179544.

111. Gong W., Xiong X., Qi Y., Jiao J., Duan C., Wen B. Identification of novel surface-exposed proteins of Rickettsia rickettsii by affinity purification and proteomics. *PLoS ONE.* 2014;9:e100253. doi: 10.1371/journal.pone.0100253.

112. Renesto P., Azza S., Dolla A., Fourquet P., Vestris G., Gorvel J.P., Raoult D. Proteome analysis of Rickettsia conorii by two-dimensional gel electrophoresis coupled with mass spectrometry. *FEMS Microbiol. Lett.* 2005;245:231–238. doi: 10.1016/j.femsle.2005.03.004.

113. Osterloh A. The neglected challenge: Vaccination against rickettsiae. *PLoS Negl. Trop. Dis.* 2020;14:e0008704. doi: 10.1371/journal.pntd.0008704.

114. He Y., Racz R., Sayers S., Lin Y., Todd T., Hur J., Li X., Patel M., Zhao B., Chung M., et al. Updates on the web-based VIOLIN vaccine database and analysis system. *Nucleic Acids Res.* 2014;42:D1124–D1132. doi: 10.1093/nar/gkt1133.

115. Doytchinova I.A., Flower D.R. VaxiJen: A server for prediction of protective antigens, tumour antigens and subunit vaccines. *BMC Bioinform.* 2007;8:4. doi: 10.1186/1471-2105-8-4.

116. Magnan C.N., Zeller M., Kayala M.A., Vigil A., Randall A., Felgner P.L., Baldi P. High-throughput prediction of protein antigenicity using protein microarray data. *Bioinformatics.* 2010;26:2936–2943. doi: 10.1093/bioinformatics/btq551.

117. Saha S., Raghava G.P. Prediction of continuous B-cell epitopes in an antigen using recurrent neural network. *Proteins.* 2006;65:40–48. doi: 10.1002/prot.21078.

118. Schlessinger A., Ofran Y., Yachdav G., Rost B. Epitome: Database of structure-inferred antigenic epitopes. *Nucleic Acids Res.* 2006;34:D777–D780. doi: 10.1093/nar/gkj053.

119. Nielsen M., Lundegaard C., Worning P., Lauemoller S.L., Lamberth K., Buus S., Brunak S., Lund O. Reliable prediction of T-cell epitopes using neural networks with novel sequence representations. *Protein Sci.* 2003;12:1007–1017. doi: 10.1110/ps.0239403.

120. Nielsen M., Lundegaard C., Worning P., Hvid C.S., Lamberth K., Buus S., Brunak S., Lund O. Improved prediction of MHC class I and class II epitopes using a novel Gibbs sampling approach. *Bioinformatics.* 2004;20:1388–1397. doi: 10.1093/bioinformatics/bth100.

121. Lin H.H., Ray S., Tongchusak S., Reinherz E.L., Brusic V. Evaluation of MHC class I peptide binding prediction servers: Applications for vaccine research. *BMC Immunol.* 2008;9:8. doi: 10.1186/1471-2172-9-8.

122. Rammensee H., Bachmann J., Emmerich N.P., Bachor O.A., Stevanovic S. SYFPEITHI: Database for MHC ligands and peptide motifs. *Immunogenetics.* 1999;50:213–219. doi: 10.1007/s002510050595.

123. Reche P.A., Reinherz E.L. Prediction of peptide-MHC binding using profiles. *Methods Mol. Biol.* 2007;409:185–200. doi: 10.1007/978-1-60327-118-9_13.

124. Zhang G.L., Srinivasan K.N., Veeramani A., August J.T., Brusic V. PREDBALB/c: A system for the prediction of peptide binding to H2d molecules, a haplotype of the BALB/c mouse. *Nucleic Acids Res.* 2005;33:W180–W183. doi: 10.1093/nar/gki479.

125. Zhang G.L., Petrovsky N., Kwoh C.K., August J.T., Brusic V. PRED(TAP): A system for prediction of peptide binding to the human transporter associated with antigen processing. *Immunome Res.* 2006;2:3. doi: 10.1186/1745-7580-2-3.

126. Guan P., Doytchinova I.A., Zygouri C., Flower D.R. MHCPred: Bringing a quantitative dimension to the online prediction of MHC binding. *Appl. Bioinform.* 2003;2:63–66.

127. Imai K., Asakawa N., Tsuji T., Akazawa F., Ino A., Sonoyama M., Mitaku S. SOSUI-GramN: High performance prediction for sub-cellular

localization of proteins in gram-negative bacteria. *Bioinformation.* 2008;2:417–421. doi: 10.6026/97320630002417.

128. Gardy J.L., Spencer C., Wang K., Ester M., Tusnady G.E., Simon I., Hua S., deFays K., Lambert C., Nakai K., et al. PSORT-B: Improving protein subcellular localization prediction for Gram-negative bacteria. *Nucleic Acids Res.* 2003;31:3613–3617. doi: 10.1093/nar/gkg602.
129. Almagro Armenteros J.J., Tsirigos K.D., Sonderby C.K., Petersen T.N., Winther O., Brunak S., von Heijne G., Nielsen H. SignalP 5.0 improves signal peptide predictions using deep neural networks. *Nat. Biotechnol.* 2019;37:420–423. doi: 10.1038/s41587-019-0036-z.
130. Bendtsen J.D., Jensen L.J., Blom N., Von Heijne G., Brunak S. Feature-based prediction of non-classical and leaderless protein secretion. *Protein Eng. Des. Sel. PEDS.* 2004;17:349–356. doi: 10.1093/protein/gzh037.
131. Gazi M., Caro-Gomez E., Goez Y., Cespedes M.A., Hidalgo M., Correa P., Valbuena G. Discovery of a protective Rickettsia prowazekii antigen recognized by CD8+ T cells, RP884, using an in vivo screening platform. *PLoS ONE.* 2013;8:e76253. doi: 10.1371/journal.pone.0076253.
132. Caro-Gomez E., Gazi M., Goez Y., Valbuena G. Discovery of novel cross-protective Rickettsia prowazekii T-cell antigens using a combined reverse vaccinology and in vivo screening approach. *Vaccine.* 2014;32:4968–4976. doi: 10.1016/j.vaccine.2014.06.089.
133. Dzul-Rosado K., Balam-Romero J., Valencia-Pacheco G., Lugo-Caballero C., Arias-Leon J., Peniche-Lara G., Zavala-Castro J. Immunogenicity of OmpA and OmpB antigens from Rickettsia rickettsii on mononuclear cells from Rickettsia positive Mexican patients. *J. Vector Borne Dis.* 2017;54:317–327. doi: 10.4103/0972-9062.225836.
134. Sumner J.W., Sims K.G., Jones D.C., Anderson B.E. Protection of guinea-pigs from experimental Rocky Mountain spotted fever by immunization with baculovirus-expressed Rickettsia rickettsii rOmpA protein. *Vaccine.* 1995;13:29–35. doi: 10.1016/0264-410X(95)80007-Z.
135. Jiao Y., Wen B., Chen M., Niu D., Zhang J., Qiu L. Analysis of immunoprotectivity of the recombinant OmpA of Rickettsia heilongjiangensis. *Ann. N. Y. Acad. Sci.* 2005;1063:261–265. doi: 10.1196/annals.1355.042.

136. Vishwanath S., McDonald G.A., Watkins N.G. A recombinant Rickettsia conorii vaccine protects guinea pigs from experimental boutonneuse fever and Rocky Mountain spotted fever. *Infect. Immun.* 1990;58:646–653. doi: 10.1128/iai.58.3.646-653.1990.

137. Bourgeois A.L., Dasch G.A. The species-specific surface protein antigen of Rickettsia typhi: Immunogenicity and protective efficacy in guinea pigs. In: Burgdorfer W., Anacker R.L., editors. *Rickettsiae and Rickettsial Diseases.* Academic Press; New York, NY, USA: 1981. pp. 71–80.

138. Ching W.M., Wang H., Jan B., Dasch G.A. Identification and characterization of epitopes on the 120-kilodalton surface protein antigen of Rickettsia prowazekii with synthetic peptides. *Infect. Immun.* 1996;64:1413–1419. doi: 10.1128/iai.64.4.1413-1419.1996.

139. Sears K.T., Ceraul S.M., Gillespie J.J., Allen E.D., Jr., Popov V.L., Ammerman N.C., Rahman M.S., Azad A.F. Surface proteome analysis and characterization of surface cell antigen (Sca) or autotransporter family of Rickettsia typhi. *PLoS Pathog.* 2012;8:e1002856. doi: 10.1371/journal.ppat.1002856.

140. Wang P., Xiong X., Jiao J., Yang X., Jiang Y., Wen B., Gong W. Thl epitope peptides induce protective immunity against Rickettsia rickettsii infection in C3H/HeN mice. *Vaccine.* 2017;35:7204–7212. doi: 10.1016/j.vaccine.2017.09.068.

141. Gong W., Qi Y., Xiong X., Jiao J., Duan C., Wen B. Rickettsia rickettsii outer membrane protein YbgF induces protective immunity in C3H/HeN mice. *Hum. Vaccines Immunother.* 2015;11:642–649. doi: 10.1080/21645515.2015.1011572.

142. Qi Y., Xiong X., Duan C., Jiao J., Gong W., Wen B. Recombinant protein YbgF induces protective immunity against Rickettsia heilongjiangensis infection in C3H/HeN mice. *Vaccine.* 2013;31:5643–5650. doi: 10.1016/j.vaccine.2013.09.064.

143. Gong W., Xiong X., Qi Y., Jiao J., Duan C., Wen B. Surface protein Adr2 of Rickettsia rickettsii induced protective immunity against Rocky Mountain spotted fever in C3H/HeN mice. *Vaccine.* 2014;32:2027–2033. doi: 10.1016/j.vaccine.2014.02.057.

144. Gong W., Wang P., Xiong X., Jiao J., Yang X., Wen B. Enhanced protection against Rickettsia rickettsii infection in C3H/HeN mice by immunization with a combination of a recombinant adhesin rAdr2 and a protein fragment rOmpB-4 derived from outer membrane protein B.

Vaccine. 2015;33:985–992. doi: 10.1016/j.vaccine.2015.01.017.

145. Chen W.J., Niu D.S., Zhang X.Y., Chen M.L., Cui H., Wei W.J., Wen B.H., Chen X.R. Recombinant 56-kilodalton major outer membrane protein antigen of Orientia tsutsugamushi Shanxi and its antigenicity. *Infect. Immun.* 2003;71:4772–4779. doi: 10.1128/IAI.71.8.4772-4779.2003.
146. Seong S.Y., Park S.G., Huh M.S., Jang W.J., Kim H.R., Han T.H., Choi M.S., Chang W.H., Kim I.S. Mapping of antigenic determinant regions of the Bor56 protein of Orientia tsutsugamushi. *Infect. Immun.* 1997;65:5250–5256. doi: 10.1128/iai.65.12.5250-5256.1997.
147. Ramaiah A., Koralur M.C., Dasch G.A. Complexity of type-specific 56 kDa antigen CD4 T-cell epitopes of Orientia tsutsugamushi strains causing scrub typhus in India. *PLoS ONE.* 2018;13:e0196240. doi: 10.1371/journal.pone.0196240.
148. Chen H.W., Zhang Z., Huber E., Mutumanje E., Chao C.C., Ching W.M. Kinetics and magnitude of antibody responses against the conserved 47-kilodalton antigen and the variable 56-kilodalton antigen in scrub typhus patients. *Clin. Vaccine Immunol.* 2011;18:1021–1027. doi: 10.1128/CVI.00017-11.
149. Seong S.Y., Huh M.S., Jang W.J., Park S.G., Kim J.G., Woo S.G., Choi M.S., Kim I.S., Chang W.H. Induction of homologous immune response to Rickettsia tsutsugamushi Boryong with a partial 56-kilodalton recombinant antigen fused with the maltose-binding protein MBP-Bor56. *Infect. Immun.* 1997;65:1541–1545. doi: 10.1128/iai.65.4.1541-1545.1997.
150. Seong S.Y., Kim H.R., Huh M.S., Park S.G., Kang J.S., Han T.H., Choi M.S., Chang W.H., Kim I.S. Induction of neutralizing antibody in mice by immunization with recombinant 56 kDa protein of Orientia tsutsugamushi. *Vaccine.* 1997;15:1741–1747. doi: 10.1016/S0264-410X(97)00112-6.
151. Choi S., Jeong H.J., Ju Y.R., Gill B., Hwang K.J., Lee J. Protective immunity of 56-kDa type-specific antigen of Orientia tsutsugamushi causing scrub typhus. *J. Microbiol. Biotechnol.* 2014;24:1728–1735. doi: 10.4014/jmb.1407.07048.
152. Kim H.I., Ha N.Y., Kim G., Min C.K., Kim Y., Yen N.T.H., Choi M.S., Cho N.H. Immunization with a recombinant antigen composed of conserved blocks from TSA56 provides broad genotype protection against scrub typhus. *Emerg. Microbes Infect.* 2019;8:946–958. doi:

10.1080/22221751.2019.1632676.

153. Yu Y., Wen B., Wen B., Niu D., Chen M., Qiu L. Induction of protective immunity against scrub typhus with a 56-kilodalton recombinant antigen fused with a 47-kilodalton antigen of Orientia tsutsugamushi Karp. *Am. J. Trop. Med. Hyg.* 2005;72:458–464. doi: 10.4269/ajtmh.2005.72.458.
154. Chattopadhyay S., Jiang J., Chan T.C., Manetz T.S., Chao C.C., Ching W.M., Richards A.L. Scrub typhus vaccine candidate Kp r56 induces humoral and cellular immune responses in cynomolgus monkeys. *Infect. Immun.* 2005;73:5039–5047. doi: 10.1128/IAI.73.8.5039-5047.2005.
155. Hickman C.J., Stover C.K., Joseph S.W., Oaks E.V. Molecular cloning and sequence analysis of a Rickettsia tsutsugamushi 22 kDa antigen containing B- and T-cell epitopes. *Microb. Pathog.* 1991;11:19 31. doi: 10.1016/0882-4010(91)90090-W.
156. Ha N.Y., Kim Y., Choi J.H., Choi M.S., Kim I.S., Kim Y.S., Cho N.H. Detection of antibodies against Orientia tsutsugamushi Sca proteins in scrub typhus patients and genetic variation of sca genes of different strains. *Clin. Vaccine Immunol.* 2012;19:1442–1451. doi: 10.1128/CVI.00285-12.
157. Ha N.Y., Sharma P., Kim G., Kim Y., Min C.K., Choi M.S., Kim I.S., Cho N.H. Immunization with an autotransporter protein of Orientia tsutsugamushi provides protective immunity against scrub typhus. *PLoS Negl. Trop. Dis.* 2015;9:e0003585. doi: 10.1371/journal.pntd.0003585.
158. Ha N.Y., Shin H.M., Sharma P., Cho H.A., Min C.K., Kim H.I., Yen N.T., Kang J.S., Kim I.S., Choi M.S., et al. Generation of protective immunity against Orientia tsutsugamushi infection by immunization with a zinc oxide nanoparticle combined with ScaA antigen. *J. Nanobiotechnol.* 2016;14:76. doi: 10.1186/s12951-016-0229-2.
159. Kheirollahpour M., Mehrabi M., Dounighi N.M., Mohammadi M., Masoudi A. Nanoparticles and Vaccine Development. *Pharm. Nanotechnol.* 2020;8:6–21. doi: 10.2174/2211738507666191024162042.
160. Cho J.H., Youn J.W., Sung Y.C. Cross-priming as a predominant mechanism for inducing CD8(+) T cell responses in gene gun DNA immunization. *J. Immunol.* 2001;167:5549–5557. doi: 10.4049/jimmunol.167.10.5549.

161. Kutzler M.A., Weiner D.B. Developing DNA vaccines that call to dendritic cells. *J. Clin. Investig.* 2004;114:1241–1244. doi: 10.1172/JCI23467.

162. Crocquet-Valdes P.A., Diaz-Montero C.M., Feng H.M., Li H., Barrett A.D., Walker D.H. Immunization with a portion of rickettsial outer membrane protein A stimulates protective immunity against spotted fever rickettsiosis. *Vaccine.* 2001;20:979–988. doi: 10.1016/S0264-410X(01)00377-2.

163. Diaz-Montero C.M., Feng H.M., Crocquet-Valdes P.A., Walker D.H. Identification of protective components of two major outer membrane proteins of spotted fever group Rickettsiae. *Am. J. Trop. Med. Hyg.* 2001;65:371–378. doi: 10.4269/ajtmh.2001.65.371.

164. Ni Y.S., Chan T.C., Chao C.C., Richards A.L., Dasch G.A., Ching W.M. Protection against scrub typhus by a plasmid vaccine encoding the 56-KD outer membrane protein antigen gene. *Am. J. Trop. Med. Hyg.* 2005;73:936–941. doi: 10.4269/ajtmh.2005.73.936.

165. Maruggi G., Zhang C., Li J., Ulmer J.B., Yu D. mRNA as a Transformative Technology for Vaccine Development to Control Infectious Diseases. *Mol. Ther. J. Am. Soc. Gene Ther.* 2019;27:757–772. doi: 10.1016/j.ymthe.2019.01.020.

166. Scheel B., Teufel R., Probst J., Carralot J.P., Geginat J., Radsak M., Jarrossay D., Wagner H., Jung G., Rammensee H.G., et al. Toll-like receptor-dependent activation of several human blood cell types by protamine-condensed mRNA. *Eur. J. Immunol.* 2005;35:1557–1566. doi: 10.1002/eji.200425656.

167. Kallen K.J., Heidenreich R., Schnee M., Petsch B., Schlake T., Thess A., Baumhof P., Scheel B., Koch S.D., Fotin-Mleczek M. A novel, disruptive vaccination technology: Self-adjuvanted RNActive((R)) vaccines. *Hum. Vaccines Immunother.* 2013;9:2263–2276. doi: 10.4161/hv.25181.

168. Rauch S., Lutz J., Kowalczyk A., Schlake T., Heidenreich R. RNActive(R) Technology: Generation and Testing of Stable and Immunogenic mRNA Vaccines. *Methods Mol. Biol.* 2017;1499:89–107. doi: 10.1007/978-1-4939-6481-9_5.

169. Chen N., Xia P., Li S., Zhang T., Wang T.T., Zhu J. RNA sensors of the innate immune system and their detection of pathogens. *IUBMB Life.* 2017;69:297–304. doi: 10.1002/iub.1625.

170. Kowalczyk A., Doener F., Zanzinger K., Noth J., Baumhof P., Fotin-Mleczek M., Heidenreich R. Self-adjuvanted mRNA vaccines induce local innate immune responses that lead to a potent and boostable adaptive immunity. *Vaccine.* 2016;34:3882–3893. doi: 10.1016/j.vaccine.2016.05.046.

171. Edwards D.K., Jasny E., Yoon H., Horscroft N., Schanen B., Geter T., Fotin-Mleczek M., Petsch B., Wittman V. Adjuvant effects of a sequence-engineered mRNA vaccine: Translational profiling demonstrates similar human and murine innate response. *J. Transl. Med.* 2017;15:1. doi: 10.1186/s12967-016-1111-6.

172. Li Z., Diaz-Montero C.M., Valbuena G., Yu X.J., Olano J.P., Feng H.M., Walker D.H. Identification of CD8 T-lymphocyte epitopes in OmpB of Rickettsia conorii. *Infect. Immun.* 2003;71:3920–3926. doi: 10.1128/IAI.71.7.3920-3926.2003.

173. Saade F., Petrovsky N. Technologies for enhanced efficacy of DNA vaccines. *Expert Rev. Vaccines.* 2012;11:189–209. doi: 10.1586/erv.11.188.

174. Harro C., Sun X., Stek J.E., Leavitt R.Y., Mehrotra D.V., Wang F., Bett A.J., Casimiro D.R., Shiver J.W., DiNubile M.J., et al. Safety and immunogenicity of the Merck adenovirus serotype 5 (MRKAd5) and MRKAd6 human immunodeficiency virus type 1 trigene vaccines alone and in combination in healthy adults. *Clin. Vaccine Immunol.* 2009;16:1285–1292. doi: 10.1128/CVI.00144-09.

175. Hill A.V., Reyes-Sandoval A., O'Hara G., Ewer K., Lawrie A., Goodman A., Nicosia A., Folgori A., Colloca S., Cortese R., et al. Prime-boost vectored malaria vaccines: Progress and prospects. *Hum. Vaccin.* 2010;6:78–83. doi: 10.4161/hv.6.1.10116.

176. Radosevic K., Wieland C.W., Rodriguez A., Weverling G.J., Mintardjo R., Gillissen G., Vogels R., Skeiky Y.A., Hone D.M., Sadoff J.C., et al. Protective immune responses to a recombinant adenovirus type 35 tuberculosis vaccine in two mouse strains: CD4 and CD8 T-cell epitope mapping and role of gamma interferon. *Infect. Immun.* 2007;75:4105–4115. doi: 10.1128/IAI.00004-07.

177. McCoy K., Tatsis N., Korioth-Schmitz B., Lasaro M.O., Hensley S.E., Lin S.W., Li Y., Giles-Davis W., Cun A., Zhou D., et al. Effect of preexisting immunity to adenovirus human serotype 5 antigens on the immune responses of nonhuman primates to vaccine regimens

based on human- or chimpanzee-derived adenovirus vectors. *J. Virol.* 2007;81:6594–6604. doi: 10.1128/JVI.02497-06.

178. Pichla-Gollon S.L., Lin S.W., Hensley S.E., Lasaro M.O., Herkenhoff-Haut L., Drinker M., Tatsis N., Gao G.P., Wilson J.M., Ertl H.C., et al. Effect of preexisting immunity on an adenovirus vaccine vector: In vitro neutralization assays fail to predict inhibition by antiviral antibody in vivo. *J. Virol.* 2009;83:5567–5573. doi: 10.1128/JVI.00405-09.
179. Xiang Z., Li Y., Cun A., Yang W., Ellenberg S., Switzer W.M., Kalish M.L., Ertl H.C. Chimpanzee adenovirus antibodies in humans, sub-Saharan Africa. *Emerg. Infect. Dis.* 2006;12:1596–1599. doi: 10.3201/eid1210.060078.
180. Gong W., Liang Y., Wu X. The current status, challenges, and future developments of new tuberculosis vaccines. *Hum. Vaccines Immunother.* 2018;14:1697–1716. doi: 10.1080/21645515.2018.1458806.
181. Skinner M.A., Yuan S., Prestidge R., Chuk D., Watson J.D., Tan P.L. Immunization with heat-killed Mycobacterium vaccae stimulates CD8+ cytotoxic T cells specific for macrophages infected with Mycobacterium tuberculosis. *Infect. Immun.* 1997;65:4525–4530. doi: 10.1128/iai.65.11.4525-4530.1997.
182. Zhang L., Jiang Y., Cui Z., Yang W., Yue L., Ma Y., Shi S., Wang C., Wang C., Qian A. Mycobacterium vaccae induces a strong Th1 response that subsequently declines in C57BL/6 mice. *J. Vet. Sci.* 2016;17:505–513. doi: 10.4142/jvs.2016.17.4.505.
183. Safar H.A., Mustafa A.S., Amoudy H.A., El-Hashim A. The effect of adjuvants and delivery systems on Th1, Th2, Th17 and Treg cytokine responses in mice immunized with Mycobacterium tuberculosis-specific proteins. *PLoS ONE.* 2020;15:e0228381. doi: 10.1371/journal.pone.0228381.
184. Meng Y., Xiong X., Qi Y., Duan C., Gong W., Jiao J., Wen B. Protective immunity against Rickettsia heilongjiangensis in a C3H/HeN mouse model mediated by outer membrane protein B-pulsed dendritic cells. *Sci. China Life Sci.* 2015;58:287–296. doi: 10.1007/s11427-014-4720-4.

CHAPTER 11

Can Digital Tools Be Used for Improving Immunization Programs

Alberto E. Tozzi[1], Francesco Gesualdo[1], Angelo D'Ambrosio[1], Elisabetta Pandolfi[1], Eleonora Agricola[1], and Pierluigi Lopalco[2]

[1]Unit of Telemedicine, IRCCS, Bambino Gesù Children's Hospital, Rome, Italy
[2]European Centre for Disease Prevention and Control, Stockholm, Sweden

ABSTRACT

In order to successfully control and eliminate vaccine-preventable infectious diseases, an appropriate vaccine coverage has to be achieved and maintained. This task requires a high level of effort as it may be compromised by a number of barriers. Public health agencies have issued specific recommendations to address these barriers and therefore improve immunization programs. In the present review, we characterize issues and challenges of immunization programs for which digital tools are a potential

---

Citation: Tozzi AE, Gesualdo F, D'Ambrosio A, Pandolfi E, Agricola E and Lopalco P (2016) Can Digital Tools Be Used for Improving Immunization Programs? Front. Public Health 4:36. doi: 10.3389/fpubh.2016.00036

solution. In particular, we explore previously published research on the use of digital tools in the following vaccine-related areas: immunization registries, dose tracking, and decision support systems; vaccine-preventable diseases surveillance; surveillance of adverse events following immunizations; vaccine confidence monitoring; and delivery of information on vaccines to the public. Subsequently, we analyze the limits of the use of digital tools in such contexts and envision future possibilities and challenges.

Keywords: vaccination strategies, vaccination, immunization, immunization programs, registries, communication, vaccine confidence, information dissemination

INTRODUCTION

Maintaining a high performance of an immunization program is one of the most challenging public health objectives. The history of immunization has shown several cases of success at a global level, including the eradication of smallpox and the control or elimination of several other vaccine-preventable diseases (1–3). Significant efforts and resources are constantly dedicated to supporting, maintaining, and improving immunization strategies, in order to achieve the goals set by national and international health agencies (2).

Several studies have been conducted to identify actions associated with an improvement of the vaccination coverage (4). In this regard, a collection of evidence-based recommendations has been issued by The Community Preventive Services Task Force (5).

Nevertheless, at a global level, a number of issues have become real threats for maintaining high immunization coverage, running an effective surveillance, and allowing immunization programs to timely react to new issues.

A large number of these issues could benefit from the adoption of digital tools (see Table Table1).1). The objective of this review is to explore the role of potential digital solutions to problems of vaccination programs.

Table 1: Potential uses of digital tools in immunization programs

Challenges	Addressed issues	Actions based on digital tools
Digitalization of immunization data	Need of data-driven activities in immunization programs	Implementation of immunization registries Integration of immunization registries with clinical information and other data
Improvement of logistics and dose tracking	Simplification of logistics in vaccine management, reduction of errors, and improvement of safety	Barcodes for dose tracking Integration of barcode scanning tech in EHRs
Decision support through appropriate algorithms for final users	Vaccination delay and vaccine hesitancy	Electronic decision support systems for health-care professionals Adoption of personal health records
Timely detection of epidemiologic signals	Delay in disease incidence reporting Low specificity of signals	Use of digital traces on the web for surveillance purposes Epidemic intelligence based on aggregation of different information sources Participatory surveillance
Improvement of vaccine safety evaluation	Underreporting and underrecognition of adverse events following immunizations	Integration and analysis of information on adverse events following immunizations obtained from EHRs and web signals
Timely assessment of the public's vaccine confidence	Reduction of public confidence in vaccinations	Acquirement and interpretation of data on vaccine confidence from web sources and social networks
Effective delivery of vaccine information to the public	Reduction of public confidence in vaccinations Vaccine hesitancy Lack of effectiveness of common communication strategies for vaccine promotion	Addressing information gaps and misconceptions through analysis of digital traces left by users Integration of different digital tools for information delivery to the public

Data Collection and Management

In most countries, especially in the developing world, the logistics of vaccination systems are paper-based, thus limiting timely update and information accessibility. Limited accessibility to vaccination information has a crucial impact on vaccination strategies, which cannot be timely and comprehensively informed by data. Monitoring vaccine safety and effectiveness can also be affected by lack of data accessibility. Moreover, in several countries, data on vaccinations are stored at the local level; therefore, citizens may have difficulties in accessing their vaccination record.

Regarding vaccination logistics, paper-based dose accountability has clear limits regarding safety and timely administration of doses according to schedules.

Digital solutions for data collection and management may streamline vaccination activities and provide important information to tailor immunization programs.

Immunization Registries

Immunization information systems (IISs) are confidential, population-based, computerized databases designed to record all immunization doses administered to a population, providing health operators with tools for maintaining a high vaccination coverage (6). IIS programs should provide solutions for (a) identification of at-risk individuals and groups; (b) management, storage, and integration of immunization data; (c) data protection; (d) facilitation in the engagement of families and individuals for timely vaccination receipt; (e) clinical decision support for health providers; and (f) framework for data sharing among health providers, at regional, national, and international levels (6).

Though several efforts have been made in some countries for adopting IIS, their use is far from being universal.

In Europe, only Denmark, Iceland, Malta, the Netherlands, and Norway have a national, fully implemented digital IIS, while six other countries have subnational IISs (7). Among the systems adopted by these countries, there is a high variability in methods, frequency of data acquisition, geographic coverage, data storage, and distribution (7, 8).

Initial efforts are being made to build digital registries at the international level, in order to facilitate global surveillance of vaccination programs and sharing of good practices. A system with such characteristics has been implemented by WHO and UNICEF: the Centralized Information System for Infectious Diseases (8).

Since 1998, Canada has had a network of regional IIS registries, which have been integrated in a national, intercommunicating network since the SARS pandemic in 2004 (9).

In the US, the CDC is working closely with electronic health record (EHR) providers to define the set of data and functionalities EHRs should have in order to be useful in the immunization setting, and to adopt a communication standard in order to allow consistent data intercommunication between health-care points and the CDC itself (10). Moreover, the American Academy of Pediatrics (AAP) is developing guidelines for tailoring immunization records to children, including parent refusal, information interchange with IISs, and clinical decision support functionalities (11).

Dose Tracking

Vaccine dose accountability may also benefit from digital solutions. The CDC vaccine tracking system (VTrckS) allows web-based ordering and tracking of publicly funded vaccines (12). The use of two-dimensional barcodes to store vaccine information (vaccine product identification, expiration date, and lot number) has been allowed by the FDA in 2011 (13, 14), is being experimented in a pilot study by the AAP and CDC (15, 16), and is already applied in Canada (17) and in Spain (13).

Integrating barcode scanning technology in EHRs has clear advantages in reducing errors and increasing safety (18, 19). In EHRs, an automatic link to vaccine information could also allow to easily tracking vaccine lots in case of adverse events following immunization (AEFI) (20).

Decision Support Systems

Clinical decision support systems help health professionals to correctly manage immunizations through the following functionalities: they suggest the appropriate immunization schedule for children based on birth date and vaccine history (21), automatically integrate changes in regulations (22, 23), proactively remind physicians of vaccinations for their patients (24), automatically recognize AEFIs (11), and suggest tailored immunizations in at-risk groups (25).

Personal health records (PHR) managed by patients and families can also improve adherence to immunization schedules through automatic notifications (26). A study performed on an adult population showed that the use of immunization PHRs correlates to a higher chance of receiving influenza immunization (27).

Developing Countries

Due to the lack of infrastructures and to high costs, adopting IISs in developing countries is a challenging task. Nevertheless, information systems enabling digital recording and transmission of immunization data are being implemented in Guatemala (28) and South Africa (29).

On the other hand, based on the observation that mobile phones are widely used in developing countries, mobile-based approaches may be promising in such contexts. For example, mobile technologies and advanced algorithms are being used to digitalize old paper-based immunization registries in low resource settings, e.g., Mozambique (30).

In Haiti, a cholera vaccination campaign has been carried out through house-by-house visits by operators equipped with wireless tablets. Children's immunization status was assessed and recorded using a family-specific bar code; data were geolocalized and sent to a central system, which provided the program staff with a real-time map of vaccination coverage (31). A similar approach has been used in China, with a mobile app for facilitating immunization data recording, tracking unimmunized children, appointment booking (32) and in Thailand, with an app for recording data during antenatal and immunization visits (33).

Tools for Vaccine-Preventable Diseases Surveillance

Traditionally, surveillance is defined as "the systematic collection, consolidation, analysis, and dissemination of data on specific diseases" (34). Classically, the main outputs of traditional surveillance systems have been indicators focused on individuals (34). More recently, surveillance activities have been aiming at rapidly capturing information about events that may represent a threat for public health and are referred to as event-based surveillance. Epidemiological surveillance is therefore extending from individual-based to event-based data (34, 35).

Traditional surveillance systems have a number of limitations:

- information is collected through health-care providers, not directly from individuals; therefore, traditional surveillance systems fail to catch signals from sick people who do not go to the doctor;
- traditional systems are based on case definitions, and therefore may miss emerging diseases with unexpected combinations of symptoms;

- there is a consistent time lag between signals of disease and production and dissemination of aggregated incidence figures.

In the context of event-based surveillance, information can be collected from news, reports, or other sources transmitted both through institutional and informal channels. Web-based data have been used to support public health in Canada since the 1990s, with the Global Public Health Intelligence Network (36), a service that automatically retrieves information about potential public health emergencies from news feed aggregators and distributes this information to public health agencies, including the WHO Global Outbreak Alert and Response Network (37).

Several studies explored the use and interpretation of spontaneous digital traces left by Internet users as a convenient and timely strategy to detect signals of trend variations in diseases and immunizations. The idea of using digital traces for syndromic surveillance was proposed by Eysenbach, who tracked demand for health information on the Internet using keyword-triggered ads for influenza (38). Subsequently, on the basis of the hypothesis that the interest of web users may correlate to disease incidence, a number of studies have focused on measuring the occurrence of specific health-related and disease-related search keywords.

Search volumes on web search engines may represent a surrogate of frequency of specific health events. A correlation between search volumes and disease trends has been shown (39). Search queries were useful to track dengue activity (40), and one study showed a correlation between search terms and laboratory confirmed cases of rotavirus infections (41). In 2008, a Google service (Google Flu Trends) has been developed to estimate and predict influenza activity by aggregating Google search query volumes (42). One study investigated the possibility of applying this approach to vaccinations, showing that search activity for HPV and H1N1 correlated to immunization coverage (43).

This "demand based" approach for surveillance has been subsequently integrated with the "supply based" approach, investigating communication contents and patterns in discussion groups, blogs, and microblogs (38), thus focusing on what users say, rather than on what users search for.

In particular, Twitter, a social network based on the sharing of short messages (up to 140 characters), which are available to the public without restrictions, has often been investigated as a source of information for infectious disease surveillance. Twitter posts are rich in data, allowing to follow disease trends both temporally and geographically. Many studies have explored whether monitoring information flow and networks on Twitter could help following the emergence of health conditions, their evolution, and the public's interest around them. Influenza surveillance has been one of the main topics in Twitter research (44), with international (45) and local scale studies (46, 47). This approach has been recently used to study the incidence trends of other infectious diseases, namely, pertussis (47), dengue (48), and cholera (49).

On the other hand, vaccine-preventable disease surveillance may benefit from systems based on an active input of information by users. Such methods, grouped under the definition of "participatory surveillance," are based on platforms (both web- and smartphone-based) that allow users to directly provide information about their health status. Typical examples of this kind of surveillance are platforms dedicated to crowdsourced influenza surveillance (e.g., Flu Near You or Influnet) (50–52), providing a powerful and precise tool for epidemic assessment.

Complex biosurveillance systems aggregate data from a variety sources: news sites, social media, crowdsourcing platforms, official resources (e.g., WHO), audio, and video sources. Information acquired by such platforms may provide geographical details both at the local and at the international level (53). One of the best examples of Internet biosurveillance systems is represented by Healthmap, which provides information about emerging and re-emerging public health threats by aggregating information from various structured and non-structured data sources (54).

GeoChat (55) is another example of an open source platform that can enable the easy deployment of crowdsourced interactive mapping applications for surveillance with web forms/e-mail, short message service (SMS), and Twitter support. This application has been used in Cambodia for disease reporting and outbreak alerts (56).

Epidemic intelligence activities can also integrate traditional surveillance systems for the detection of vaccine failure and lack of effectiveness. Detection of outbreak clusters in vaccinated populations may trigger a signal of potential vaccine failure. Such signals must be verified and properly assessed using "traditional" epidemiological techniques in order to get estimates of vaccine effectiveness and either confirm or reject the signal (57).

Tools for Surveillance of Adverse Events Following Immunizations

Assessment of vaccine safety is a priority for public health and may have a powerful impact on the success of an immunization program. AEFI may be studied before marketing authorization, in the context of phase I-III clinical trials, which, though, may not have a sufficient magnitude for adequately detecting rare AEFIs. Subsequently to licensure, AEFIs are monitored through *ad hoc*, formal studies, or, more frequently, through passive surveillance by health-care workers. Limits of such systems are under-reporting and biased reporting (8).

Innovative ways to monitor AEFIs and capture signals of vaccine safety have been explored extensively (58).

In particular, large databases of EHRs have been used to set up systems able to capture signals in a semi-automated way. The US Vaccine Safety Datalink (VSD) is the oldest of such infrastructures, established by the CDC in 1990 for population-based, post-marketing monitoring of vaccine safety (59). The VSD project includes a population of nearly 9 million individuals, including 2.1 million children, attending various participating organizations. In addition, the FDA has recently established the Post-Licensure Rapid Immunization Safety Monitoring (PRISM) system, built on the VSD model, linking information from different health insurance databases (60). Such systems are extremely efficient in addressing specific questions related to vaccine safety in a timely manner. On the other hand, they are not specifically designed to detect safety signals.

Clusters of mass psychogenic illness following pandemic influenza vaccination were detected and investigated in Taiwan in 2009 using a combination of institutional data sources (the passive AEFI surveillance system and the web-based Emergency Medical Management System) (61). Association between narcolepsy and H1N1 pandemic vaccine Pandemrix in Finland was initially flagged up by an informal digital network of neurologists and was afterward confirmed by epidemiological studies (62). The same signal would have been hardly detected through the routine AEFI passive reporting system, being an unexpected and rare event.

Surveillance of AEFIs through a participatory approach has been a matter of debate. In many AEFI surveillance systems, patients do report health events. Post-marketing surveillance of AEFI has been implemented through mobile devices (63) and SMS (64).

Monitoring Vaccine Confidence

The media resonate with uncontrolled, scaring information on vaccine safety, and the general public is exposed to conflicting information on the balance between risks and benefits of immunizations. As a matter of fact, vaccine hesitancy, due to lack of confidence in vaccinations, may have a crucial impact on vaccine uptake (65, 66).

Vaccine confidence is traditionally measured through classic surveys or interviews (67, 68).

As information regarding vaccinations is largely acquired on the Internet, web data mining may enlighten various aspects of vaccine confidence. Parents seeking vaccine information on the Internet, compared to those using other information sources, are less likely to agree with accepted principles of vaccine science and less likely to recognize the benefits of vaccinations (69). Side effects, ingredients, and immunization policies are the most searched and discussed topics on the web (70–72).

Web monitoring of vaccine confidence may allow to timely intercept negative trends and, therefore, to set up immediate actions. This objective may be achieved by collection and interpretation of heterogeneous data on vaccines derived from various web sources (73), including social networks (74).

Delivering Information on Immunizations to the Public

Vaccine information campaigns are mostly based on paper material and delivered through traditional media channels. This approach has several limitations. First, a large time lag may exist between the detection of information needs and information delivery. Second, the tailoring potential of this approach is limited, as information campaigns usually target the general population.

Digital tools and new media can be exploited as means for accurately identifying information needs and effectively delivering vaccination campaigns. A number of studies have investigated the presence of vaccine information on the web, often starting from an analysis of Google search results obtained through the use of specific keywords (72, 75–78). This approach can be useful for addressing information trends and gaps on available web pages.

Information on vaccinations published on the web, as well as most common information gaps and misconceptions, have often been assessed through an analysis of vaccine-related websites. A number of studies have analyzed both websites promoting vaccinations and anti-vaccination websites (79–81). To this regard, the World Health Organization conducts the Vaccine Safety Network, an initiative aimed at crediting web sites meeting certain quality criteria, which are periodically reviewed (82).

For a complete assessment of information trends on vaccinations, social networks cannot be neglected, as they are a fertile medium for anti-vaccine sentiments (83). Blogs commonly support variable positions on vaccinations (84); nevertheless, negative messages may have a greater influence on the decision to vaccinate, compared to pro-vaccine messages (85). Twitter has been studied as a source of information on vaccines (86), with pro-vaccine contents being more prevalent on Twitter compared to other social media (87). Moreover, the use of social media and, specifically, of Facebook, seems promising for the implementation of communication strategies targeting adolescents, who are particularly engaged in this kind of media. In this regard, a number of studies investigated patients' enrollment and HPV vaccination promotion on Facebook (88–91).

Recently, some attempts have been conducted to pilot tailored communication toward vaccine-hesitant parents (92, 93). A successful tailored education on HPV vaccine has been experienced by Chinese health professionals who collaborated with popular websites for women to gear up vaccination uptake (94).

A large part of the literature on digital tools aimed at improving immunization uptake focuses on electronic reminders. Reminders can be included in more complex interventions to improve immunization coverage (95). Several authors have documented the impact of text messaging and computerized reminders in increasing adherence to schedules, especially if integrated with educational interventions (27, 96–99). On the other hand, the evidence supporting these interventions, as for others digital tools for public health, seems insufficient (24).

COMMENTS AND DISCUSSION

Only small progresses have been made to integrate digital tools into immunization programs. Moreover, research aimed at assessing the effectiveness of digital technologies as potential responses to problems of immunization programs is scarce.

Electronic data management is not fully applied to traditional components of immunization programs yet. Electronic data on vaccinations should be integrated at the international level, in order to facilitate cross collaboration and coordination among health services and research institutes. Moreover, easy access to large amounts of immunization records may give a robust contribution to vaccine research, facilitating the design of timely and inexpensive epidemiological studies to assess vaccine efficacy and safety (9).

With regard to surveillance, crowdsourcing, mobile phones, personal computing devices, and geolocalization of information promise to become stable pillars of public health strategies. Combining these resources with other tools, such as SMS and social networks, may generate innovative instruments to support surveillance of health events and other public health activities (56). This observation extends to monitoring vaccine coverage and AEFI occurrence, where social media can enhance traditional information sources.

The analysis of search queries and social media information is a powerful approach, although it may suffer from information biases, as interest in specific diseases may be amplified by media attention, independently from the disease incidence (100, 101). Nevertheless, refined algorithms and the use of text mining techniques for sense disambiguation, topic filtering, and mood analysis (102, 103) are allowing to isolate actual signals of disease from information noise with increasing precision. The use of natural language processing techniques and of algorithms allowing an automatic or semi-automatic classification of contents may also allow to overcome the manual evaluation of contents, which is an expensive and energy-consuming task. One step toward a more accurate surveillance may be represented by participatory surveillance. A recent review showed a large potential benefit of participatory surveillance, although specificity may be limited and participating bias can affect its performance (104).

Parallel to surveillance of medical events, the use of digital tools for monitoring vaccine confidence and information needs may greatly enhance the performance of immunization programs. Vaccine hesitancy and vaccine opposition are increasingly worrying phenomena. The detection of a drop in vaccine confidence may anticipate a more severe event, such as a decrease in vaccine coverage, allowing the immunization program to react. Monitoring information needs may allow to identify issues which are potentially linked with vaccine hesitancy or opposition. Moreover, data gathering and analysis of information found on the web has the advantage to shortcut traditional channels and to allow continuous monitoring and interpretation of data (73). An interesting approach for investigating information needs may be represented by the analysis of search behaviors through the study of search terms and queries, which are commonly used by Internet users (105, 106), using specific platforms such as Google Adwords.

On the other hand, monitoring vaccine confidence and information needs on the Internet does not allow to consider segments of the public that do not have access to the Internet or that do not leave digital traces on the web. Therefore, information obtained from the Internet is not sufficient to describe the entire population and should be used to integrate other traditional sources of information. Nevertheless, web-derived information may allow to finely profile specific populations of Internet users, which may become the target of tailored immunization campaigns delivered through multiple web channels.

Business marketing strategies are based on rules for translating information delivery into monetary return of investments. The same approach may be used to rapidly spread vaccine information to different segments of the public to maintain vaccine confidence and increase immunization coverage. Information strategies should exploit more comprehensively the speed and pervasiveness of digital tools. The web community will likely become increasingly demanding, switching from a passive acceptance of static, limited contents to an active request of detailed information. Vaccine information delivery should therefore move from classical, static web sites to real interactions with the users, especially through the use of social networks. The future scenarios of immunization policies will possibly be characterized by a direct participation of the public in designing appropriate information strategies and even efficacious approaches to maintaining immunization programs.

All the potentials hitherto described can be greatly enhanced when the tool used for interaction is a mobile phone. Indeed, people who access the Internet through a smartphone rather than a computer are more likely to interact with the technology than simply consuming information (107). Apps dedicated to vaccination might be highly impacted in this population group. On the other hand, one of the issues in using electronic reminders based on cellular phone may be that the global penetration of these devices is not universal (108).

At present, a number of barriers limit the adoption of digital tools in immunization programs. First, ability to use digital tools may be hampered by cultural background and infrastructure availability. This implies an urgent need for educational activities aimed at empowering health professionals and patients for the use of such tools. Institutional stakeholders should drive political decisions toward the use of digital tools both in research and clinical activities. Second, the adoption of digital tools requires a relatively large, initial investment in human and financial resources. This may represent a limit, in particular, in developing countries or in countries suffering from financial crisis. Nevertheless, the return of such investments is high in terms of increased quality of immunization programs and subsequent cost savings. Third, the use of Internet for managing health data is subject to security and privacy issues. Strategies to maintain anonymity and preserve confidentiality are difficult to implement. Moreover, relying on proprietary resources, such as those offered by Google, may be problematic, since algorithms used by this company are not explicit (109).

Research studies assessing the impact of digital tools in immunization programs are still rare and do not follow the rapid pace of development of technology and digital tools. This discrepancy is common to other domains and should be rapidly filled up to improve immunization programs.

In conclusion, despite digital tools may greatly enhance the effectiveness of immunization programs, only few examples of implementation are available at present. The use of digital tools can favor the intersection of three crucial dimensions of immunization programs: immunization registries, surveillance of vaccine-preventable diseases, and surveillance of AEFI. The fourth dimension, represented by monitoring confidence in immunization programs, can be easily integrated through the use of digital instruments, which would allow implementation of data-driven vaccine information strategies. An expanded use of digital tools is expected to ultimately increase immunization coverage, reduce vaccine-preventable disease incidence, reduce AEFIs, and increase the active participation of the public to immunization strategies through informed decisions.

AUTHOR CONTRIBUTIONS

AET, FG and AED conceived the project and wrote the article. EP and EA performed the literature review. PL critically revised the article and gave the final approval for publication.

REFERENCES

1. World Health Organization. *The Global Eradication of Smallpox: Final Report of the Global Commission for the Certification of Smallpox Eradication, Geneva, December 1979*. Geneva: World Health Organization; (1980).
2. World Health Organization. *Global Vaccine Action Plan 2011–2020*. Geneva: World Health Organization; (2013).
3. Global polio eradication initiative: 10th meeting of the Independent Monitoring Board. *Wkly Epidemiol Rec* (2014) 89:361–7.
4. Briss PA, Rodewald LE, Hinman AR, Shefer AM, Strikas RA, Bernier RR, et al. Reviews of evidence regarding interventions to improve vaccination coverage in children, adolescents, and adults. The Task Force on Community Preventive Services. *Am J Prev Med* (2000) 18:97–140. 10.1016/S0749-3797(99)00118-X
5. *Guide to Community Preventive Services. Increasing Appropriate Vaccination* (2015). Available from: www.thecommunityguide.org/vaccines/index.html
6. Centers for Disease Control and Prevention (CDC). *Increasing Appropriate Vaccination* (2012). Available from: http://www.cdc.gov/vaccines/programs/iis/about.html
7. VENICE II Consortium. *Vaccination Coverage Assessment in EU/EEA, 2011* (2012). Available from: http://venice.cineca.org/Final_Vaccination_Coverage_Assesment_Survey_2011_1.pdf
8. Lopalco PL, Carrillo Santisteve P. Actual immunization coverage throughout Europe: are existing data sufficient? *Clin Microbiol Infect* (2014) 20(Suppl 5):7–11. 10.1111/1469-0691.12453
9. Writing team for the Public Health Agency of Canada/Canadian Institutes of Health Research Influenza Research Network Vaccine Coverage Theme Group. Why collect individual-level vaccination data? *CMAJ* (2010) 182:273–5. 10.1503/cmaj.091515
10. Centers for Disease Control and Prevention (CDC). Progress in immunization information systems – United States, 2012. *MMWR Morb Mortal Wkly Rep* (2013) 62:1005–8.
11. Spooner SA, Council on Clinical Information Technology, American Academy of Pediatrics . Special requirements of electronic health record systems in pediatrics. *Pediatrics* (2007) 119:631–7. 10.1542/peds.2006-3527

12. Centers for Disease Control and Prevention (CDC). Progress in immunization information systems – United States, 2010. *MMWR Morb Mortal Wkly Rep* (2012) 61:464–7.
13. Pebody R. Vaccine registers – experiences from Europe and elsewhere. *Euro Surveill* (2012) 17:20159.
14. O'Connor AC, Kennedy ED, Loomis RJ, Haque SN, Layton CM, Williams WW, et al. Prospective cost-benefit analysis of a two-dimensional barcode for vaccine production, clinical documentation, and public health reporting and tracking. *Vaccine* (2013) 31:3179–86. 10.1016/j.vaccine.2013.04.073
15. American Academy of Pediatrics. *Vaccine Barcoding* (2013). Available from: http://www2.aap.org/immunization/pediatricians/barcoding.html
16. Centers for Disease Control and Prevention (CDC). *Two-Dimensional (2D) Vaccine Barcodes* (2015). Available from: http://www.cdc.gov/vaccines/programs/iis/2d-vaccine-barcodes/
17. Pereira JA, Quach S, Hamid JS, Quan SD, Diniz AJ, Van Exan R, et al. The integration of barcode scanning technology into Canadian public health immunization settings. *Vaccine* (2014) 32:2748–55. 10.1016/j.vaccine.2013.11.015
18. Kaushal R, Barker KN, Bates DW. How can information technology improve patient safety and reduce medication errors in children's health care? *Arch Pediatr Adolesc Med* (2001) 155:1002–7. 10.1001/archpedi.155.9.1002
19. Larrabee S, Brown MM. Recognizing the institutional benefits of bar-code point-of-care technology. *Jt Comm J Qual Saf* (2003) 29:345–53.
20. Supply and safety issues surrounding an H1N1 vaccine. *Lancet* (2009) 374:358. 10.1016/S0140-6736(09)61395-7
21. Zhu VJ, Grannis SJ, Tu W, Rosenman MB, Downs SM. Evaluation of a clinical decision support algorithm for patient-specific childhood immunization. *Artif Intell Med* (2012) 56:51–7. 10.1016/j.artmed.2012.04.004
22. Miller PL, Frawley SJ, Brandt C, Sayward FG. Tools for immunization guideline knowledge maintenance. II. Automated Web-based generation of user-customized test cases. *Comput Biomed Res* (1998) 31:190–208. 10.1006/cbmr.1998.1471

23. Miller PL, Frawley SJ, Sayward FG. Issues in accommodating national changes and local variation in a computer-based guideline for childhood immunization and in related knowledge maintenance tools. *Proc AMIA Symp* (1998):563–7.

24. Stockwell MS, Fiks AG. Utilizing health information technology to improve vaccine communication and coverage. *Hum Vaccin Immunother* (2013) 9:1802–11. 10.4161/hv.25031

25. Fiks AG, Hunter KF, Localio AR, Grundmeier RW, Bryant-Stephens T, Luberti AA, et al. Impact of electronic health record-based alerts on influenza vaccination for children with asthma. *Pediatrics* (2009) 124:159–69. 10.1542/peds.2008-2823

26. Tom JO, Mangione-Smith R, Solomon C, Grossman DC. Integrated personal health record use: association with parent-reported care experiences. *Pediatrics* (2012) 130:e183–90. 10.1542/peds.2011-1786

27. Lau AY, Sintchenko V, Crimmins J, Magrabi F, Gallego B, Coiera E. Impact of a web-based personally controlled health management system on influenza vaccination and health services utilization rates: a randomized controlled trial. *J Am Med Inform Assoc* (2012) 19:719–27. 10.1136/amiajnl-2011-000433

28. World Health Organization. *An Integrated Approach to Health Information Systems in Guatemala* (2015). Available from: http://www.who.int/immunization/programmes_systems/supply_chain/optimize/guatemala/en/

29. *Integrated Health Information Systems for Vaccination in Developing Countries* (2015). Available from: https://www.idi.ntnu.no/~ericm/gavi.pdf

30. Dell N, Breit N, Chaluco T, Crawford J, Borriello G. Digitizing paper forms with mobile imaging technologies. Proceedings of the 2nd ACM Symposium on Computing for Development New York (2012).

31. Teng JE, Thomson DR, Lascher JS, Raymond M, Ivers LC. Using mobile health (mHealth) and geospatial mapping technology in a mass campaign for reactive oral cholera vaccination in rural Haiti. *PLoS Negl Trop Dis* (2014) 8:e3050. 10.1371/journal.pntd.0003050

32. Chen L, Wang W, Du X, Rao X, van Velthoven MH, Yang R, et al. Effectiveness of a smart phone app on improving immunization of children in rural Sichuan Province, China: study protocol for a paired

cluster randomized controlled trial. *BMC Public Health* (2014) 14:262. 10.1186/1471-2458-14-262

33. Kaewkungwal J, Singhasivanon P, Khamsiriwatchara A, Sawang S, Meankaew P, Wechsart A. Application of smart phone in "Better Border Healthcare Program": a module for mother and child care. *BMC Med Inform Decis Mak* (2010) 10:69. 10.1186/1472-6947-10-69
34. Langmuir AD. The surveillance of communicable diseases of national importance. *N Engl J Med* (1963) 268:182–92. 10.1056/NEJM196301242680405
35. World Health Organization. *A Guide to Establishing Event-Based Surveillance*. Manila: WHO Regional Office for the Western Pacific; (2008).
36. Mykhalovskiy E, Weir L. The Global Public Health Intelligence Network and early warning outbreak detection: a Canadian contribution to global public health. *Can J Public Health* (2006) 97:42–4.
37. Heymann DL, Rodier GR, Operational Support WHO Team to the Global Outbreak Alert and Response Network: hot spots in a wired world: WHO surveillance of emerging and re-emerging infectious diseases. *Lancet Infect Dis* (2001) 1:345–53. 10.1016/S1473-3099(01)00148-7
38. Eysenbach G. Infodemiology and infoveillance: framework for an emerging set of public health informatics methods to analyze search, communication and publication behavior on the Internet. *J Med Internet Res* (2009) 11:e11. 10.2196/jmir.1157
39. Polgreen PM, Chen Y, Pennock DM, Nelson FD. Using internet searches for influenza surveillance. *Clin Infect Dis* (2008) 47:1443–8. 10.1086/593098
40. Althouse BM, Ng YY, Cummings DA. Prediction of dengue incidence using search query surveillance. *PLoS Negl Trop Dis* (2011) 5:e1258. 10.1371/journal.pntd.0001258
41. Desai R, Lopman BA, Shimshoni Y, Harris JP, Patel MM, Parashar UD. Use of Internet search data to monitor impact of rotavirus vaccination in the United States. *Clin Infect Dis* (2012) 54:e115–8. 10.1093/cid/cis121
42. Ginsberg J, Mohebbi MH, Patel RS, Brammer L, Smolinski MS, Brilliant L. Detecting influenza epidemics using search engine query data. *Nature* (2009) 457:1012–4. 10.1038/nature07634

43. Kalichman SC, Kegler C. Vaccine-related internet search activity predicts H1N1 and HPV vaccine coverage: implications for vaccine acceptance. *J Health Commun* (2015) 20(3):259–65. 10.1080/10810730.2013.852274
44. Gesualdo F, Stilo G, Agricola E, Gonfiantini MV, Pandolfi E, Velardi P, et al. Influenza-like illness surveillance on Twitter through automated learning of naive language. *PLoS One* (2013) 8:e82489. 10.1371/journal.pone.0082489
45. Chew C, Eysenbach G. Pandemics in the age of Twitter: content analysis of Tweets during the 2009 H1N1 outbreak. *PLoS One* (2010) 5:e14118. 10.1371/journal.pone.0014118
46. Broniatowski DA, Paul MJ, Dredze M. National and local influenza surveillance through Twitter: an analysis of the 2012-2013 influenza epidemic. *PLoS One* (2013) 8:e83672. 10.1371/journal.pone.0083672
47. Nagar R, Yuan Q, Freifeld CC, Santillana M, Nojima A, Chunara R, et al. A case study of the New York City 2012-2013 influenza season with daily geocoded Twitter data from temporal and spatiotemporal perspectives. *J Med Internet Res* (2014) 16:e236. 10.2196/jmir.3416
48. Gomide J, Veloso A, Meira W, Jr, Almeida V, Benevenuto F, Ferraz F, et al. Dengue surveillance based on a computational model of spatio-temporal locality of Twitter. Proceedings of the 3rd International Web Science Conference New York (2011).
49. Chunara R, Andrews JR, Brownstein JS. Social and news media enable estimation of epidemiological patterns early in the 2010 Haitian cholera outbreak. *Am J Trop Med Hyg* (2012) 86:39–45. 10.4269/ajtmh.2012.11-0597
50. Paolotti D, Carnahan A, Colizza V, Eames K, Edmunds J, Gomes G, et al. Web-based participatory surveillance of infectious diseases: the Influenzanet participatory surveillance experience. *Clin Microbiol Infect* (2014) 20:17–21. 10.1111/1469-0691.12477
51. Cantarelli P, Debin M, Turbelin C, Poletto C, Blanchon T, Falchi A, et al. The representativeness of a European multi-center network for influenza-like-illness participatory surveillance. *BMC Public Health* (2014) 14:984. 10.1186/1471-2458-14-984
52. *Flu Near You* (2015). Available from: https://flunearyou.org/
53. Hartley DM, Nelson NP, Arthur RR, Barboza P, Collier N, Lightfoot N, et al. An overview of internet biosurveillance. *Clin Microbiol Infect* (2013) 19:1006–13. 10.1111/1469-0691.12273

54. Schwind JS, Wolking DJ, Brownstein JS, PREDICT Consortium1. Mazet JA, Smith WA. Evaluation of local media surveillance for improved disease recognition and monitoring in global hotspot regions. *PLoS One* (2014) 9:e110236. 10.1371/journal.pone.0110236
55. Innovative Support to Emegencies Diseases and Disasters (InSTEDD). *Geochat* (2015). Available from: http://instedd.org/technologies/geochat/
56. Kamel Boulos MN, Resch B, Crowley DN, Breslin JG, Sohn G, Burtner R, et al. Crowdsourcing, citizen sensing and sensor web technologies for public and environmental health surveillance and crisis management: trends, OGC standards and application examples. *Int J Health Geogr* (2011) 10:67. 10.1186/1476-072x-10-67
57. Yen C, Figueroa JR, Uribe ES, Carmen-Hernandez LD, Tate JE, Parashar UD, et al. Monovalent rotavirus vaccine provides protection against an emerging fully heterotypic G9P[4] rotavirus strain in Mexico. *J Infect Dis* (2011) 204:783–6. 10.1093/infdis/jir390
58. Lopalco PL, DeStefano F. The complementary roles of phase 3 trials and post-licensure surveillance in the evaluation of new vaccines. *Vaccine* (2014) 33:1541–8. 10.1016/j.vaccine.2014.10.047
59. Baggs J, Gee J, Lewis E, Fowler G, Benson P, Lieu T, et al. The Vaccine Safety Datalink: a model for monitoring immunization safety. *Pediatrics* (2011) 127(Suppl 1):S45–53. 10.1542/peds.2010-1722H
60. Yih WK, Lee GM, Lieu TA, Ball R, Kulldorff M, Rett M, et al. Surveillance for adverse events following receipt of pandemic 2009 H1N1 vaccine in the Post-Licensure Rapid Immunization Safety Monitoring (PRISM) System, 2009-2010. *Am J Epidemiol* (2012) 175:1120–8. 10.1093/aje/kws197
61. Huang WT, Hsu CC, Lee PI, Chuang JH. Mass psychogenic illness in nationwide in-school vaccination for pandemic influenza A(H1N1) 2009, Taiwan, November 2009-January 2010. *Euro Surveill* (2010) 15:19575.
62. Partinen M, Saarenpaa-Heikkila O, Ilveskoski I, Hublin C, Linna M, Olsen P, et al. Increased incidence and clinical picture of childhood narcolepsy following the 2009 H1N1 pandemic vaccination campaign in Finland. *PLoS One* (2012) 7:e33723. 10.1371/journal.pone.0033723
63. Cashman P, Moberley S, Dalton C, Stephenson J, Elvidge E, Butler M, et al. Vaxtracker: active on-line surveillance for adverse events

following inactivated influenza vaccine in children. *Vaccine* (2014) 32:5503–8. 10.1016/j.vaccine.2014.07.061

64. Baron S, Goutard F, Nguon K, Tarantola A. Use of a text message-based pharmacovigilance tool in Cambodia: pilot study. *J Med Internet Res* (2013) 15:e68. 10.2196/jmir.2477
65. Atwell JE, Salmon DA. Pertussis resurgence and vaccine uptake: implications for reducing vaccine hesitancy. *Pediatrics* (2014) 134:602–4. 10.1542/peds.2014-1883
66. Roberts JR, Thompson D, Rogacki B, Hale JJ, Jacobson RM, Opel DJ, et al. Vaccine hesitancy among parents of adolescents and its association with vaccine uptake. *Vaccine* (2015) 33:1748–55. 10.1016/j.vaccine.2015.01.068
67. Kennedy A, Lavail K, Nowak G, Basket M, Landry S. Confidence about vaccines in the United States: understanding parents' perceptions. *Health Aff (Millwood)* (2011) 30:1151–9. 10.1377/hlthaff.2011.0396
68. Opel DJ, Mangione-Smith R, Taylor JA, Korfiatis C, Wiese C, Catz S, et al. Development of a survey to identify vaccine-hesitant parents: the parent attitudes about childhood vaccines survey. *Hum Vaccin* (2011) 7:419–25. 10.4161/hv.7.4.14120
69. Jones AM, Omer SB, Bednarczyk RA, Halsey NA, Moulton LH, Salmon DA. Parents' source of vaccine information and impact on vaccine attitudes, beliefs, and nonmedical exemptions. *Adv Prev Med* (2012) 2012:932741. 10.1155/2012/932741
70. Witteman HO, Zikmund-Fisher BJ. The defining characteristics of web 2.0 and their potential influence in the online vaccination debate. *Vaccine* (2012) 30:3734–40. 10.1016/j.vaccine.2011.12.039
71. Bianco A, Zucco R, Nobile CG, Pileggi C, Pavia M. Parents seeking health-related information on the internet: cross-sectional study. *J Med Internet Res* (2013) 15:e204. 10.2196/jmir.2752
72. Ruiz JB, Bell RA. Understanding vaccination resistance: vaccine search term selection bias and the valence of retrieved information. *Vaccine* (2014) 32:5776–80. 10.1016/j.vaccine.2014.08.042
73. Larson HJ, Smith DM, Paterson P, Cumming M, Eckersberger E, Freifeld CC, et al. Measuring vaccine confidence: analysis of data obtained by a media surveillance system used to analyse public concerns about vaccines. *Lancet Infect Dis* (2013) 13:606–13. 10.1016/S1473-3099(13)70108-7

74. Salathe M, Khandelwal S. Assessing vaccination sentiments with online social media: implications for infectious disease dynamics and control. *PLoS Comput Biol* (2011) 7:e1002199. 10.1371/journal.pcbi.1002199

75. Madden K, Nan X, Briones R, Waks L. Sorting through search results: a content analysis of HPV vaccine information online. *Vaccine* (2012) 30:3741–6. 10.1016/j.vaccine.2011.10.025

76. Harmsen IA, Doorman GG, Mollema L, Ruiter RA, Kok G, de Melker HE. Parental information-seeking behaviour in childhood vaccinations. *BMC Public Health* (2013) 13:1219. 10.1186/1471-2458-13-1219

77. Oncel S, Alvur M. How reliable is the internet for caregivers on their decision to vaccinate their child against influenza? Results from googling in two languages. *Eur J Pediatr* (2013) 172:401–4. 10.1007/s00431-012-1889-z

78. Pias-Peleteiro L, Cortes-Bordoy J, Martinon-Torres F. Dr. Google: what about the human papillomavirus vaccine? *Hum Vaccin Immunother* (2013) 9:1712–9. 10.4161/hv.25057

79. Nundy S, Surati M, Nwadei I, Singal G, Peek ME. A web-based patient tool for preventive health: preliminary report. *J Prim Care Community Health* (2012) 3:289–94. 10.1177/2150131911436011

80. Shropshire AM, Brent-Hotchkiss R, Andrews UK. Mass media campaign impacts influenza vaccine obtainment of university students. *J Am Coll Health* (2013) 61:435–43. 10.1080/07448481.2013.830619

81. Starling R, Nodulman JA, Kong AS, Wheeler CM, Buller DB, Woodall WG. Beta-test results for an HPV information web site: GoHealthyGirls.org – increasing HPV vaccine uptake in the United States. *J Consum Health Internet* (2014) 18:226–37. 10.1080/15398285.2014.931771

82. World Health Organization. *Vaccine Safety Net* (2015). Available from: http://www.who.int/vaccine_safety/initiative/communication/network/vaccine_safety_websites/en/

83. Wilson K, Keelan J. Social media and the empowering of opponents of medical technologies: the case of anti-vaccinationism. *J Med Internet Res* (2013) 15:e103. 10.2196/jmir.2409

84. Keelan J, Pavri V, Balakrishnan R, Wilson K. An analysis of the human papilloma virus vaccine debate on MySpace blogs. *Vaccine* (2010) 28:1535–40. 10.1016/j.vaccine.2009.11.060

85. Nan X, Madden K. HPV vaccine information in the blogosphere: how positive and negative blogs influence vaccine-related risk perceptions, attitudes, and behavioral intentions. *Health Commun* (2012) 27:829–36. 10.1080/10410236.2012.661348

86. Love B, Himelboim I, Holton A, Stewart K. Twitter as a source of vaccination information: content drivers and what they are saying. *Am J Infect Control* (2013) 41:568–70. 10.1016/j.ajic.2012.10.016

87. Keelan J, Pavri-Garcia V, Tomlinson G, Wilson K. YouTube as a source of information on immunization: a content analysis. *JAMA* (2007) 298:2482–4. 10.1001/jama.298.21.2482

88. Raviotta JM, Nowalk MP, Lin CJ, Huang HH, Zimmerman RK. Using Facebook to recruit college-age men for a human papillomavirus vaccine trial. *Am J Mens Health* (2016) 10(2):110–9. 10.1177/1557988314557563

89. Remschmidt C, Walter D, Schmich P, Wetzstein M, Delere Y, Wichmann O. Knowledge, attitude, and uptake related to human papillomavirus vaccination among young women in Germany recruited via a social media site. *Hum Vaccin Immunother* (2014) 10:2527–35. 10.4161/21645515.2014.970920

90. Stratton SL, Spencer HJ, Greenfield WW, Low G, Hitt WC, Quick CM, et al. A novel use of a statewide telecolposcopy network for recruitment of participants in a phase I clinical trial of a human papillomavirus therapeutic vaccine. *Clin Trials* (2015) 12(3):199–204. 10.1177/1740774514566333

91. Gunasekaran B, Jayasinghe Y, Brotherton JM, Fenner Y, Moore EE, Wark JD, et al. Asking about human papillomavirus vaccination and the usefulness of registry validation: a study of young women recruited using Facebook. *Vaccine* (2015) 33:826–31. 10.1016/j.vaccine.2014.11.002

92. Gowda C, Schaffer SE, Kopec K, Markel A, Dempsey AF. A pilot study on the effects of individually tailored education for MMR vaccine-hesitant parents on MMR vaccination intention. *Hum Vaccin Immunother* (2013) 9:437–45. 10.4161/hv.22821

93. Williams SE, Rothman RL, Offit PA, Schaffner W, Sullivan M, Edwards KM. A randomized trial to increase acceptance of childhood vaccines by vaccine-hesitant parents: a pilot study. *Acad Pediatr* (2013) 13:475–80. 10.1016/j.acap.2013.03.011

94. Zhang C, Gotsis M, Jordan-Marsh M. Social media microblogs as an HPV vaccination forum. *Hum Vaccin Immunother* (2013) 9:2483–9. 10.4161/hv.25599
95. Community Preventive Services Task Force. *Guide to Community Preventive Services. Increasing Appropriate Vaccination: Client Reminder and Recall Systems* (2015). Available from: www.thecommunityguide.org/vaccines/universally/clientreminder.html
96. Peck JL, Stanton M, Reynolds GE. Smartphone preventive health care: parental use of an immunization reminder system. *J Pediatr Health Care* (2014) 28:35–42. 10.1016/j.pedhc.2012.09.005
97. Free C, Phillips G, Felix L, Galli L, Patel V, Edwards P. The effectiveness of M-health technologies for improving health and health services: a systematic review protocol. *BMC Res Notes* (2010) 3:250. 10.1186/1756-0500-3-250
98. Hartzler A, Wetter T. Engaging patients through mobile phones: demonstrator services, success factors, and future opportunities in low and middle-income countries. *Yearb Med Inform* (2014) 9:182–94. 10.15265/IY-2014-0022
99. Odone A, Ferrari A, Spagnoli F, Visciarelli S, Shefer A, Pasquarella C, et al. Effectiveness of interventions that apply new media to improve vaccine uptake and vaccine coverage. *Hum Vaccin Immunother* (2015) 11:72–82. 10.4161/hv.34313
100. Batler D. *When Google Got Flu Wrong. US Outbreaks Foxes a Leading Web-Based Method for Tracking Seasonal Flu* (2013). Available from: http://www.nature.com/news/when-google-got-flu-wrong-1.12413
101. Eberth JM, Kline KN, Moskowitz DA, Montealegre JR, Scheurer ME. The role of media and the Internet on vaccine adverse event reporting: a case study of human papillomavirus vaccination. *J Adolesc Health* (2014) 54:289–95. 10.1016/j.jadohealth.2013.09.005
102. Cohen AM, Hersh WR. A survey of current work in biomedical text mining. *Brief Bioinform* (2005) 6:57–71. 10.1093/bib/6.1.57
103. Berendt B. Text mining for news and blogs analysis. In: Sammut C, Webb GI, editors. *Encycolpedia of Machine Learning*. New York: Springer; (2010).
104. Wojcik OP, Brownstein JS, Chunara R, Johansson MA. Public health for the people: participatory infectious disease surveillance in the digital age. *Emerg Themes Epidemiol* (2014) 11:7. 10.1186/1742-7622-11-7

105. Baazeem M, Abenhaim H. Google and women's health-related issues: what does the search engine data reveal? *Online J Public Health Inform* (2014) 6:e187. 10.5210/ojphi.v6i2.5470

106. Lewis SP, Mahdy JC, Michal NJ, Arbuthnott AE. Googling self-injury: the state of health information obtained through online searches for self-injury. *JAMA Pediatr* (2014) 168:443–9. 10.1001/jamapediatrics.2014.187

107. Fox S. *Peer-to-Peer Healthcare*. Washington, DC: Pew Research Center's Internet & American Life Project; (2011).

108. Clark SJ, Butchart A, Kennedy A, Dombkowski KJ. Parents' experiences with and preferences for immunization reminder/recall technologies. *Pediatrics* (2011) 128:e1100–5. 10.1542/peds.2011-0270

109. Milinovich GJ, Williams GM, Clements AC, Hu W. Internet-based surveillance systems for monitoring emerging infectious diseases. *Lancet Infect Dis* (2014) 14:160–8. 10.1016/S1473-3099(13)70244-5

INDEX

D

E

F

G

H

I